SpringerWienNewYork

CISM COURSES AND LECTURES

The series presents lecture notes, monographs, edited works and proceedings in the field of Mechanics, Engineering, Computer Science and Applied Mathematics.
Purpose of the series is to make known in the international scientific and technical community results obtained in some of the activities organized by CISM, the International Centre for Mechanical Sciences.

INTERNATIONAL CENTRE FOR MECHANICAL SCIENCES

COURSES AND LECTURES - No. 531

ENVIRONMENTAL WIND ENGINEERING AND DESIGN OF WIND ENERGY STRUCTURES

EDITED BY

CHARALAMBOS C. BANIOTOPOULOS
ARISTOTLE UNIVERSITY OF THESSALONIKI, GREECE

CLAUDIO BORRI
UNIVERSITY OF FLORENCE, FIRENZE, ITALY

THEODORE STATHOPOULOS
CONCORDIA UNIVERSITY, MONTREAL, CANADA

SpringerWienNewYork

This volume contains 201 illustrations

SPIN 80112209

All contributions have been typeset by the authors.

ISBN 978-3-7091-0952-6 SpringerWienNewYork

PREFACE

The present volume presents topics from environmental aerodynamics, which in the last decade has significantly progressed, particularly at the computational front. The volume is divided into two parts; the first is devoted to environmental wind engineering, whereas the second addresses the design of wind energy support structures.

In the first part of the present volume, the fundamentals of both experimental and computational approaches along with examples from actual studies involving pedestrian level winds, comfort levels, relevant legislation and remedial measures. Pollutant dispersion in the building environment is also presented and discussed along with the methodologies available to deal with potentially critical design problems, which—if left unattended—could jeopardize the successful operation of new construction developments.

Considering that wind energy production involves special wind turbine support structures, the second part of this volume concerns their design, which needs some special reliability assessment. Experimental and computational approaches are analyzed in order to make the reader familiar with the uncertainties involved and the currently available design methodologies and codes framework.

The volume contains seven chapters written by wind engineering experts. Chapters 1 and 2, authored by Theodore Stathopoulos, include a broad introduction to environmental aerodynamics, their fundamentals and applications respectively. Chapter 3, by Bert Blocken, introduces the theory and applications of computational wind engineering, an area of vital contemporary interest in the evolution of wind engineering.

Chapter 4, co-authored by Alberto Zasso, Paolo Schito, Carlo Botasso and Alessandro Croce, refers to the fundamental interaction mechanism between the wind flow and the wind turbine that is the basis for the design of wind turbines applying numerical and wind tunnel modeling approaches. Chapter 5, by Peter Schaumann, Cord Böker, Anne Bechtel and Stephan Lochte-Holtgreven, introduces the different types of support structures for onshore and offshore wind energy converters including the relevant design criteria for different limit design states. Chapter 6, by Charalampos Baniotopoulos, Iakovos Lavassas, George Nikolaidis and Pantelis Zervas, deals with specific topics on the design of tubular steel wind turbine towers considering that the design is governed by extreme wind and earthquake loading when the Aeolian park is constructed in seismic hazardous regions. Finally, Chapter 7, co-authored by Claudio Borri, Paolo Biagini and Enzo Marino, addresses structural design questions for large wind turbines in seismic areas with reference to main dimensioning criteria, international Norm specifications and erection technologies. This chapter includes a thorough discussion on various structural analysis approaches and the design, construction and in-situ testing.

The editors would like to thank all contributors to this volume for the excellence of their work and to also extend their sincere appreciation to the CISM General Secretary, Professor Bernhard Schrefler, the CISM Rector, Professor Giulio Maier, the Editor of the Series Professor Paolo Serafini, as well as to the entire CISM staff in Udine for their excellent cooperation.

The produced volume will be of interest to engineers, researchers and academicians who work on relevant scientific research or design topics in research centers or institutes, universities, industry and government agencies. The volume is written to address the interests of practicing engineers and professionals as well.

C.C. Baniotopoulos, C. Borri, and T. Stathopoulos

CONTENTS

PART I

ENVIRONMENTAL WIND ENGINEERING TOPICS

Introduction to Environmental Aerodynamics

Theodore Stathopoulos

Department of Building, Civil and Environmental Engineering, Concordia University, Montreal, Canada

Abstract

A brief introduction to environmental aerodynamics, as part of wind engineering is provided. Of particular interest are the areas of pedestrian level winds and exhaust dispersion around buildings, which are introduced and discussed in the context of urban environment aerodynamics.

1. Introduction

The area of environmental aerodynamics examines those aspects of wind engineering that are closely related to the effects of wind to the built environment in conjunction to basic environmental issues. Environmental aerodynamics is complementary to the so-called structural wind engineering, which deals with wind loading on structures. Environmental aerodynamics requires fundamental knowledge of wind engineering principles (Stathopoulos and Baniotopoulos 2007) in order to study phenomena such as pedestrian level winds and their effect to humans; dispersion of pollutants; natural and hybrid ventilation; snow dispersion and accumulation; wind flows over various topographies, as well as in the urban or building environment; and the like. Particular emphasis is placed here on pedestrian level winds and dispersion of pollutants in the urban and building environment.

2. Pedestrian Level Winds

2.1 Background

The quality of open urban spaces has received a lot of attention in recent years. There is a broad recognition that microclimatic conditions contribute to the quality of life in cities, both from the economic as well as from the social viewpoint. Consequently, universities and other research organizations, municipal and other government agencies, as well as construction and architectural companies have expressed significant interest and allocated resources to examine microclimatic conditions, particularly the effect of wind, on the outdoor human comfort. This task is quite complex because, contrary to the more or less controllable indoor comfort conditions, outside human comfort in an urban climate is generally affected by a wide range of variables such as wind speed and direction, air temperature, solar radiation, possible precipitation in various forms etc.

In Europe, a 3-year (2001-2004) EU-funded project entitled *Rediscovering the Urban Realm in Urban Spaces (RUROS)* with extensive surveys carried out at different open spaces has completed approximately 10,000 interviews (http://alpha.cres.gr/ruros). The project aims to produce an urban design tool that provides architects, engineers, urban planners and other decision makers with means to assess effectively the construction of new buildings and the development of cities from the economic, psycho-physiological and sociological perspective of human comfort, air temperature, relative humidity, solar radiation, air quality, human activity.

Work has also been carried out within the auspices of the European Action C14 dealing with Impact of Wind and Storm on City Life and Built Environment, with a working group interested in the effects of wind on pedestrians, their assessment and comparisons, as well as the parameters they influence human comfort and its evaluation. Results have appeared in the 2002 Workshop in Nantes, e.g. Westbury et al. (2002) and in the International Conference in Urban Wind Engineering and Building Aerodynamics organized by Von Karman Institute for Fluid Dynamics in May 2004. In addition, the American Society of Civil Engineers has put out a pertinent state-of-the-art document (ASCE 2003), which was developed with input from the European Action C14.

The present lecture notes will describe the aerodynamics of the urban environment and the reasons causing high wind speeds at sidewalks and, consequently, potential discomfort to pedestrians; they will address the experimental and computational evaluations of the wind on people in the urban environment and will focus on the state-of-the-art of the development of human outdoor comfort criteria by considering a wide range of parameters, including wind speed, air temperature, relative humidity, solar radiation, air quality, human activity, clothing level, age and the like.

2.2 Aerodynamics of the urban environment

Strong winds are usually accelerated at the pedestrian level within the urban environment, say around tall buildings, due to particular aerodynamic configurations generally associated with tall buildings. In the case of a simple rectangular tall building, it is the boundary layer flow that causes descending flows towards the street level due to the pressure differences created by the velocity differences between higher and lower levels. This down-flow is significant due to the pressure proportionality to the square of the velocity (Bernoulli equation) and its strength increases with the building height. Figure 1 demonstrates this effect, which is termed in the literature as downwash. Clearly, downwash is diminished drastically in the absence of boundary layer flow and this explains the lack of adequate representation of wind effects in the building environment for simulationscarried out in the past using aeronautical wind tunnels for building aerodynamics applications.

In general, buildings will only induce high wind speeds at lower levels if a significant part of them is exposed to direct wind flows. It is actually the direct exposure to wind rather than building height alone, which causes the problem. This is shown diagrammatically in Figure 2.

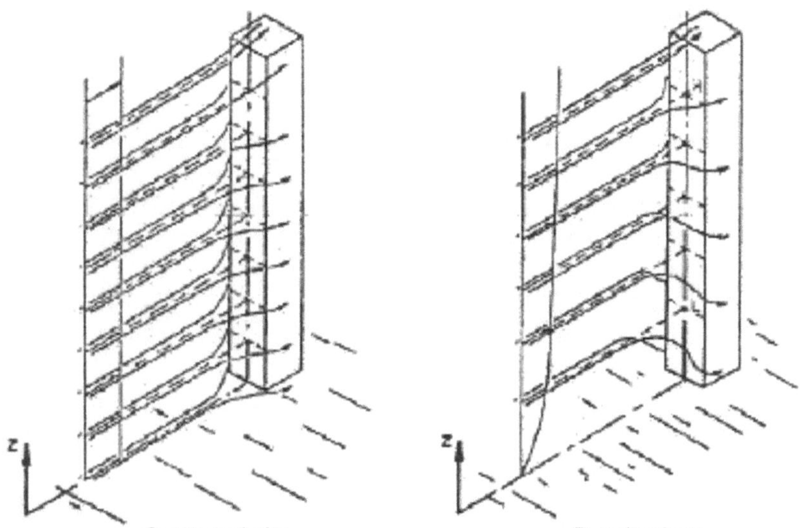

Figure 1. Uniform and boundary layer wind flow around a tall rectangular building.

Another type of pedestrian-level winds is formed when high-speed winds pass through openings between high-pressure air on the windward wall and low-pressure regions on the leeward side of a building. Once more, the fair character of nature, which does not like pressure differences, prevails and strong flow is induced to correct the problem. Pedestrians in arcades of commercial buildings can testify regarding this situation, which is unpleasant to the store-owners in these areas as well. Figure 3 shows this type of configuration along with other flow-induced mechanisms creating disturbances to the urban environment in the vicinity of buildings. These include but there are not limited to the effects of the large standing vortex in front of a building, the vortex flows generated after the flow separates and accelerates along the building front edges and the wake-induced disturbances via the interaction of the flow coming from the building side faces and the re-circulation flow regime created by the shear layer flow above the building. Clearly, wind direction is a significant factor here, in addition to the magnitude of the oncoming wind speed.

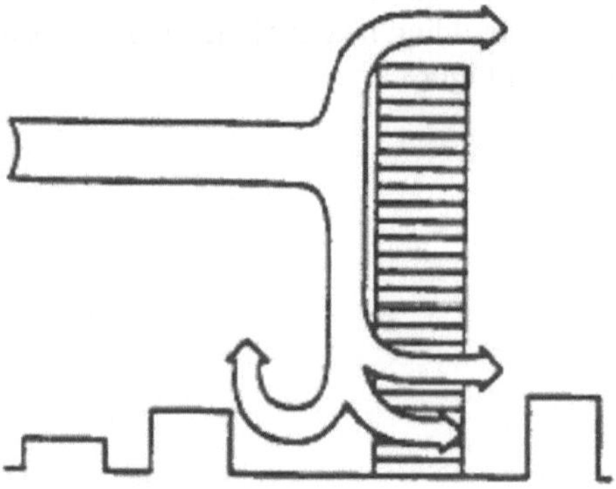

Figure 2. Wind flow around buildings significantly taller than their surroundings, after Cochran (2004).

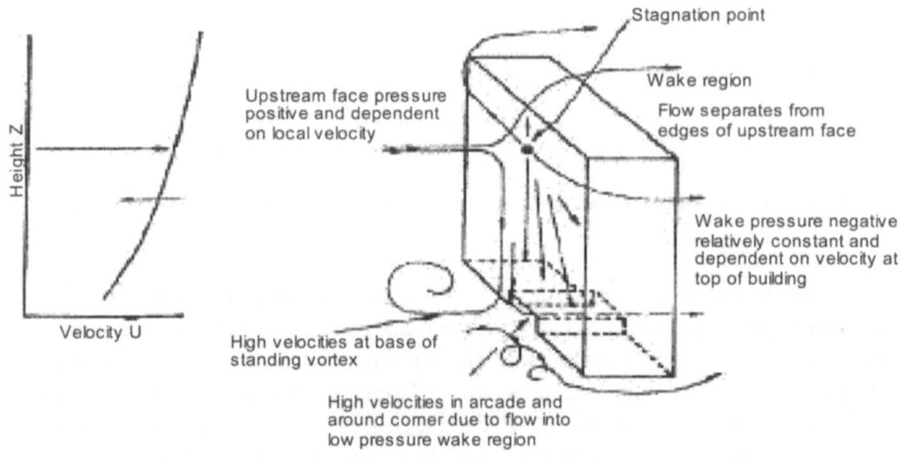

Figure 3. Interaction of building with oncoming wind flow.

Additional common building configurations and potential influences on pedestrian-level winds are shown in Figure 4 taken from Cochran (2004). These configurations include the effects of canopies, which may act as deterrents to the strong down-flow prior to impacting on sidewalks or other pedestrian free access areas around the building. However, such measures may create other problems by deflecting the wind from, say, a building entrance to another area around the building corners or across the street. Setbacks on the building surfaces or penthouses are elements generally remediating the pedestrian-level winds and are used rather extensively. Furthermore, a podium not intended for long-term pedestrian activities or vegetation in terms of bushes and coniferous-evergreen trees can also be used as a positive measure to amend harsh wind conditions at pedestrian level. Porous screens are also successful in deflecting winds without relocating the adverse conditions on other places. Entrance alcoves, as well as balconies, diminish sidewalk winds in the cities. However, high winds may be transferring on balconies themselves, particularly those near the edges of the building facades.

Previous discussion is really about isolated and mainly rectangular buildings. Curved buildings such as cylindrical shapes generally promote lateral flow, so they behave better as far as effects of pedestrian-level winds are concerned. Channeling effects appearing in the case of two or more buildings are generally critical, particularly if the wind direction is along the street or corridor formed between the buildings. This is a result of the quasi-Venturi effect, which can be critical in some cases.

If the wind conditions with one or two simple-shaped buildings in place can become so complex, one can easily imagine what would really happen with buildings of complex shapes interacting with the wind flow passing amongst them, particularly when the effect of ground topography and all adjacent buildings are taken into account. The problem becomes really difficult and for a number of years could only be solved experimentally via appropriate simulation in a boundary layer wind tunnel. Only recently, more specifically during the last few years with the significant progress in computational technology, attempts were made to address the problem of pedestrian-level winds in the urban environment computationally. More detailed discussion on the state-of-the-art of this approach will be presented in a subsequent section.

Regardless of the approach used to determine the impact of wind flows at the pedestrian level, the previous comments have demonstrated that the direction of the oncoming wind together with its magnitude, i.e. speed, will be of paramount importance. If the wind climate in a city is distinctly directional, i.e. strong winds come always from a particular narrow fetch, it is clear that this set of directions should be really scrutinized because, in all likelihood, critical results will occur when the wind comes from these particular directions.

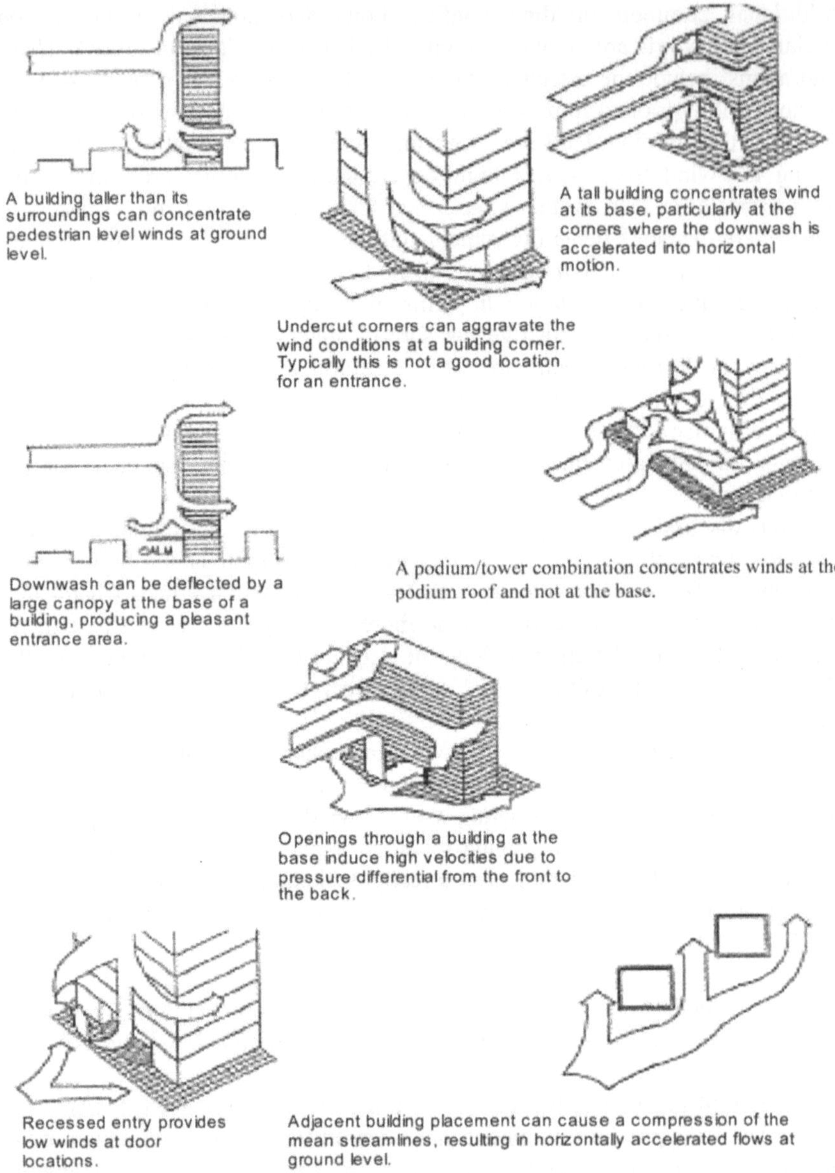

Figure 4. Design features to change and/or ameliorate pedestrian wind conditions, after Cochran (2004).

As an example, the basic wind environment of Montreal in terms of wind speeds and probabilities of exceedance from different directions is presented in Figure 5. As clearly shown, westerly and southwesterly winds dominate while north and northeasterly winds may also be high. Note that these are upper level winds and significant changes may occur near the ground areas. In addition, differences exist between summer and winter wind data. Maximum summer winds are dominant from west, while winter winds are certainly higher and they blow primarily from southwest. In the great majority of pedestrian wind studies carried out for tall buildings in Montreal, it has been found that winds for west / southwest and, to a lesser extent from northeast have produced the most critical adverse conditions.

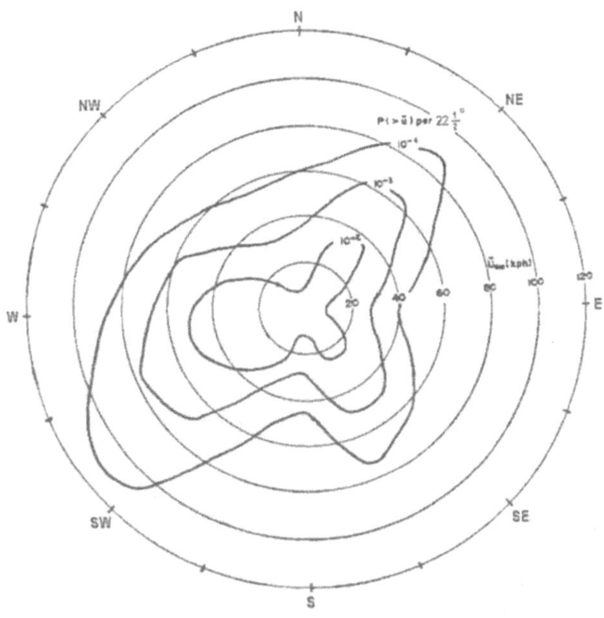

Figure 5. Probability distributions of hourly mean wind speed at 300 m over Montreal for daylight hours during the winter (derived from 10 year record of wind data obtained at a height of 10 m at Trudeau – previously Dorval - Airport)

In summary, there are two main flow types causing high pedestrian-level winds in the urban environment: downwash flows and horizontally accelerated flows. The former are diminished by podia, architectural features such as setbacks, balconies and the like; the latter are ameliorated by alcoves, chamfered corners, landscaping (vegetation) or porous screens.

2.3 Wind comfort criteria

Several criteria have been developed in the wind engineering community for evaluating only the wind-induced mechanical forces on the human body and the resulting pedestrian comfort and safety. There are significant differences among the criteria used by various countries and institutions to establish threshold values for tolerable and unacceptable wind conditions even if a single parameter, such as the wind speed is used as criterion. These differences range from the speed averaging period (mean or gust) and its probability of exceedance (frequency of occurrence) to the evaluation of its magnitude (experimental or computational).

Table 1 shows the traditional Beaufort scale used in ship navigation in a modified version applicable to land regions and for heights representative of pedestrians. This table provides an idea of the mechanical effects of wind of different speeds on the human body. Physiological effects are more complex since they depend on additional factors and their interactions.

Table 1. Extended Land Beaufort Scale showing wind effects on people (Lawson and Pendwarden 1975; Isyumov and Davenport 1975).

Beaufort number	Description	Wind speed at 1.75 m height (m/s)	Effect
0	Calm	0.0 – 0.1	
1	Light air	0.2 – 1.0	No noticeable wind
2	Light breeze	1.1 – 2.3	Wind felt on face
3	Gentle breeze	2.4 – 3.8	Hair disturbed, clothing flaps, newspaper difficult to read
4	Moderate breeze	3.9 – 5.5	Raises dust and loose paper, hair disarranged
5	Fresh breeze	5.6 – 7.5	Force of wind felt on body, danger of stumbling when entering a windy zone
6	Strong breeze	7.6 – 9.7	Umbrellas used with difficulty, hair blown straight, difficult to walk steadily, sideways wind force about equal to forward walking force, wind noise on ears unpleasant
7	Near gale	9.8 – 12.0	Inconvenience felt when walking
8	Gale	12.1 – 14.5	Generally impedes progress, great difficulty with balance in gusts
9	Strong gale	14.6 – 17.1	People blown over by gusts

A simple rule of thumb has been provided by Wise (1970) and Pendwarden (1973). This is based on mean speeds (V) assuming the following effects:

- V = 5 m/s or 18 km/h; onset of discomfort
- V = 10 m/s or 36 km/h; definitely unpleasant
- V = 20 m/s or 72 km/h; dangerous

Conditions for pedestrians are considered acceptable if V > 5 m/s is less than 20% of the time (Pendwarden and Wise 1975).

Recognizing the importance of frequency of occurrence along with the magnitude of wind speeds, Figures 6, 7, and 8 provide threshold mean wind speeds for various types of activity as functions of theaverage annual number of storm occurrences. Naturally the mean wind speed threshold level drops significantly as the yearly average number of occurrences increases.

Utilization of mean wind speeds as comfort criteria for pedestrian-level winds has been questioned by the wind engineering community. In fact, the most prevailing opinion seeks an effective wind speed, which is related to the gustiness of the wind, to be used for that purpose. Such effective speeds can be derived from the following equation expressing their outcome in terms of the mean and a number (ranging from 1 to 3) of standard deviations of the wind speed:

$$V^e = \overline{V} \cdot \left(1 + \kappa \cdot \frac{\overline{v'^2}^{1/2}}{\overline{V}}\right) \tag{1}$$

where $\overline{v'^2}^{1/2}$ is the rms of longitudinal velocity fluctuations and κ is a constantequal to 1~3.

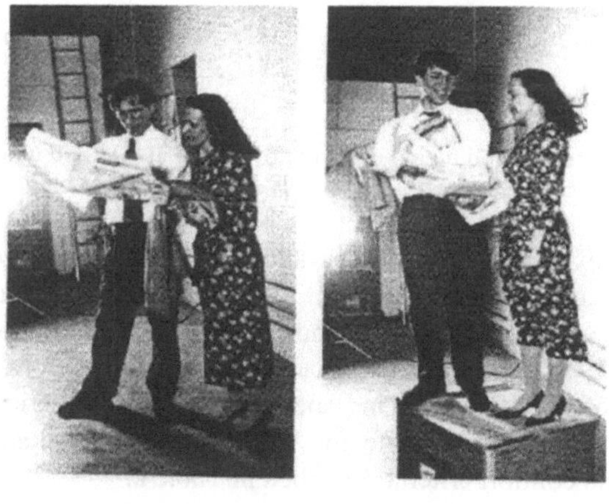

Figure 6. Wind tunnel exposure of people at 10-15 km/h winds.

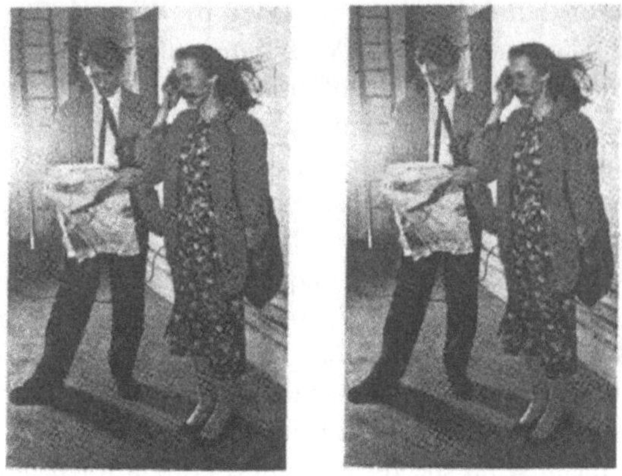

Figure 7. Wind tunnel exposure of people at 20 (left) and 40 (right) km/h winds.

Figure 8. Wind tunnel exposure of people at 70 km/h winds.

Wind tunnel experiments and observations of pedestrian performance suggest that $\kappa = 3$ is the most appropriate value. Figure 9 shows acceptance criteria for wind speeds for various annual frequencies of occurrence proposed by Isyumov and Davenport (1975). Note that these criteria are different from previous criteria in that, instead of specifying a wind speed for various activities, frequencies of occurrence are specified for different wind speeds. Murakami et al. (1986) produced the wind comfort criteria described in Table 2.

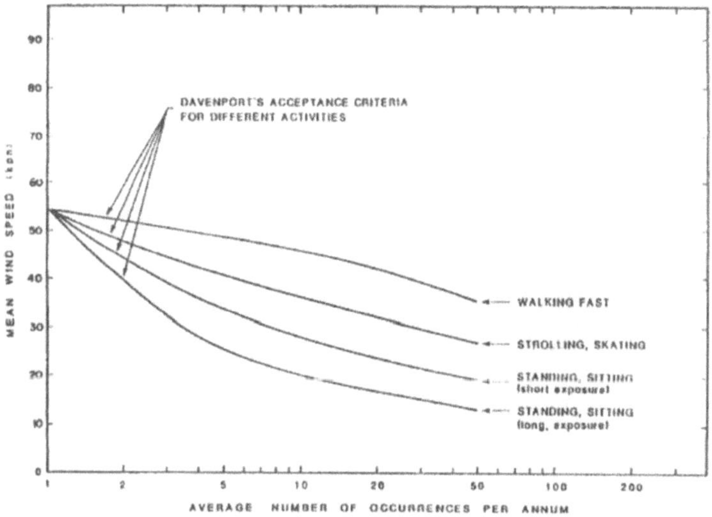

Figure 9. Acceptance criteria for wind speeds for various annual frequencies of occurrence, after Isyumov and Davenport (1975).

Table 2. Wind environment criteria of Murakami et al. (1986).

Activity	**Probability of Exceedance (P(>u))**		
	$\hat{u}$= 36 km/h	**$\hat{u}$ = 54 km/h**	**$\hat{u}$ = 72 km/h**
Long-term and short-term stationary exposure	0.10	0.008	0.0008
Strolling	0.22	0.036	0.006
Walking	0.35	0.07	0.015

Melbourne (1978) has produced separate criteria based on mean and gust speeds. He proposed their application only for daylight hours and on the assumption that the max 2-sec gust speed will be roughly twice as large as the mean speed, he produced the curves shown in Figure 10. These curves identify threshold wind speed criteria for different types of activity similar to those shown in Table 2. Criteria for dangerous wind conditions were also specified.

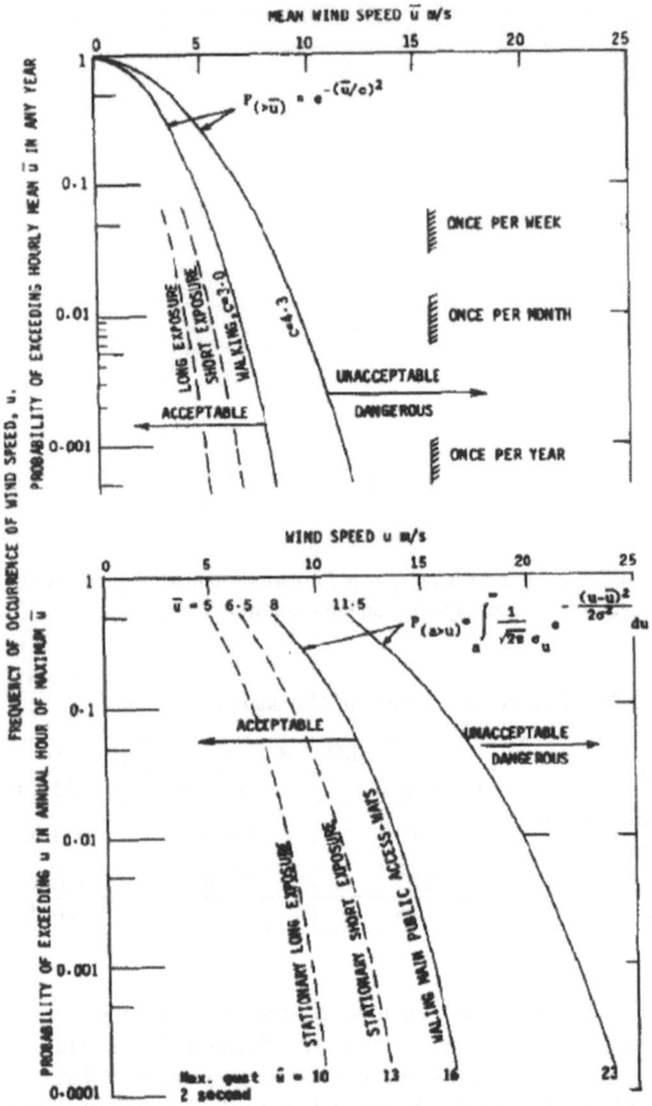

Figure 10. Probability distributions of Melbourne's criteria for environmental wind conditions for daylight hours for a turbulence intensity of 30% and $\hat{u} = 2\bar{u}$, after Melbourne (1978).

Such conditions are particularly important for cities with harsh winter conditions where icy sidewalks become source of frequent accidents when combined with high winds. Several cases of this nature have been reported, most involving accidents happened on elderly people.Liability issues are also interesting for such cases and courts have always a hard time dealing with them.

On the basis of experience over a number of projects and wind tunnelstudies, it has been concluded that Melbourne's criteria are on the strict side, i.e. if prevailing conditions abide by the prescribed limits, most sets of other criteria available in the literature or included in ordinances of various municipalities will be satisfied. Consequently, these criteria can be used as upper limits for pedestrian-level winds and, in this regard, are indeed valuable.

Wind ordinances in major cities. There is great variation regarding wind ordinances in various cities around the world. In some cases, specific legislation has passed and new building permits are not provided until the developers/owners demonstrate that the project will not generate dangerous or even uncomfortable and undesirable pedestrian-level wind conditions. In other cases, this is expected to happen as part of assumed good engineering and architectural practice. In general, the following points can be made:

- Most major cities (Montreal, Toronto, Sydney, etc) have some guidelines addressing the problem at the approval stage for new construction projects.
- San Francisco has adopted a very strict wind ordinance; they use $V^e = 42$ km/h with P ($>V^e$) = 0.01% as safety criterion; this is significantly lower than that proposed in most of the current literature.
- New York has strict air pollution standards, which tend to work against guidelines for the pedestrian wind environment; about 30% of new developments have to go through a review process.
- Boston Planning Department specifies that a wind tunnel study is required to assess wind environmental conditions near new developments for the following cases:
  i) for any new building taller than 30 m (100 ft) and at least two times taller than its adjacent buildings
  ii) forother buildings in special circumstances

As it is always the case with any adoption of code provisions or changes, passing legislation regarding pedestrian wind conditions is always problematic. It is worth mentioning that a new wind ordinance has been approved in the Netherlands only recently after several years of intense efforts by several experts, architects and engineers.

2.4 Experimental procedure: wind tunnel approach

General. As mentioned previously, the flows around buildings even in simple surrounding environments, let alone in complex urban settings are still extremely difficult to predict by computational methods. However, the testing of scale models in a boundary layer wind tunnel capable of simulating the mean-velocity profile and turbulence of the natural wind has been shown to be a very effective method of prediction by comparison with respective full-scale data. The wind-tunnel model typically includes all buildings in the surrounding landscape; thus, their effect is automatically included. Both existing conditions and those with the new building(s) in place can be readily measured, thus allowing the impact of the new building(s) to be identified. Furthermore, the effects of changes to the building itself, or to landscaping, can also be studied, particularly where undesirable wind conditions are found.

A typical set up of a wind-tunnel model in a boundary-layer wind tunnel is illustrated in Figure 11. The building itself and the model of its surroundings are mounted on the wind-tunnel turntable, which can be rotated to allow various wind directions to be simulated. Typical model scales for large buildings are in the range of 1:200 to 1:500. Larger scales have been used for smaller buildings. The model of surroundings enables the complex flows created by other buildingsnear the study building to be automatically included in the tests. However, it is also essential to create a proper simulation of the natural wind approaching the modeled area. The requirements for modeling the natural wind in a wind tunnel are described in the ASCE Manual of Practice (1999). In typical wind tunnel tests, the airflow speed above the boundary layer is in the range 10 to 30 m/s.

Figure 11. Typical wind tunnel set-up for a pedestrian wind assessment study.

The process followed in the experimental approach consists of the following steps:

1. Meteorological records
2. Wind tunnel testing
3. Combination of (1) and (2)
4. Comparison with comfort criteria
5. Remedial measures

Details of this process will be presented in a case study described in Chapter 2.

2.5 Computational procedure: CFD

In Computational Wind Engineering (CWE) the computer essentially replaces the physical simulation in the boundary layer wind tunnel, at least in principle. CFD methods involve very large amounts of computation even for relatively simple problems and their accuracy is often difficult to assess when applied to a new problem where prior experimental verification has not been done. Castro and Graham (1999) summarized the concerns expressed with respect to these issues. However, there have been cases for which the application of CFD methodologies appears to give somewhat satisfactory responses. These are cases requiring the determination of mean flow conditions and pressures, i.e. those related primarily with environmental issues. Typical problems of this category include but are not limited to pedestrian level winds, snow dispersion and accumulation, dispersion of pollutants in the near-building and/or urban environment, ventilation and the like. There is increasing evidence that for such problems CFD-based techniques may provide adequate responses – see Stathopoulos (2002).

Pedestrian-level winds can be described quite adequately in terms of mean velocities in the presence and absence of a new building within a specific urban environment. Although it can be argued that pedestrians are mostly affected by gust effects and mean wind speeds may not be sufficient to produce satisfactory results, the fact remains that several major cities require only the satisfaction of certain mean (sustainable) speeds with a specified probability of exceedance. A number of recent computational studies for the evaluation of pedestrian level winds and the comparison of their results with respective experimental data are described in Chapter 2. The process of comparison between computational and experimental results has already been challenged and appears problematic on its own. For instance, is it more meaningful to carry out point-by-point comparisons or does it make more sense to examine pedestrian-level wind speeds affecting a particular zone or area of influence for a specific activity within the urban environment? Furthermore, and after due consideration to the fact that pedestrian level wind speeds measured in the proximity of buildings, i.e. in areas of high turbulence, are not very accurate, it may be conceivable that "errors" in the results might be better described in terms of their impact on design decisions. Clearly, this may be more reasonable, at least in the context of engineering perspective.

2.6 Outdoor comfort issues

Outdoor human comfort in an urban climate depends on a wide range of weather and human factors. Studies have shown integrated effects of wind speed, air temperature, relative humidity and solar radiation on the human perception, preference and overall comfort in an urban environment. Some analysis of these issues has been presented in the ASCE SOA Report (2003). Furthermore, the studies by Nicolopoulou et al. (2001; 2002) also address the influence of microclimatic characteristics in outdoor urban spaces and the comfort implications for the people using them. A significant characteristic is the psychological adaptation, which has also been addressed. An equivalent temperature has been defined and related to the outdoor human comfort by considering acclimatization and other bio-meteorological principles (Stathopoulos et al. 2004; Zacharias et al. 2001). However, the implications of this approach are far fetching and the overall assessment problems are still quite intriguing. Some basic ideas are presented in these notes.

Temperature and relative humidity. Both can have a significant impact on a person's comfort, since sensation of comfort in cold conditions is linked to the heat balance of the human body, i.e. the balance of heat generated by metabolic processes and heat lost by conduction, convection, radiation and evaporation. In convective and evaporative losses, the effects of temperature and humidity are closely linked with the wind conditionsand cannot be treated in isolation from wind speed. This is why, for example, in the colder regions of Europe and North America, the wind chill equivalent temperature is used to provide a more meaningful description of how cold weather will really feel, rather than simply giving air temperature. The equivalent temperature is obtained by calculating the temperature in standard wind (set at 1.8 m/s = 4 mph) that would give the same rate of heat loss from exposed skin at 33°C as occurs in the actual wind and temperature conditions. Generally, in cold conditions, humidity is low and has little direct effect on thermal comfort, although there may be indirect effects, such as humidity changing the insulation value of clothing. In hot conditions, the human body needs to increase heat losses to maintain thermal comfort. This is largely achieved by reducing clothing and through sweating and the corresponding heat losses associated with the latent heat of evaporation. Since the efficiency of evaporation is decreased as the relative humidity of the air increases, the relative humidity becomes a much more important parameter in hot climates. Also, since the efficiency of evaporation is increased with wind speed, in cold climates it is often desirable to reduce wind speeds but the opposite is sometimes the case in hot climates. The well-known Humidex is an effective temperature, combining the temperature and humidity into one number to reflect the perceived temperature and to quantify human discomfort due to excessive heat and humidity. In general, almost everyone will feel uncomfortable when the Humidex ranges from 40 to 45, and many types of labor must be restricted when the Humidex is 46 and higher. The incorporation of relative humidity effects into the overall assessment of thermal comfort is discussed in Stathopoulos et al. (2004).

Solar radiation. Any assessment of thermal comfort must account for the effects of sun/shade conditions. The angle of the sun, the amount of radiation absorbed by clouds, dust and particles in the atmosphere, and the sun light absorbed and reflected by buildings need to be taken into account.

Precipitation. In heavy rain conditions, people are less likely to be outside, thus their wind and thermal comfort will usually be less critical compared with other microclimate factors. However, it may be of interest to evaluate how far under a sheltering canopy roof the precipitation will infiltrate and how often this will happen. Dampness of clothes may also be of interest because it will affect thermal comfort.

A working group of the International Society of Biometeorology is in the process of developing a new standardized Universal Thermal Climate Index (UTCI), which can also be used in the development of a criterion for human outdoor comfort (Hoppe, 2002). An example of application of such an approach is shown in Figure 12 taken from Stathopoulos et al. (2004). The dependence of the overall comfort is expressed on the basis of a group of survey respondents as a function of the difference of two equivalent temperatures: one based on the weather norm, $T_{e,n}$ and the other based on the actual outdoor conditions, $T_{e,a}$. Equivalent temperatures take into account the effect of relative humidity and solar radiation as well. It should be noted that $(T_{e,a} - T_{e,n})$ is the most influential factor on the overall comfort of the respondents.

Figure 12 shows that (1) most comfortable conditions occur when the equivalent temperature difference is about 5°C, which may be attributed to the preference of local residents for higher air temperature as well as the temperature difference between an urban environment downtown and the airport; (2) lower comfort occurs with a negative temperature difference, or when the actual equivalent temperature islower than the norm; and (3) if the temperature difference is beyond a certain limit, say greater than 10°C, less comfortable (overall comfort < 1) outdoor conditions may be perceived, although more field data are necessary to confirm this observation. At present, it is still considered premature to draw a curve for a definite mathematical relationship of overall comfort and equivalent temperature difference.

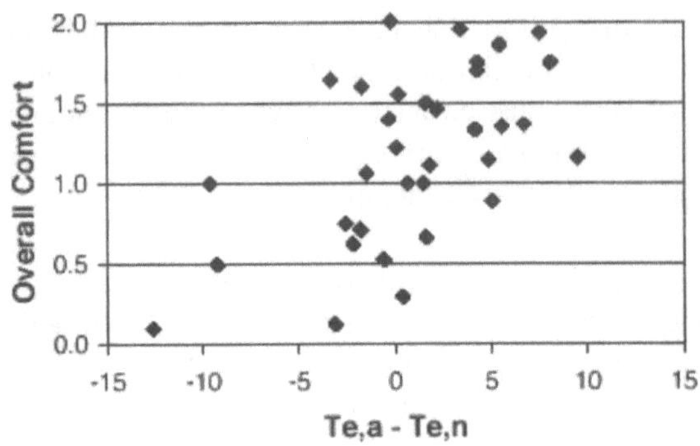

Figure 12. Overall comfort example in terms of equivalent temperature difference, after Stathopoulos et al. (2004).

3. Exhaust Dispersion around Buildings

3.1 Background

Current standards for building ventilation systems recommend that rooftop stacks from industrial, laboratory or hospital buildings be designed such that their emissions do not contaminate fresh air intakes of the emitting building or nearby buildings. This may require extending the height of the stack or increasing its exit velocity. Unfortunately, the state-of-the-art has not been sufficiently advanced to allow building engineers to apply appropriate design criteria to avoid this problem for new construction or to help alleviate it for existing buildings. Consequently, numerous incidents of poor air quality have been recorded and documented.

In most previous studies, e.g. Schulman and Scire (1991),plume dispersion was evaluated for a single isolated building model in the wind tunnel. Wilson et al. (1998) investigated the effect of stack height and exhaust speed on the dispersion of building exhaust in a series of water channel experiments. In addition to an isolated low-rise building, tests were also performed for cases in which an adjacent building was upwind or downwind of the emitting building. A lower adjacent building upwind of the emitting building tends to increase dilution on the emitting building roof. Results obtained with a taller upwind building showed that the leeward wall of this upwind building may experience low dilution values (high concentrations), depending on stack location, stack height, exhaust velocity and the distance between the two buildings. In this case, increasing exhaust velocity was found more beneficial than increasing stack height.

Fluid modeling studies have demonstrated the benefits of high exit velocities and increasing stack height in reducing pollutant concentrations at critical receptors. However, field studies have shown that even with high exit velocities and moderately high stacks, pollutant concentrations may be unacceptably high at particular locations (Wilson and Lamb, 1994, Georgakis et al., 1995, Saathoff et al., 2002). Several factors may account for the occasional poor performance of rooftop stacks. These factors include the location of the stack relative to regions of flow separation and flow re-attachment, the presence of rooftop irregularities such as penthouses and high upstream turbulence. It is important to validate the results of fundamental fluid modeling studies with full-scale data. Although most flow features (e.g. wake size, reattachment lengths etc.) under neutral atmospheric conditions can be accurately simulated in wind tunnels and water channels, it is necessary to determine the limitations of fluid modeling with respect to plume dispersion. Relatively few studies have compared wind tunnel concentration data with field data for near-field diffusion cases, i.e. receptors within – say - 50 m of a stack. This is one of the most difficult fluid modeling applications, since the plume characteristics may be sensitive to a number of local factors such as building wake effects, the position of the stack relative to rooftop recirculation zones, possible delta-wing vortices, stack Reynolds number etc. On the other hand, for far-field applications, plume characteristics are much less sensitive to these factors. Higson et al. (1996) conducted field tracer gas experiments with a stack at varying distances upwind of a small building and found that the maximum concentrations were generally overestimated in the wind tunnel tests while the minimum concentrations were underestimated. This suggests that the wind tunnel plume was narrower than the field plume due to the absence of large-scale turbulence in the wind tunnel. Several studies at Concordia University have evaluated the accuracy of wind tunnel dispersion measurements (Stathopoulos et al., 2002, Saathoff et al., 2002) and found generally good agreement between wind tunnel and field data. The wind tunnel concentration values were usually within a factor of two of the field values. In general, the accuracy of the wind tunnel increases as stack-receptor distance increases.

Various models have been developed for estimating near-field dilution of plumes emitted from rooftop stacks for open fetch situations. Two such models are recommended in ASHRAE (1999) and ASHRAE (2003). For the case of a tall building upwind of an emitting building, dilution estimates are required for receptors on the adjacent building leeward wall, as well as the roof of the emitting building. To date, an acceptable dilution model for such cases has not been developed (e.g. see Wilson et al., 1998). In addition to dilution models that provide quantitative estimates of plume dispersion, ASHRAE (2003)also provides a geometric method to predict the likelihood of a plume making contact with a critical rooftop receptor. The accuracy of these models is evaluated using some of the field data obtained in the present study, in which the emission source was either exposed to the approaching flow (open fetch) or it was in the wake of a tall building. ASHRAE (2007) is easier to use due to a simplification in the definition of initial plume height.

Finally, one of the most widely used models is the Atmospheric Dispersion Modeling System (ADMS), developed in England, which is also endorsed by the United States Environmental Protection Agency (EPA). In a recent study to assess the quality and applicability of various dispersion models for near-field dispersion, ADMS was the only model producing good comparisons with ASHRAE and wind tunnel results, as discussed by Stathopoulos et al (2008).Acomprehensive comparative study of ADMS, ASHRAE (2003 and 2007 versions) and wind tunnel results,has been carried out by Hajra et al. (2010).

3.2 ASHRAE Dispersion Models

The ASHRAE geometric design method. ASHRAE (2003) provides a geometric stack design method for estimating the minimum stack height to avoid plume entrainment in the flow re-circulation zones of a building and its rooftop structures. Dimensions of the re-circulation zones are expressed in terms of scaling length, R, which is defined as:

$$R = B_s^{0.67} B_L^{0.33} \tag{2}$$

where B_s is the smaller of upwind building height or width and B_L is the larger of these dimensions. The dimensions of flow re-circulation zones that form on the building and rooftop structures are:

$$H_c = 0.22R \tag{3}$$

$$X_c = 0.5R \tag{4}$$

$$L_c = 0.9R \tag{5}$$

$$L_r = 1.0R \tag{6}$$

whereH_c is the maximum height of the roof re-circulation zone, X_c is the distance from the leading edge to H_c, L_c is the length of the roof re-circulation zone, and $L_r$is the length of the building wake zone. Note that the height of the wake zone is equal to the height of the structure. Figure 13 shows the re-circulation zones for a typical building.

The design method assumes that the boundary of the high turbulence region is defined by a line with a slope of 10:1 extending from the top of the leading edge separation bubble. The location of the plume relative to the re-circulation zones is determined by taking into account plume rise due to exhaust momentum and assuming a conical plume with a slope of 5:1.

It should be noted that the geometric design method is applicable for wind directions that are approximately normal to the windward wall of the building. This is the critical case for flat-roof buildings (without significant rooftop structures), since depending on the geometry of the building, the rooftop recirculation zone may encompass most of the roof (Saathoff et al., 2003). On the other hand, for oblique winds, building-generated turbulence is normally not significant in the central part of the roof (where stacks are usually located). For such winds, the building-generated turbulence is confined to the leading edges of the roof where the familiar delta-wing vortices are formed.

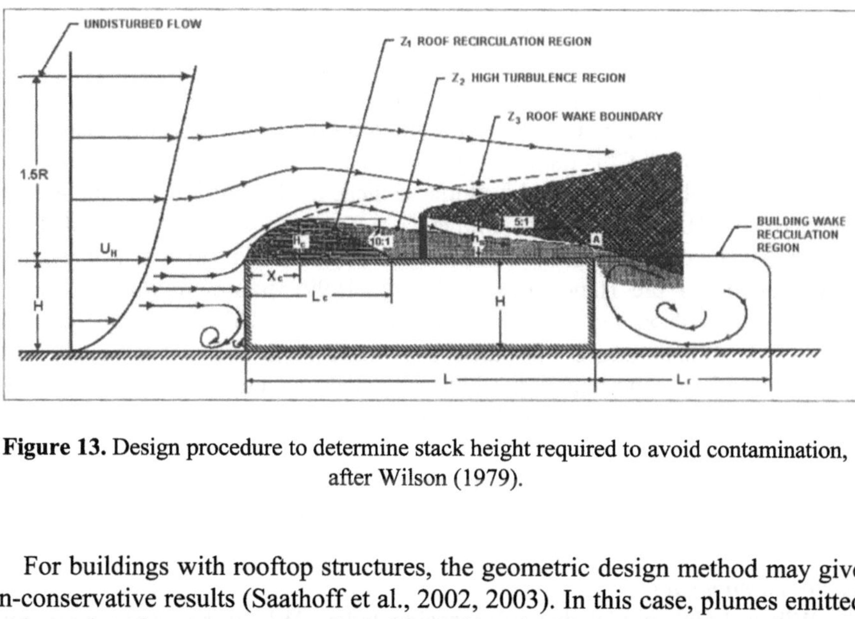

Figure 13. Design procedure to determine stack height required to avoid contamination, after Wilson (1979).

For buildings with rooftop structures, the geometric design method may give un-conservative results (Saathoff et al., 2002, 2003). In this case, plumes emitted in the wake of a structure may be subjected to significant downwash, which, in turn, will cause high roof-level concentrations. It is important to note that for buildings with rooftop structures, the oblique wind case may be critical since such winds tend to produce the strongest downwash. The effective height of the plume above the roof or rooftop structure is:

$$h = h_s + h_r - h_d \quad (7)$$

whereh_s is stack height, h_r is plume rise and h_d is the reduction in plume height due to entrainment into the stack wake during periods of strong winds. It should be noted that h_s is the height of the stack tip above the roof minus the height of rooftop obstacles (including their re-circulation zones) that are in the path of the plume.

Plume rise, which is assumed to occur instantaneously, is calculated using the formula of Briggs (1984):

$$h_r = 3\beta d_e\, (w_e/U_H) \tag{8}$$

where d_e is the stack diameter, w_e is the exhaust velocity, U_H is the wind speed at building height and β is the stack capping factor. The value of β is 1 for uncapped stacks and 0 for capped stacks. To account for the stack downwash caused by low exit velocities, when $w_e/U_H < 3.0$, Wilson et al. (1998) recommended a stack wake downwash adjustment h_d, which is defined as:

$$h_d = d_e\, (3.0 - \beta w_e/U_H) \tag{9}$$

For $w_e/U_H > 3.0$ there is no stack downwash ($h_d = 0$).

Dilution models for an open fetch. A number of semi-empirical models have been developed for predicting minimum dilution ($D_{min} = C_e/C_{max}$) of exhaust from rooftop stacks, where C_e is the exhaust concentration and C_{max} is the concentration at a roof or wall receptor on the plume centerline. The ASHRAE Applications Handbook (ASHRAE, 2003) recommends a Gaussian dilution model that was developed using data from water channel experiments of Wilson et al. (1998). In an earlier version of the Handbook (ASHRAE, 1999), minimum dilution models formulated by Wilson and Lamb (1994) and Halitsky (1963) were recommended. Of these, the Wilson and Lamb model provides a more accurate lower bound of dilution, based on wind tunnel and field case studies (Petersen and Wilson, 1989, Stathopoulos et al., 2002). The Halitsky model has been shown to be overly conservative in most cases.

The accuracy of the Wilson-Lamb and ASHRAE (2003) dilution models are evaluated using data obtained in field tests conducted with an open fetch.

Wilson-Lamb model

The Wilson-Lamb model, hereafter designated as WL, is based on a previous dilution model derived from wind tunnel data obtained with isolated building models (Wilson and Chui, 1985, 1987, Chui and Wilson, 1988). In this model, minimum dilution along the plume centerline is given by:

$$D_{min} = (D_o^{0.5} + D_d^{0.5})^2 \tag{10}$$

where D_o is the initial dilution at the location and D_d is the distance dilution which is produced by atmospheric and building-generated turbulence. The formulas for D_o and D_d recommended in ASHRAE (1999) are:

$$D_o = 1 + 13M \tag{11}$$

$$D_d = B_1 S^2 / MA_e \tag{12}$$

where B_1 is the distance dilution parameter, S is the stretched string distance from stack to receptor and M is the ratio of exhaust gas velocity, w_e, to the mean wind speed at the building height, U_H. The parameter, B_1, is set at a constant value with the magnitude dependent on the location of the receptors.

Dilution data obtained in a field study (Wilson and Lamb, 1994) and a wind tunnel study (Wilson and Chui, 1987) indicate that B_1 is strongly affected by the level of atmospheric turbulence in the approaching flow. The effect of upstream turbulence on the distance dilution parameter is approximated by the following formula:

$$B_1 = 0.027 + 0.0021\sigma_\theta \tag{13}$$

whereσ_θ is the standard deviation of wind direction fluctuations in degrees and varies between 0° and 30°. The model suggests that distance dilution has two components -- the dilution due to building-generated turbulence and that due to atmospheric turbulence. It assumes that D_d is significantly enhanced by atmospheric turbulence. For an urban environment, ASHRAE (1997) recommends a typical value of σ_θ =15°, which gives a value of 0.032 for the atmospheric component of the distance dilution parameter, B_1=0.059. Thus, more than 50% of D_d is assumed to be due to upstream turbulence.

Gaussian dilution model (ASHRAE 2003)

The Gaussian dilution model recommended in ASHRAE (2003) is based on a series of experiments carried out in a water flume by Wilson et al. (1998). The model predicts worst-case dilution at roof-level, D_r, assuming that the plume has a Gaussian (bell-shaped) concentration profile in both the vertical and horizontal directions, as shown in Figure 14. It should be noted that D_r is the predicted dilution on the plume centre-line and thus, corresponds to D_{min} obtained using the WL model.

The roof-level dilution for a plume at height, h, at a receptor distance, X, from the stack is given as:

$$D_r = 4 \frac{U_H}{w_e} \frac{\sigma_y}{d_e} \frac{\sigma_z}{d_e} \exp\left[\frac{h^2}{2\sigma_z^2}\right] \tag{14}$$

where U_H is the wind speed at the building height, d_e is stack diameter, w_e is the exhaust speed and σ_y and $\sigma_z$are the plume spreads in the horizontal and vertical directions, respectively. The height of plume above the roof or rooftop structure, h, is determined using Equation7.

The equations for σ_y and $\sigma_z$are used in the ISCST dispersion model, which was developed by EPA (1995), adjusted from a 60-minute averaging time to a 2-minute averaging time using the 0.2 power law applied to both vertical and crosswind spreads. The plume spread formulas are based on the water channel data of Wilson et al. (1998), which are assumed to have a full-scale equivalent averaging time of 2 minutes. The cross-wind and vertical spreads are given by the equations:

$$\frac{\sigma_y}{d_e} = 0.071\left(\frac{t_{avg}}{2.0}\right)^{0.2}\frac{X}{d_e} + \frac{\sigma_o}{d_e} \tag{15}$$

$$\frac{\sigma_z}{d_e} = 0.071\frac{X}{d_e} + \frac{\sigma_o}{d_e} \tag{16}$$

wheret_{avg} is the concentration averaging time in minutes, and σ_o is the initial source size that accounts for stack diameter and for dilution due to jet entrainment during plume rise.

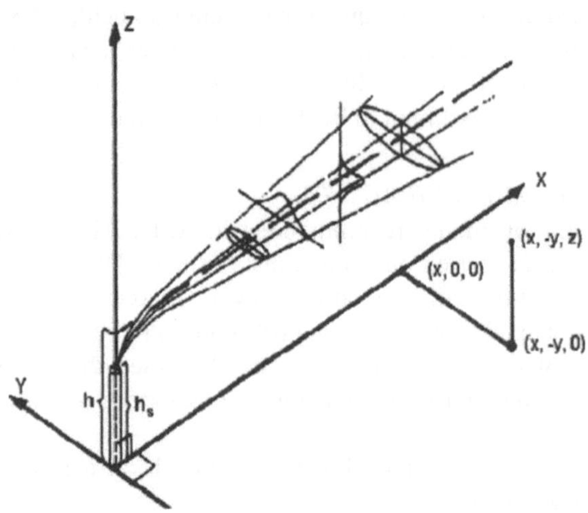

Figure 14. Coordinate system showing Gaussian distributions in the horizontal and vertical direction, after Turner (1994).

The formula for σ_o/d_e is:

$$\frac{\sigma_o}{d_e} = [0.125\beta \frac{w_e}{U_H} + 0.911\beta \left(\frac{w_e}{U_H}\right)^2 + 0.250]^{0.5} \tag{17}$$

The Gaussian dilution model (Equation14) should not be used when the plume height, h, is less than the maximum height of the roof recirculation zones that are in the path of the plume. This critical height is referred to as h_{top} and is shown in Figure 15. For cases in which the plume height is greater than h_{top} but less than the height required to escape all critical recirculation zones (h_{valid}in Figure 15), the physical stack height should be set at 0 when calculating h, as per ASHRAE (2003).

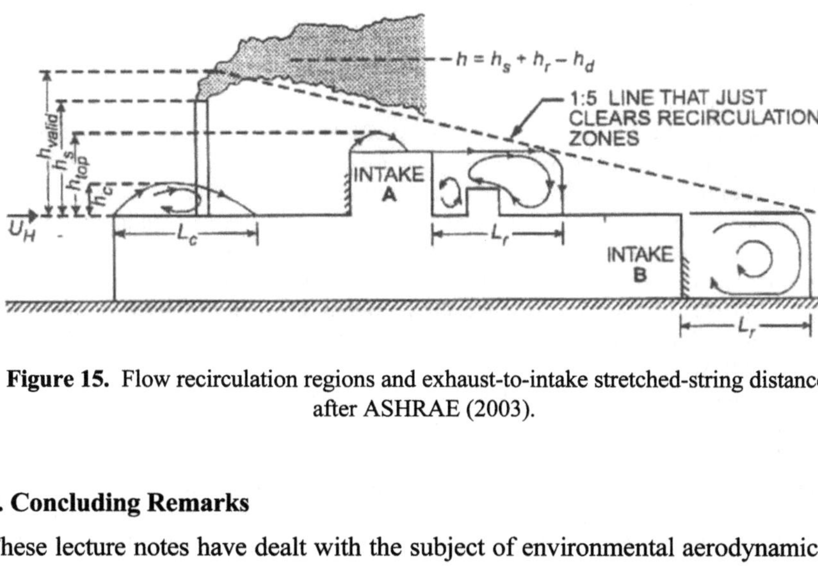

Figure 15. Flow recirculation regions and exhaust-to-intake stretched-string distances, after ASHRAE (2003).

4. Concluding Remarks

These lecture notes have dealt with the subject of environmental aerodynamics. In particular, pedestrian level wind conditions, their origin, their experimental and computational assessment in the urban environment, as well as the criteria used for outside human comfort have been discussed. Special attention has been paid on the state-of-the-art and the current capabilities of Computational Wind Engineering to determine at least mean values of wind speeds in the vicinity of buildings in urban areas. An approach towards the establishment of an overall comfort index taking into account, in addition to wind speed, the temperature and relative humidity in the urban area under consideration was presented.

The area of exhaust dispersion around buildings has also been introduced as part of environmental aerodynamics. Evaluatingairflow around buildings and the transportation ofpollutants released in that area is a very complex problem affecting building design and performance. The general accuracy of the wind tunnel data to represent actual design situations and the limitations of ASHRAE and other models to predict actual dilutions for particular building configurations and stack locations are discussed.

References

ASHRAE (1997). Chapter 15 - Airflow around buildings, *ASHRAE Handbook*. American Society of Heating, Refrig. and Air-Cond. Eng., Inc., Atlanta.

ASHRAE (1999). Chapter 43 - Building Air Intake and Exhaust Design. *ASHRAE Applications Handbook*, American Society of Heating, Refrig. and Air-Cond. Eng., Inc., Atlanta.

ASHRAE (2003). Chapter 44 - Building Air Intake and Exhaust Design. *ASHRAE Applications Handbook*, American Society of Heating, Refrig. and Air-Cond. Eng., Inc., Atlanta.

ASHRAE (2007). Chapter 44 - Building Air Intake and Exhaust Design. *ASHRAE Applications Handbook*, American Society of Heating, Refrig. and Air-Cond. Eng., Inc., Atlanta.

American Society of Civil Engineers - ASCE (1999). Wind tunnel studies of buildings and structures. *ASCE Manuals and Reports on Engineering Practice No. 67*, Aerospace Division of the American Society of Civil Engineers, Reston, VA.

American Society of Civil Engineers - ASCE. (2003). Aerodynamics Committee, Outdoor human comfort and its assessment. *State of the Art Report*, Task Committee on Outdoor Human Comfort, American Society of Civil Engineers, Reston, VA.

Briggs, G. (1984). Plume rise and buoyancy effects. *Atmospheric Science and Power Production*. Randerson Edition, U.S. Department of Energy, D.O.E./TIC-27601 (DE 84005177), Washington, DC.

Castro, I.P. and Graham, J.M.R. (1999). Numerical wind engineering: the way ahead?. *Proceedings Institute Civil Engineers. Structures & Buildings*, 134.

Chui, E.H. and Wilson, D.J. (1988). Effects of varying wind direction on exhaust gas dilution. *Journal of Wind Engineering and Industrial Aerodynamics*, 31, 87-104.

Cochran, L.S. (2004). Design Features to Change and/or Ameliorate Pedestrian Wind Conditions. *Proceedings of the ASCE Structures Congress*, Nashville, Tennessee, May 2004.

EPA (1995). A user guide for the Industrial Source Complex (ISC3) dispersion models: Volume III - Description of model algorithms. *EPA-454/B-95-003,* U.S. Environmental Protection Agency, Research Triangle Park, NC.

Georgakis, K., Smith, J., Goodfellow, H. and Pye, J. (1995). Review and evaluation of models estimating the minimum atmospheric dilution of gases exhausted near buildings. *Journal of the Air & Waste Management Assoc.*, 45, 722-729.

Hajra, B., Stathopoulos, T. and Bahloul, A. (2010). Assessment of Pollutant Dispersion from Rooftop Stacks: ASHRAE, ADMS and Wind Tunnel Simulation. *Building and Environment*, 45, 2768-2777.

Halitsky, J. (1963). Gas diffusion near buildings. *ASHRAE Transactions*. 69, 464-484.

Higson, H.L., Griffiths, R.F., Jones, C.D. and Hall, D.J. (1996). Flow and dispersion around an isolated building. *Atmospheric Environment*, 30 (16), 2859-2870.

Hoppe, P. (2002). Different aspects of assessing indoor and outdoor thermal comfort. *Energy and Buildings*, 34, 661-665.

Isyumov, N. and Davenport A.G. (1975). The ground level wind environment in built-up areas. *4th International Conf. on Wind Effects on Buildings and Structures*, London.

Lawson, T.V. and Penwarden, A.D. (1975). Effect of wind on people in the vicinity of buildings. *in Proc. 4th Int'l Conf. on Wind Effects on Buildings and Structures*, Heathrow, UK, 605-622.

Melbourne, W.H. (1978). Criteria for environmental wind conditions. *Journal of Wind Engineering and Industrial Aerodynamics*, 3, 241-249.

Murakami, S., Iwasa, Y. and Morikawa, Y. (1986). Study on Acceptable Criteria for Assessing Wind Environment on Ground Level Based on Residents' diaries. *Journal of Wind Engineering and Industrial Aerodynamics*, 24, 1-18.

Nicolopoulou, M., Baker, N. and Steemers, K. (2001). Thermal comfort in outdoor urban spaces: understanding the human parameter. *Solar Energy*, 70 (3), 227-235.

Nicolopoulou, M. and Lykoudis, S. (2002). Thermal comfort in open spaces: the human factor. *6th Panhellenic Congress of Meteorology, Climatologic and Atmospheric Physics*, September 25-28, Ioannina, Greece.

Pendwarden, A.D. (1973). Acceptable wind speeds in towns. *Building Science,* 8, 259-267.

Pendwarden, A.D. and Wise, A.F.E. (1975). Wind environment around buildings. *Building Research Establishment Digest.*

Petersen, R. and Wilson, D.J. (1989). Analytical versus wind tunnel determined concentrations due to laboratory exhaust. *ASHRAE Transactions,* 95 (2), 729-736.

Saathoff, P., Lazure, L., Stathopoulos, T. and Peperkamp, H. (2002). The influence of a rooftop structure on the dispersion of exhaust from a rooftop stack. *ASHRAE Transactions*, 108 (2), 1029-1038.

Saathoff P., Gupta A., Stathopoulos T. and Lazure L. (2003). Effect of rooftop structures on the plume from a nearby stack. *11th International Conference on Wind Engineering,* June 2-5, Lubbock, Texas.

Schulman and Scire, J. (1991). The effect of stack height, exhaust speed, and wind direction on concentrations from a rooftop stacks. *ASHRAE Transactions*, 97 (2), 573-582.

Stathopoulos, T. (2002). The numerical wind tunnel for industrial aerodynamics: real or virtual in the new millennium?, *Wind and Structures*, 5 (2-4), 193-208.

Stathopoulos, T., Lazure, L., Saathoff, P. and Wei, X. (2002). Dilution of exhaust from a rooftop stack on a cubical building in an urban environment. *Atmospheric Environment*, 36 (29), 4577-4591.

Stathopoulos, T., Wu, H. and Zacharias, J. (2004). Outdoor human comfort in an urban climate, *Building and Environment*, 39 (3), 297-305.

Stathopoulos, T. and Baniotopoulos, C.C. (2007). Wind effects on buildings and design of wind-sensitive structures. *CISM Courses and Lectures*, No. 493, International Centre for Mechanical Sciences.

Stathopoulos, T., Bahloul, A. and Hajra, B. (2008). Analytical evaluation of dispersion of exhaust from rooftop stacks on buildings. *IRSST research report R-576*, Institut de recherche Robert-Sauvé en santé et en sécurité du travail, Montreal, Canada.

Turner, D.B. (1994). *Workbook of Atmospheric Dispersion Estimates*. 2nd Ed., CRC Press.

Westbury, P.S., Miles, S. D. and Stathopoulos, T. (2002). CFD application on the evaluation of pedestrian-level winds. Workshop on Impact of Wind and Storm on City Life and Built Environment, *Cost Action C14*, CSTB, June 3-4, Nantes, France.

Wilson, D. (1979). Flow patterns over flat-roofed buildings and application to exhaust stack design. *ASHRAE Transactions*, 85 (2), 284-295.

Wilson D.J. and Chui E.H. (1985). Influence of exhaust velocity and wind incidence angle on dilution from roof vents. *ASHRAE Transactions*, 91 (2B), 1693-1706.

Wilson D.J. and Chui E.H. (1987). Effect of turbulence from upwind buildings on dilution of exhaust gases. *ASHRAE Transactions*, 93 (2), 2186-2197.

Wilson, D.J. and Lamb, B. (1994). Dispersion of exhaust gases from roof level stacks and vents on a laboratory building. *Atmospheric Environment*, 28, 3099-3111.

Wilson, D.J., Fabris, I., Chen, J. and Ackerman, M. (1998). Adjacent building effects on laboratory fume hood exhaust stack design. *ASHRAE Research Report 897*, Atlanta, GA.

Wise, A. F. E. (1970). Wind effects due to groups of buildings. *BRS Current Paper CP 23/70*, Royal Society Symposium 'Architectural Aerodynamics'.

Zacharias, J., Stathopoulos, T. and Wu, H. (2001). Microclimate and downtown open Space activity. *Environment and Behavior*, 33 (2), 296-315.

Applications of Environmental Aerodynamics

Theodore Stathopoulos

Department of Building, Civil and Environmental Engineering, Concordia University, Montreal, Canada

Abstract

Applications of environmental aerodynamics, namely pedestrian level winds and dispersion of pollutants in the building environment, are discussed with particular emphasis to suitable engineering design issues and practices. Both experimental and computational approaches are presented with reference to particular case studies.

1. Pedestrian Level Winds

1.1 Experimental procedure

The case illustrated is an actual study in Montreal but names have been withheld and sample results presented are coming from another unnamed study for reasons of confidentiality. The study was conducted in the Building Aerodynamics Laboratory of Concordia University using a 1:500 model of the buildings and their surroundings. Wind environmental conditions at the sidewalks around the new buildings were assessed in terms of peak (gust) and mean speeds.

The following configurations were considered in the study:

1. Current conditions
2. Proposed development

The architects of the project provided the details of the design and the building models were placed on a maquette with the surroundings on a full-scale equivalent radius of 300 m. Environmental wind conditions on sidewalks around the buildings were assessed for the critical south-westerly and north-easterly wind directions. Measurements were obtained at the locations shown in Figures 1 and 2.

The 1:500-scaled model of the proposed development and the surroundings was mounted on the turntable of the boundary layer wind tunnel. Peak wind velocities were measured at 31 ground-level locations but results for only 6 points are shown herein. Five of the sampling points (3, 5, 6, 7 and 8) were located on S street and one point (15) was located in the courtyard. Pedestrian wind data were obtained with a hot film sensor located at a full-scale height of 2 m above ground.

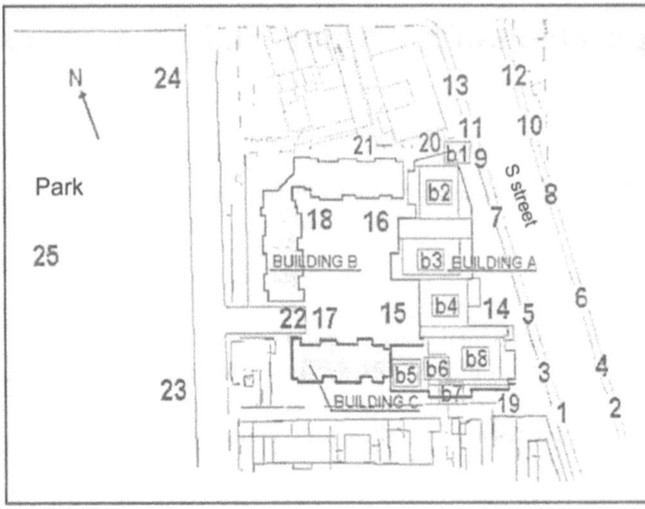

Figure 1. Plan view of proposed development showing measurement locations.

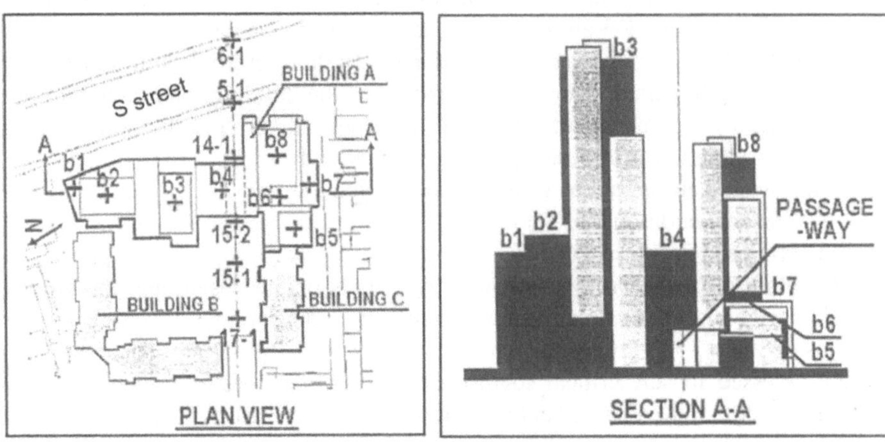

Figure 2. Plan and elevation view of proposed development showing measurement locations near and on buildings.

The reference mean velocity in the wind tunnel at a full-scale equivalent height of 300 m, U_{300}, was also measured. The ratio $\hat{U}_{local}/U_{300}$ was determined for each wind direction, where $\hat{U}_{local}$ is the gust velocity at ground level. The test results were then combined with meteorological data recorded at the P.E. Trudeau (Dorval) International Airport for a 10-year period in order to predict wind speeds with a particular return period for each test location. The Trudeau data, consisting of mean hourly wind speeds obtained in open country terrain at a height of 10 m, have been used to derive probability distributions of wind speed and direction for Montreal. These wind speeds were adjusted to an urban exposure at a height of 300 m using the following formulas:

$$U_{300,open\ country} = U_{10,\ Dorval}\,(300/10)^{0.15} \tag{1}$$

and

$$U_{300,urban} = U_{300,open\ country}(300/450)^{0.34} \tag{2}$$

since $U_{450,urban} = U_{300,open\ country}$. Note that the exposure category characterizes the terrain several kilometers upwind of the site.

Probability distributions of wind speed and direction for a suburban exposure at a height of 300 m have been evaluated for summer and winter, respectively. These distributions, which are based on data for daylight hours (07:00 – 19:00), indicate that strong winds occur most often from W, WSW and SW directions and occur more frequently during the winter than in the summer. Data also show that the probability of strong winds from the NE is relatively high. The probability of exceedance [P(>)] corresponding to once per month and once per year are approximately 10^{-2} and 10^{-3}, respectively. For instance, a westerly wind at a height of 300 m is expected to exceed 48 km/h approximately once per month during the winter. Although the proposed development will be highly exposed to winds from the east quadrant, the probability of strong winds from the east is relatively low. It should be noted that the influence of Mount Royal on Montreal's wind climate is not evident in the meteorological data used, since the probability curves were derived from data obtained at Trudeau (Dorval) Airport. It is anticipated that the presence of Mount Royal should reduce the probability of strong winds at the site for westerly and south-westerly winds. Therefore, the results presented in this study are expected to be generally conservative.

Melbourne's criteria shown in Table 1 have been used for the characterization of wind comfort conditions at the pedestrian level. The peak wind speed acceptable for walking was reduced to 32 km/h for the winter months to take into account the effect of temperature on pedestrian comfort. As suggested in ASCE (2003), the summer wind speed should be reduced by one Beaufort number for every 20°C reduction in temperature. The second criterion indicates that wind conditions are hazardous if the gust velocity exceeds 83 km/h more than once per year. At this wind speed, people can be blown over.

Table 1. Melbourne's pedestrian comfort criteria, after Melbourne 1978).

Wind Condition	$\hat{U}_{local}$	$P(>\hat{U}_{local})$
Acceptable for walking:		
summer	48 kph	0.01 (once per month)
winter	32 kph	0.01 (once per month)
Hazardous	83 kph	0.001 (once per year)

Sample test results. Data obtained under the current conditions and with the proposed development are plotted in polar form in Figure 3 for a typical set of four measurement points. As suggested in Westbury et al. (2002), the data are plotted in terms of the velocity pressure ratio $(\hat{U}_{local}/U_{300})^2$. Melbourne's wind environment criteria for winter conditions have been included on the diagrams, in terms of curves determined by incorporating the probability distributions. These curves show the limiting wind pressure values for walking comfort and for extreme hazard.

Figure 3 shows whether the proposed development does adversely affect the wind climate at different locations. For instance, the building causes an increase in gust speed at locations 6 and 8 for westerly and south-westerly winds, and at location 6 for north-easterly winds. For these cases, Melbourne's comfort criteria were exceeded. However, it should be noted that the proposed development will improve wind conditions at all points for at least some wind directions. It can also be noted that at location 7, wind conditions were improved for all of the tested wind directions.

The results of the study indicate that the proposed buildings will have relatively little adverse effect on the pedestrian wind environment. Winter winds, which are presented in Table 2, are more critical than those occurring during the summer. For the proposed development, the maximum wind gust expected to occur once per month in the winter is approximately 43 km/h. Under current conditions (i.e. without the new building), the peak monthly gusts in the winter are approximately 33 km/h at the same location.

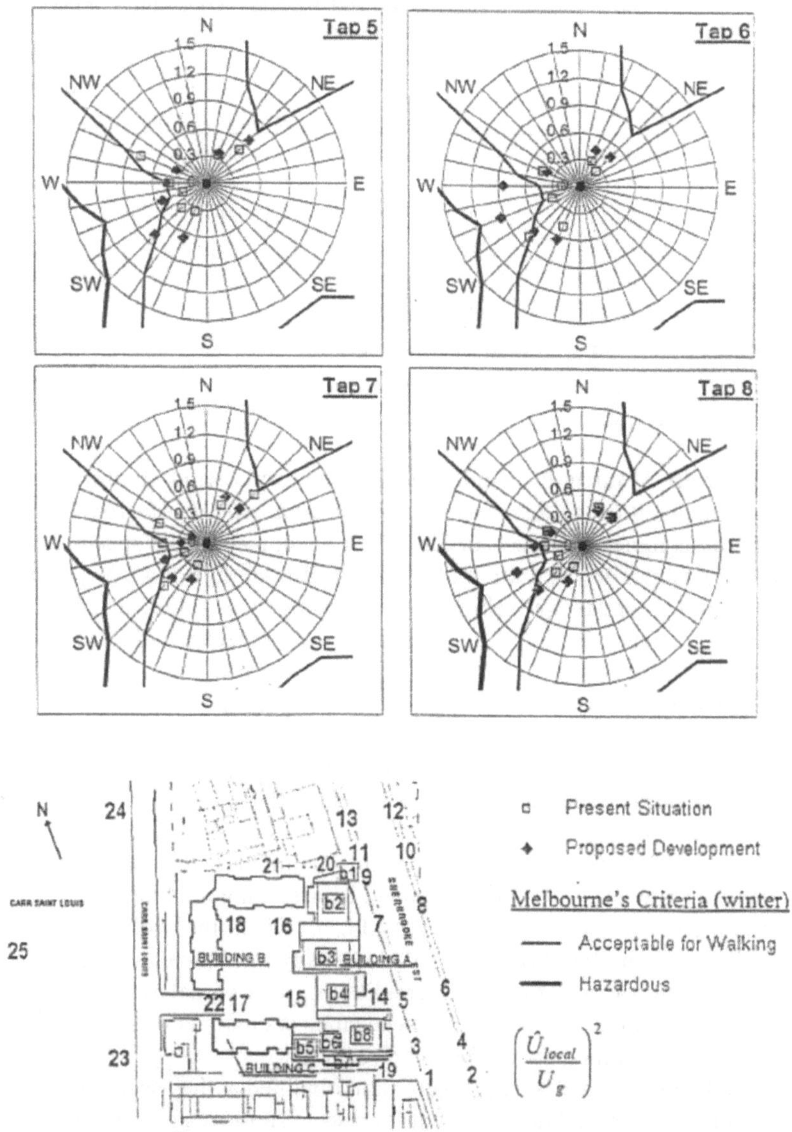

Figure 3. Measurement of velocity pressure ratio at locations 5, 6, 7 and 8.

Table 2. Comparison with Montreal's wind criteria and expected probabilities of exceeding the Montreal winter criterion (U_{local}>14.4 km/h) at location 6.

Wind direction	$(\hat{U}_{local}/U_{300})^2$	$\hat{U}_{local}/U_{300}$	U_{local}/U_{300}	U_{300} (kph) at U_{local}=14.4 kph	Probability U_{local}>14.4 kph
NNE	0.421	0.649	0.32	45.0	0.0016
NE	0.456	0.675	0.34	42.4	0.0040
SSW	0.643	0.802	0.40	36.0	0.0008
SW	0.701	0.837	0.42	34.3	0.00095
WSW	0.935	0.967	0.48	30.0	0.0360
W	0.834	0.913	0.46	31.3	0.0300
WNW	0.381	0.617	0.31	46.3	0.0018
Other directions					0.0100
				Total	0.0937

Comparison with Montreal's wind criteria. The city's wind comfort criteria, specified in Article 39 of the *Règlements refondus de la ville de Montréal,* refer to mean wind speeds rather than gust speeds. The critical mean wind speeds, U_{local}, for winter and summer are 14.4 km/h (4 m/s) and 21.6 km/h (6 m/s), respectively, and the maximum acceptable probabilities of exceeding these values are as follows:

Location	Probability of exceedance of Ulocal
Main Streets	15%
Secondary Streets	25%
Parks	10%

As previously mentioned, the wind tunnel data are expressed in terms of the local peak velocity pressure ratio, which is given as $(\hat{U}_{local}/U_{300})^2$, where $\hat{U}_{local}$ is the local gust velocity and U_{300} is the mean wind speed measured at the reference full-scale height of 300 m for an urban exposure. Using these data, the local mean velocity pressure ratio $(U_{local}/U_{300})^2$ is obtained by assuming that the peak velocity is two times the mean value, U_{local} (see Art. 38.5). Given the probability of exceedance of U_{300}, the probability of exceedance of any local mean wind speed - in this case, 14.4 km/h (4 m/s) for winter conditions - can be determined for each wind direction. The total probability of exceedance is obtained by simply adding the probabilities determined for each wind direction.

For example, the gust velocity pressure ratio obtained at the assumed windiest location 6 with the proposed development for a west-southwesterly wind is 0.935 (see Figure 3). Taking the square root and dividing by 2 gives the local mean velocity ratio of U_{local}/U_{300} = 0.48. Setting U_{local} equal to 14.4 km/h gives the critical reference velocity of U_{300} = 30 km/h for this wind direction. Based on wind speed data from Trudeau (Dorval) Airport, the probability of exceedance of this wind speed during daylight hours (07:00-19:00) in the winter months (November-April) is approximately 0.036 (3.6%). Following this procedure for the other tested wind directions gives a total probability of occurrence of 0.0837 (8.37%). For the wind directions not tested, the probability of U_{local} >14.4 km/h is expected to be very small – of the order of 1%. Thus, the total probability of U_{local} >14.4 km/h at location 6 is approximately 9.37%, as shown in Table 2. Consequently, the Montreal winter criterion for wind comfort on main streets (U_{local} >14.4 km/h less than 15 % of the time) is satisfied at location 6.

Furthermore, since location 6 is the windiest point in the project area, the Montreal standard is satisfied at all other locations and for all configurations tested in the study. This includes locations in an adjacent park, tested previously, even though the criterion is stricter for park areas (U_{local} >14.4 km/h less than 10 % of the time).

The preceding example illustrates that strong winds at ground level are not likely to occur for a given wind direction unless $(\hat{U}_{local}/U_{300})^2$ is large and the probability of strong winds from that direction is also large.

Regarding the wind comfort criterion for the summer months, analysis of the data shows that the criterion will also be met at all locations. Due to the larger acceptable wind speed (21.6 km/h) and the reduction in the frequency of high winds for the summer months, the probability of exceeding the summer criterion is less than that for the winter criterion at each location.

With respect to the city's criterion concerning hazardous wind conditions (Art. 40, *Règlements refondus de la ville de Montréal*), the wind conditions around the proposed development were found to be acceptable. The wind tunnel results show that predicted gust speeds with a probability of exceedance of 0.01 (once per month) are well below the criterion of 72 km/h (20 m/s) at all measurement locations. For all design options, the maximum monthly gust in the winter is expected to be approximately 43 km/h at locations 6 (on S street) and 15 (on passageway).

Finally, it should be recalled that the anticipated wind speeds are based on statistical expectations and actual wind conditions during a particular storm may be different. Future building developments in the surrounding area may also affect the pedestrian wind environment, but this has not been considered in the present study.

1.2 Computational procedure: CFD

Case studies. Bottema (1993) has attempted the evaluation of pedestrian level wind conditions in the vicinity of an isolated building by using the CWE approach and a simple turbulence model but with only limited success. Studies published by Stathopoulos and Baskaran (1996) have demonstrated that by using a simple version of k-ε model one may obtain computational results very comparable with corresponding values originating from respective wind tunnel studies. Figure 4 shows the cluster of Buildings A, B, C, D, E and F around the building of interest X. The non-uniform grid distribution used for this arrangement contains nearly 300,000 nodes. In spite of inadequacies in the vector plot of computed velocity field around this group of buildings, the comparison of computed and measured velocity ratios at 2 m above the ground level in the presence and absence of building X is generally satisfactory with a maximum discrepancy of the order of 30%. It is noteworthy that the maximum discrepancies between the experimental and numerical data appear in highly complex re-circulating flow regions (Locations 3 and 8 in Figure 4), for which neither the measured nor the computed values can be considered very accurate.

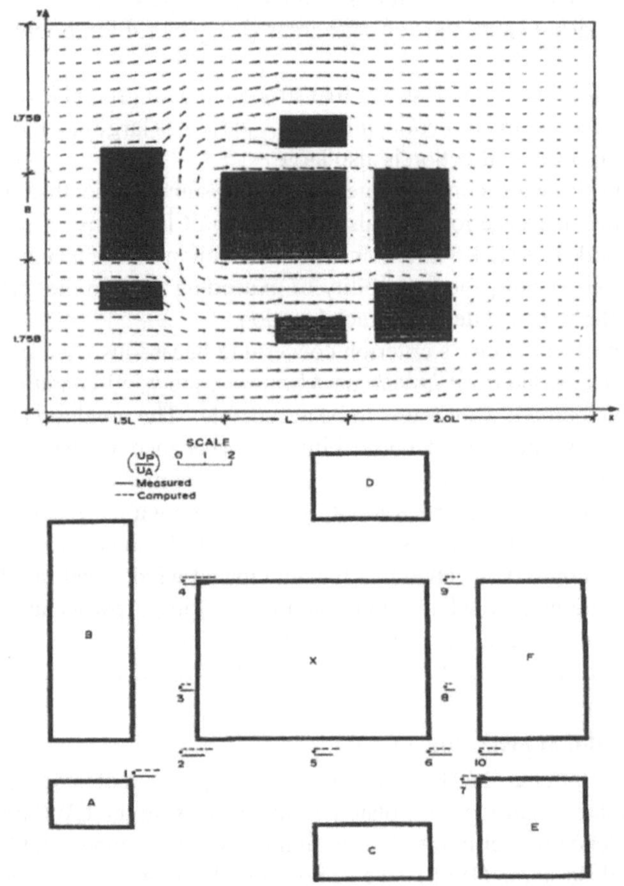

Figure 4. Vector plot of computed velocity field around a group of buildings and comparison of computed and measured velocity ratios at 2 m from ground level, after Stathopoulos and Baskaran (1996).

This type of data response has been found and confirmed by several authors. For instance, Timofeyef (1998) evaluated the wind flow around a five-storey high development in Kazakhstan and produced full-scale results, wind tunnel data and numerical results by using the discrete vortex method (two-dimensional flow). Surprisingly enough, computational results compare better with corresponding full-scale data than the latter with wind-tunnel results! This means that this rather crude computational approach provides more representative results than wind tunnel testing, at least for this particular case. In a cooperative study between Portugal and Canada, Ferreira et al. (1999) produced wind flow around a group of low-rise buildings (Expo' 98 – Lisbon). Both wind tunnel and field data were compared with numerical results obtained with the standard formulation and the Re-Normalization Group (RNG) extension of the k-ε turbulence model. By and large the comparisons are satisfactory, at least for engineering design purposes.

Hu and Wang (2005) have also attempted the evaluation of street-level winds in a built-up area by using CFD. Computational velocity ratios obtained through the commercially available code "PHOENICS" agree reasonably well with the experimental results for surrounding blocks of similar height. However, the comparison results deteriorate when the surrounding blocks consist of buildings with different heights. Such trends are possibly due to the more significant interaction of horizontal separated flow with the down-flow originating from the pressure difference at different heights.

Wind tunnel and CFD data have been compared for a complex development in central London, which includes a 40 m high apartment block adjacent to an 80 m tower block, and a number of other buildings – see Miles and Westbury (2002). This study forms part of a research program at BRE, which aims to assess whether CFD is 'fit for purpose' for use during the design process, and to determine the major sources of inaccuracy associated with user's modeling decisions.

CFD simulations were performed using CFX-5, a general-purpose RANS solver employing an unstructured mesh of tetrahedral and prism elements. A practical, but reasonably fine, numerical mesh was employed, containing approximately 1.2 million elements. Each simulation required about six hours to converge on a dual-Pentium desktop computer with 1 GB of RAM, representing a realistic computing effort for a practical application. The simulations were carried out at a scale of 1:200 in order to be consistent with the BRE environmental wind tunnel scale. Both mean and gust wind speed measurements were carried out at an equivalent full-scale height of 1.2 m but only mean wind speeds were used for comparison with the CFD results. Figure 5 presents the physical model and the wind directions for which simulations were carried out. Figure 6 shows the locations where comparisons of pedestrian-level wind speeds were made.

The inlet mean velocity profile measured in the wind tunnel, representing an urban boundary layer, has been used in the CFD simulation. Turbulence kinetic energy and dissipation rate profiles have been derived using boundary layer parameters obtained from fitting the logarithmic law to the measured velocity profile, combined with expressions from Richards and Hoxey (1992). Wall boundary conditions were defined using a rough-wall turbulent wall function applied at the ground, and a smooth-wall function applied to the building envelopes. A number of turbulence models have been tested – see details in Westbury et al. (2002).

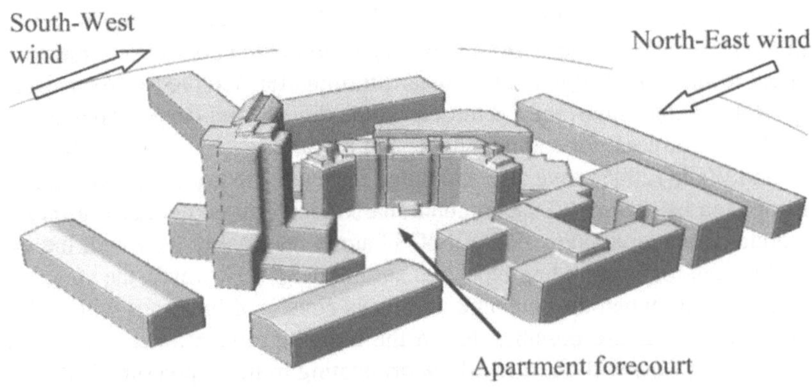

Figure 5. Building geometry in CFD model, after Westbury et al. (2002).

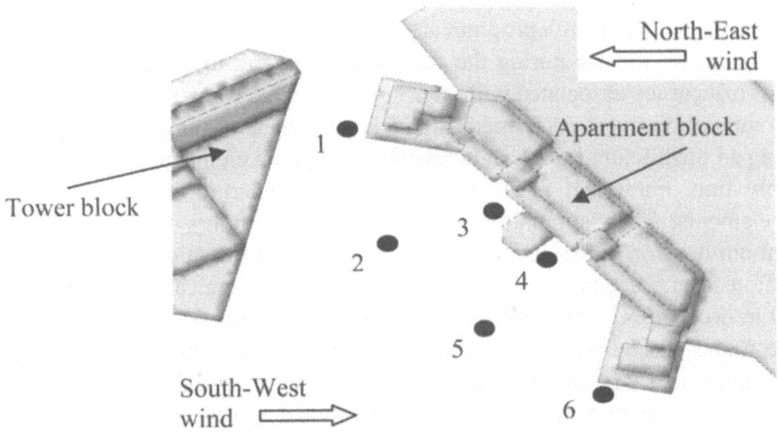

Figure 6. CFD model showing measurement locations, after Westbury et al. (2002).

The results in Table 3 show comparisons between wind tunnel and CFD wind speed predictions for the south-west and north-east winds, respectively. Mean wind speeds from the CFD simulation have been extracted from what was considered to be the nearest cell to the wind tunnel measurement location. The error bands reflect the variation in wind speed in the adjacent cells, and are shown due to the possibility that wind tunnel and CFD locations are not exactly coincident. Table 3 indicates that the discrepancies of the results are of a similar order of magnitude for the two wind directions studied. The maximum discrepancy occurs for the north-east wind direction, and is greater than 200% (Location 2). For both wind directions, the discrepancies are within 50% for 3 out of the 6 locations for which comparisons were made. The highest discrepancies tend to occur within the courtyard area (Locations 2, 3, 4 and 5) while the lowest discrepancies occur close to building corners (Locations 1 and 6). It is difficult at this stage to distinguish whether the discrepancies in the courtyard area are due to the limitations of CFD in representing complex wind flow patterns such as re-circulating regions, or are due to measurement errors associated with the use of hot-wire anemometers in highly turbulent and complex flows.

Table 3. Mean pedestrian wind speeds with south-west and north-east wind

Location	South-west wind speed (m/s)		North-east wind speed (m/s)	
	Wind tunnel	CFD	Wind tunnel	CFD
1	1.5	2.75 +/- 0.5	1.9	1.00 +/- 0.5
2	1.9	3.25 +/- 0.5	1.2	3.75 +/- 0.5
3	1.5	1.50 +/- 0.5	0.6	0.50 +/- 0.5
4	0.7	2.00 +/- 0.5	1.0	1.75 +/- 0.5
5	2.1	2.50 +/- 0.5	2.0	3.75 +/- 0.5
6	1.4	1.75 +/- 0.5	1.6	1.50 +/- 0.5

The sensitivity of the results to the turbulence model used, the geometric complexity of the model, the flow angle and the boundary condition specifications has been investigated. The choice of turbulence model was found to have little effect on the results. Also, the inclusion of finer geometrical details did not lead to any significant variation of the results. Simulations were run at angles of 15° measured clockwise and anti-clockwise from north-east, in order to test the sensitivity of the results to flow direction. The results were found to be very similar for the three wind directions, indicating that the discrepancy between prediction and measurement was not a result of the fluid flow structure being sensitive to the precise wind direction.

Finally, it is worth mentioning an intensive Japanese effort to compare CWE results in terms of pedestrian level wind speeds with respective experimentally measured values in an actual environment – see Figure 7, taken from Tominaga et al. (2004) Although CFD simulations have been performed for the different sixteen wind directions, only the wind distributions for wind directions of NNE and W, which are the prevailing wind directions in the region are shown. Various computational grids (structured, Code M; unstructured, Code T; and overlapping structured, Code O) have been used. No clear differences among the horizontal distributions of wind speeds near ground surface (z = 2 m) were observed, as shown in Figure 8, in which results are presented in terms of speed ratios normalized by their respective values at the same height at inflow boundary. High-speed regions appear in the area around the corner of the north side and east side of building A and the strong wind blows into the space between buildings with the wind direction from NNE. On the other hand, a high-speed region is observed in the area around the corner of the south side of building A with the wind direction from W. The wind speed values in the street in the case of NNE wind direction are smaller than those in the case of W wind direction. Figure 8 shows the correlation between the normalized speed values obtained by each code and those of wind tunnel experiment. Black circles indicate the speeds of the measurement points in the wake region. A similar tendency is observed in all results shown in Figure 8, i.e. the wind speed predicted by all CFD codes tested here tends to be smaller in the wake region compared to the experimental value. Excluding the speeds determined in the wake region, the correspondence between the CFD results and the experimental results is fairly good.

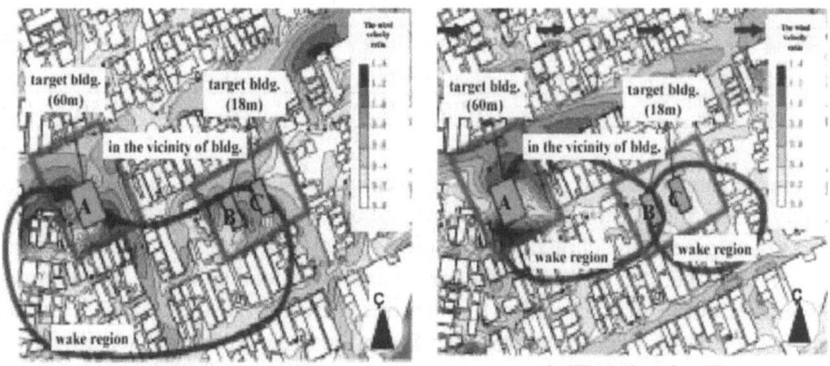

(1) Wind direction:NNE (2) Wind direction:W

Figure 7. Distributions of normalized scalar velocity near ground surface (z = 2 m) (code T), after Tominaga et al. (2004).

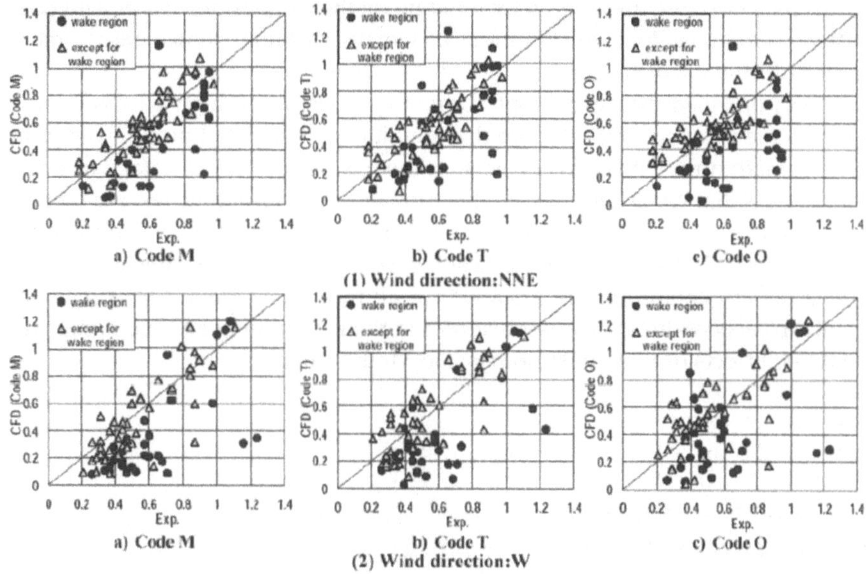

Figure 8. The correlation between the normalized velocity values predicted by each code and wind tunnel experiment, after Tominaga et al. (2004).

2. Exhaust Dispersion around Buildings

2.1 Experimental case study

General. Full-scale tracer gas experiments were performed on the roof of a 3-storey building (BE) in downtown Montreal. Rows of buildings of similar height are connected to the BE building on its northeast and southeast sides. A detailed drawing of the roof of the BE building is shown in Figure 9. The height of the main roof is 12.5 m. Several small structures are located on the roof and these vary in height from 2.2 m to 4 m. Four stack locations (SL1-SL4) were used in the study. The most significant nearby structure is the 12-storey Faubourg building, which is 50m in height and is located across the street on the southwest side – see Figure 10. A number of high-rise apartment buildings are located to the west and northwest of the BE building at distances between 100 and 300 m. Mount Royal is a 233 m tall hill whose summit is located approximately 1 km northwest of the BE building. The wind climate in the vicinity of the BE building is similar to that of Pierre Elliot Trudeau (Dorval) Airport, located 20 km west of Montreal. The predominant wind directions are west-southwest and northeast. In the city center, the frequency and magnitude of westerly winds is reduced due to the sheltering effect of Mount

Royal. Data obtained on a 14-storey building on the campus of McGill University, located approximately 1 km northeast of the BE building, shows clearly the influence of Mount Royal.

It was intended that tests be performed in moderate to high winds to ensure neutral atmospheric stability, since this condition is simulated in the wind tunnel and assumed in the ASHRAE dispersion models. A further advantage to neutral conditions is that variations in wind direction during a 50-minute test are typically small, compared to those obtained in stable or unstable conditions. Although low wind speeds (< 2 m/s) were measured at the BE building during some tests, neutral or near neutral conditions are assumed for all tests since the airport wind speed usually exceeded 4 m/s. Low wind speeds at the BE building are due primarily to sheltering effects of upwind buildings and Mount Royal.

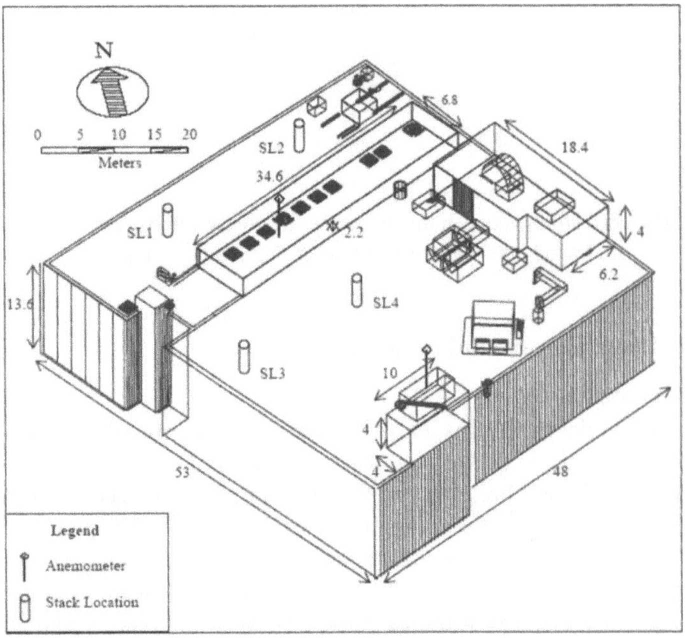

Figure 9. Detailed view of the BE building showing stack locations, anemometers and various rooftop structures (dimensions in meters).

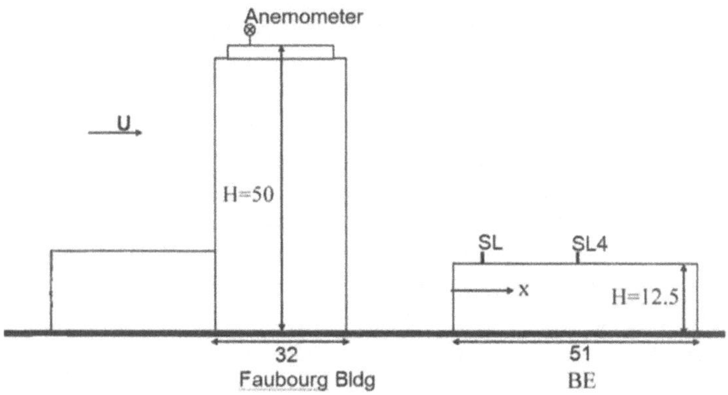

Figure 10. Elevation view of BE and Faubourg buildings (dimensions in meters).

Results of field and wind tunnel experiments are organized separately for

- tests performed with an open fetch (stack locations 1, 2 and 4); and
- tests performed with a tall upwind adjacent building (stack locations 3 and 4).

The former are suitable for comparison with the Gaussian dispersion models recommended in ASHRAE (2003) since the plume traveled downwind. Receptor concentrations in these tests are largely dependent on their distance from the source. In contrast, in the latter tests the plume was trapped in the near-wake region of the upwind building and consequently, travelled upwind initially before being dispersed by the wake turbulence. Due to the complexity of the flow in building wakes, Gaussian dispersion models are not applicable for the prediction of dilution profiles on the roof of the emitting building and leeward wall of the adjacent building. A complex Gaussian wake model developed by Wilson et al (1998) had relatively poor predictive capability but it was successful in demonstrating the trend of dilution associated with changes in M. The field results presented can be compared to the available dispersion models and with wind tunnel results. The concentration data will be expressed in terms of normalized concentration (k) as generally used in the scientific literature (e.g. Snyder, 1994) or in terms of dilution (D), as used in the ASHRAE (1999), ASHRAE (2003) and ASHRAE (2007) dispersion models.

A total of thirteen 50-minute tests were performed with open fetch conditions. Table 4 provides meteorological and stack parameters for the selected tests presented here. The wind direction was generally southwesterly ($228° < \theta < 260°$) for stack 1 (SL1), northwesterly ($280° < \theta < 355°$) for stack 2 (SL2) and southeasterly

($\theta \sim 160^{o}$) for stack 4 (SL4). In each case, it is assumed that upwind buildings do not significantly influence the plume trajectory but may enhance plume spread.

Table 4. Wind and stack data for selected open fetch tests

Date	Test No.	Stack Location	Stack Height (m)	M	U_H (m/s)	θ (deg)
10/12/00	1	SL1	1	5.4	3.3	240
	2	SL1	1	2.5	3.0	242
11/15/00	1	SL1	1	6.3	2.8	248
	2	SL1	3	7.5	2.4	252
05/15/02	1	SL1	3	2.5	3.0	267
10/30/01	1	SL2	1	3.4	2.3	305
	2	SL2	3	3.6	2.1	316
11/21/02	1	SL4	1	5.6	1.5	160
	2	SL4	1	10.5	1.7	160

Comparison of field dilutions with ASHRAE model predictions. ASHRAE provides a number of models for the prediction of plume dilution at rooftop and wall receptors. In ASHRAE (1999), formulas are given for minimum dilution (D_{min}) for flush stacks and critical dilution (D_{crit}) for stacks with $h_s > 0$. ASHRAE (2003) recommends another dilution formula, based on the Gaussian dispersion model, to predict roof-level dilution, D_r. As in the previous models, the ASHRAE (2003) model estimates dilution on the plume centre-line. Descriptions of the ASHRAE dilution models have been provided in a previous section (see "Dilution models for an open fetch"). Further details and assumptions used in this study are given below.

D_r ASHRAE (2003)

Assumption No. 1: stack height, h_s=0

In applying the D_r model (Eqs. 15-18), the final rise plume height h, must be specified (Equation 7 in Chapter 1). According to ASHRAE (2003), proper stack design suggests that h should be greater than the smallest height required to avoid a critical receptor, assuming a 5:1 slope of the plume (see Figure 15 in Chapter 1). This smallest height is referred to as h_{small}. ASHRAE (2003) specifies that "*If the plume height is less* than h_{small} *but higher than any rooftop obstacle or recirculation zone* (h_{top} - see Figure 15 – Chapter 1*), then only the physical height above* h_{top} *should be used to compute plume height rather than the full physical stack* *height.* " If the plume height does not reach h_{top}, ASHRAE (2003) recommends the use of another dilution model (D_s), which does not consider plume rise.

In a design situation, the value of h_{small} depends on the location of the critical receptor (e.g. fresh air intake). It also depends on the exhaust velocity and the design wind speed, since these parameters determine the minimum plume rise (see Equation 8 in Chapter 1). However, in the present study, it is simply assumed that h_{small} is large, as would be the case for a leeward wall intake, and thus, h is always less than h_{small}. It is also assumed that the momentum plume rise always exceeds the maximum height of the roof re-circulation zones, h_{top}. Consequently, since $h_{top} < h < h_{small}$, stack height has not been included in the calculation of h.

The predicted re-circulation zones corresponding to tests carried out with stack locations SL1 and SL2 are shown in Figure 11. Estimated values of h_{small} are indicated for intakes A, B and C, located on the leeward walls of the skylight structure, penthouse and main building, respectively. Note that this figure is strictly applicable only for winds that are nearly normal to the northwest wall of the building. This condition was generally not satisfied for the SL1 tests (Oct. 12, 2000, Nov. 15, 2000), in which oblique winds likely produced conical vortices at leading edges of the building and roof structures. Nevertheless, Figure 11 will be used for the analysis of SL1 tests as well as SL2 tests, based on the assumption that the dimensions of re-circulation zones for SL1 tests are roughly similar to those indicated.

Assumption No. 2: no averaging time correction

The D_r model assumes an averaging time of 2 minutes. Values of lateral plume spread are adjusted for other averaging times using the 0.2 power law (see Equation 15 in Chapter 1). However, if the stack tip and receptor are in the same re-circulation zone dilution is not expected to be sensitive to averaging time (ASHRAE, 2003). In the present case, since some samplers were in a roof re-circulation zone with the stack, no averaging time correction has been made. Furthermore, the correction would be small anyway since the actual sampling time was only 5 minutes.

D_{min} ASHRAE (1999)

Assumption No.1: stack height, h_s=0

The D_{min} model is applicable for exhaust vents that are flush with the roof and for stacks that do not exceed the height of rooftop structures (ASHRAE, 1999). In the present case study, although the 3 m stack was slightly higher than the skylight rooftop structure, the effective stack height is considered to be zero for both 1 m and 3 m stacks.

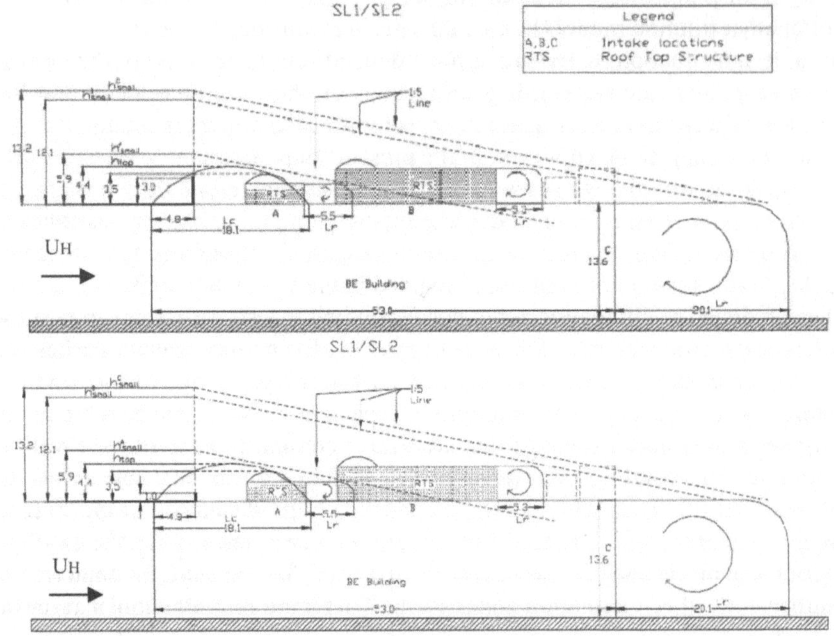

Figure 11. ASHRAE geometric design method for stack locations 1 and 2: stack height 3m and 1m.

Assumption No.2: no averaging time correction

The D_{min} model assumes an averaging time of 10 minutes. However, if the stack tip and receptor are in the same re-circulation zone, dilution is not expected to be sensitive to averaging time (ASHRAE, 1999). In the present case study, since some samplers were in a roof re-circulation zone with the stack, no averaging time correction has been made.

Assumption No.3: distance dilution parameter, B_1 = 0.059

The distance dilution parameter, B_1, is a function of the amount of turbulence in the approaching flow, as shown in Equation 15 (Chapter 1). The recommended design value of B_1 for an urban area is 0.059, based on the assumption that $\sigma_\theta = 15°$. Although σ_θ was not directly measured on the BE building, high values of longitudinal and lateral turbulence intensity measured during the tests indicate that $\sigma_\theta > 15°$. Nevertheless, the default value of B_1 was chosen to provide conservative estimates of dilution.

ASHRAE dilution model comparisons. Field data from the tests at stack locations SL1, SL2 and SL4 have been used to evaluate the ASHRAE models. Recall that the ASHRAE models are not applicable for the SL3 tests, which were performed with the Faubourg building upwind. Except for the SL4 test (Nov. 21, 2002), the selected tests were generally performed in strong wind conditions.
Figure 12a shows 5-min dilution data plotted versus distance from the stack, located at SL1. The data were obtained with the 1 m stack and M=5.5 on Oct. 12, 2000 hr. 1 test. Also plotted are ASHRAE (2003) D_r curves for rooftop and skylight receptors and the ASHRAE (2001) D_{min} curve. Relatively low values of dilution ($100<D<200$) were measured near the stack (S~10 m), while at the most distant sampler, located on the 4 m high penthouse, the average dilution during the test was approximately 1000. Generally, the D_{min} model provides an acceptable lower bound to the data, although dilution data obtained at one rooftop sampler and one skylight sampler were less than the predicted values. On the other hand, the D_r curve for rooftop samplers significantly overestimates the measured dilution values near the stack (S<25 m). The actual plume rise in this case may have been less than that predicted by the model due to high turbulence in the leading edge re-circulation zone. Note, however, the D_r curve plotted for the skylight samplers fits the data well. The dilution indicated for the skylight samplers is lower due to the lower value of h, the height of the plume above the roof surface on which the samplers are located.

Figure 12b shows test results and model predictions for low M (SL1, M=2.5) Oct. 12, 2000 hr. 2 test. In this case, the D_r model is conservative for both rooftop and skylight samplers – under-predicting the measured dilutions by approximately a factor of two at most locations. The D_{min} curve fits the data quite well.

Figure 13 shows data obtained during the Nov. 21, 2002 tests with the 1 m stack at SL4 (M=5.6, M=10.5). Note that the 3 samplers farthest from the stack were wall samplers. Comparing the two data sets, the effect of M (exhaust velocity) is evident. Near the stack (S~10 m), the minimum dilution at a rooftop sampler was approximately 300 for M=5.6. Increasing M to a value of 10.5 increased the minimum dilution at this sampler by a factor of 3 (D~ 1000). Farther from the stack, the effect of M appears to be less significant. At S~30 m, minimum dilution obtained for M=10.5 is approximately 40% larger than that obtained for M=5.6. This appears to support the use of a two-component dilution model, like D_{min} that takes into account initial dilution near the stack and distance dilution.

The D_{min} model again provides an acceptable lower bound to the data, although it is overly conservative near the stack since it does not take into account the apparent dilution due to plume rise. The D_r curve for the skylight samplers fits the high M data well but is overly conservative for the M=5.6 data set. On the other hand, the D_r curve for the rooftop samplers fits the low M data reasonably well. However, the model is not conservative in predicting dilution at rooftop samplers for the high M case with predicted values approximately twice as large as the measured minimum dilutions at each location. This may indicate that the model overestimated the plume rise in this case.

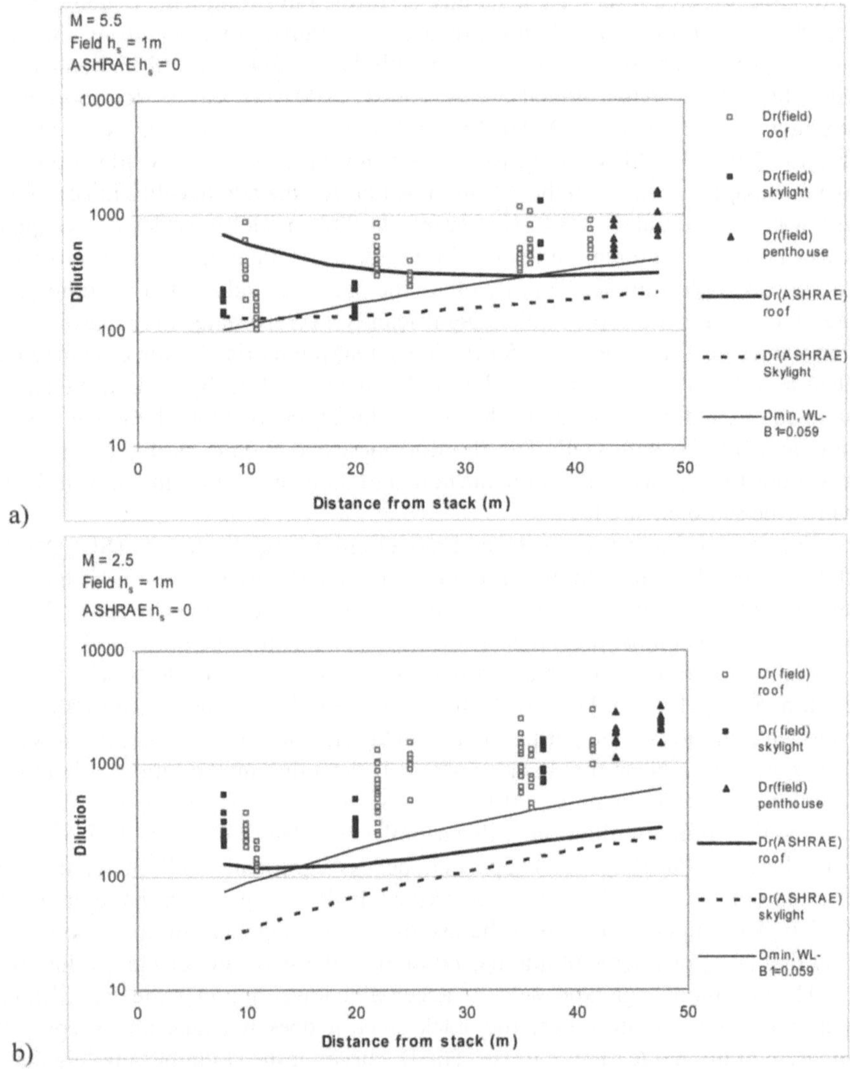

Figure 12. Comparison of field test dilution data with ASHRAE (1999) D_{min} and ASHRAE (2003) D_r provisions: Oct.12-00: hour 1 and hour 2 test.

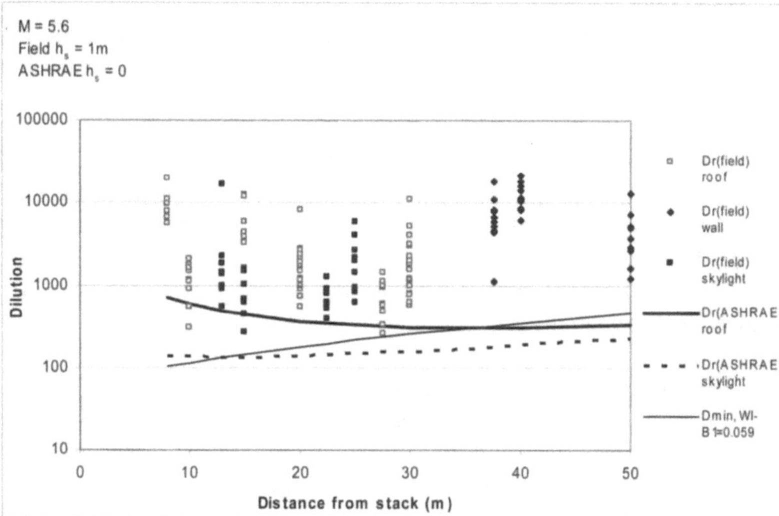

a)

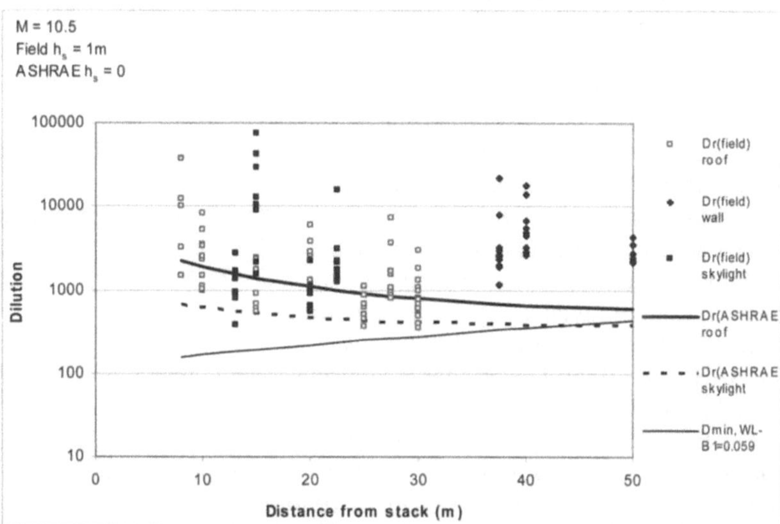

b)

Figure 13. Comparison of field test dilution data with ASHRAE (1999) D_{min} and ASHRAE (2003) Dr provisions: Nov.21-02 test: hour 1 and hour 2.

2.2 Design guidelines

The following provides a summary of various design guidelines formulated on the basis of results obtained in the study:

Stack location:
For open fetch situations, it is better to place the stack near the center of the roof. In this way, the leading edge re-circulation zone is avoided, thus, maximizing plume rise. In addition, the required plume height to avoid contact with leeward wall receptors is minimized.

For the case of a taller building upwind of the emitting building, the center of the roof may not be the optimum stack location for receptors on the emitting building. Concentrations over most of the roof can be reduced by placing the stack near the leading edge. However, this stack location will result in higher concentrations on the leeward wall of the adjacent building. Naturally, this depends on the distance between the two buildings.

Stack height:
For open fetch situations, increasing the stack height from 1 m to 3 m reduces concentrations near the stack by approximately a factor of two. Far from the stack ($x > 20$ m), the effect is negligible. A stack height of at least 5 m is required to provide significant reduction of k at such distances.

For an upwind adjacent building, small changes in stack height have little effect on concentration.

Stack exhaust speed:
Increasing stack exhaust speed by a factor of 2.5 reduces concentrations near the stack by the same factor. For distant receptors ($x>20$ m), the effect of exhaust speed depends on the M-value (the ratio of exhaust speed to wind speed). In the low M range ($1.5<M<4.5$), which is typical of wind speeds exceeding 5 m/s, increasing exhaust speed may not be beneficial for distant receptors because the plume rise may not be sufficient to avoid them. On the other hand, for light wind conditions, doubling the exhaust speed may cause M to be high enough so that concentrations are reduced over the entire roof.

ASHRAE (2003) vs ASHRAE (1999) model:
The ASHRAE (1999) D_{min} model is less conservative than the ASHRAE (2003) D_r model and significantly better for distant samplers ($S>30$m).

For the typical design situation of low M cases ($2.5<M<3.5$), the ASHRAE (2003) D_r model appears to be overly conservative, especially for distant samplers – it underestimates dilution by a factor of 10 for receptors located more than 30 m from the stack. For high M cases ($M{\sim}10$), i.e. low wind speed, the ASHRAE (2003) D_r model is not conservative for samplers near the stack.

Placement of fresh air intakes:
The case of an emitting low building in the wake of a taller building was particularly investigated. For wind coming from the direction of the taller building:

- intakes on emitting building should be placed on its leeward wall if possible.
- intakes should not be placed on leeward wall of upwind building.

3. Concluding Remarks

The results are encouraging because they demonstrate the general adequacy of the wind tunnel data to represent real design situations. In the case of pedestrian level winds, both experimental and computational approaches can produce useful data. In the case of pollutant (exhaust) dispersion, the limitations of the ASHRAE models to predict real dilutions for particular building configurations and stack locations were shown. The design guidelines provided in this chapter will be helpful to ventilation design engineers to tackle a multi-faceted complicated problem, for which codes and standards are either mute or extremely general to apply to specific cases of particular real conditions.

References

ASHRAE (1999). Chapter 43 - Building Air Intake and Exhaust Design. *ASHRAE Applications Handbook*, American Society of Heating, Refrig. and Air-Cond. Eng., Inc., Atlanta.

ASHRAE (2003). Chapter 44 - Building Air Intake and Exhaust Design. *ASHRAE Applications Handbook*, American Society of Heating, Refrig. and Air-Cond. Eng., Inc., Atlanta.

ASHRAE (2007). Chapter 44 - Building Air Intake and Exhaust Design. *ASHRAE Applications Handbook*, American Society of Heating, Refrig. and Air-Cond. Eng., Inc., Atlanta.

American Society of Civil Engineers - ASCE. (2003). Aerodynamics Committee, Outdoor Human Comfort and its assessment. *State of the Art Report*, Task Committee on Outdoor Human Comfort, American Society of Civil Engineers, Reston, VA.

Bottema, M. (1993). *Wind climate and urban geometry*. Ph.D. Dissertation, Technische Universitiet Eindhoven, Facultyeit Bouwkunde, Vakgroep Fago.

Ferreira, A.D., Viegas, D.X. and Sousa, A.C.M. (1999). Numerical and experimental study of the wind flow around a group of low-rise buildings. *Proc. 10th Intern. Conf. on Wind Engineering*, June 21-24, Copenhagen, Denmark.

Hu, Cheng-Hu and Wang, F. (2005). Using a CFD approach for the study of street level winds in a built-up area. *Building and Environment*, 40 pp. 617-631.

Melbourne, W.H. (1978). Criteria for environmental wind conditions. *Journal of Wind Engineering and Industrial Aerodynamics*, 3, 241-249.

Miles, S.D. and Westbury, P.S. (2002). Assessing CFD as a tool for practical wind engineering applications. *Proc. 5th UK Conf. on Wind Engineering.*

Richards, P.J. and Hoxey, R.P. (1992). Computational and wind tunnel modeling of mean wind loads on the Silsoe structures building. *Journal of Wind Engineering and Industrial Aerodynamics*, 41-44, 1641-1652.

Snyder, W.H. (1994). Downwash of plumes in the vicinity of buildings: a wind tunnel study. *Recent Research Advances in the Fluid Mechanics of Turbulent Jets and Plumes*, Kluwer Academic Pub., 343-356.

Stathopoulos, T. and Baskaran, A. (1996). Computer simulation of wind environmental conditions around buildings. *Engineering Structures*, 18 (11), 876-885.

Timofeyef, N. (1998). Numerical study of wind mode of a territory development. *Proc. 2nd East European Conf. on Wind Engineering*, Sept. 7-11, Prague, Czech Rep.

Tominaga, Y., Mochida, A., Shirasawa, T., Yoshie, R., Kataoka, H., Harimoto, K., Nozu, T. (2004). Cross comparisons of CFD results of wind environment at pedestrian level around a high-rise building and within a building complex. *Journal of Asian architecture and building engineering*, 3 (1), 63-70.

Westbury, P.S., Miles, S. D. and Stathopoulos, T. (2002). CFD application on the evaluation of pedestrian-level winds. Workshop on Impact of Wind and Storm on City Life and Built Environment, *Cost Action C14*, CSTB, June 3-4, Nantes, France.

Wilson, D.J., Fabris, I., Chen, J. and Ackerman, M. (1998). Adjacent building effects on laboratory fume hood exhaust stack design. *ASHRAE Research Report 897*, American Society of Heating and Refrigerating and Air-conditioning Engineers, Atlanta, GA.

Computational Wind Engineering: Theory and Applications

Bert Blocken

Building Physics and Systems, Eindhoven University of Technology, Eindhoven, the Netherlands

1. Introduction

Computational Wind Engineering (CWE) is the application of computational methods to Wind Engineering problems. While CWE is more than Computational Fluid Dynamics (CFD) alone, CFD has so far constituted the major part of CWE.

Many problems in Wind Engineering can be tackled by one of three approaches, or a combination of these: (1) on-site measurements, (2) reduced-scale wind tunnel measurements or (3) numerical simulation based on CFD. Deciding which approach is most appropriate for a given problem is not always straightforward, as each approach has specific advantages and disadvantages. An important disadvantage of on-site measurements and wind tunnel measurements is that usually only point measurements are obtained. Techniques such as Particle Image Velocimetry (PIV) and Laser-Induced Fluorescence (LIF) in principle allow planar or even full 3D data to be obtained, but the cost is considerably higher and application for complicated geometries can be hampered by laser-light shielding by the obstructions constituting the model, e.g. in case of an urban model consisting of many buildings. Another disadvantage is the required adherence to similarity criteria in reduced-scale testing. This can be a problem for, e.g., multiphase flow problems and buoyant flows. Examples are the transport and deposition of sand, dust, rain, hail, and snow, and buoyancy-driven natural ventilation and pollutant dispersion studies.

Numerical modeling with CFD can be a powerful alternative because it can avoid some of these limitations. It can provide detailed information on the relevant flow variables in the whole calculation domain ("whole-flow field data"), under well-controlled conditions and without similarity constraints. However, the accuracy of CFD is an important matter of concern. Care is required in the geometrical implementation of the model, in grid generation and in selecting proper solution strategies. The latter include choices between solving the steady Reynolds-averaged Navier-Stokes (RANS) equations, the unsteady RANS (URANS) equations, using Large Eddy Simulation (LES) or using hybrid URANS/LES, between different turbulence models, discretization schemes, etc. In addition, numerical and physical modeling errors need to be assessed by solution verification and validation studies. CFD validation in turn requires high-quality full-scale or reduced-scale measurements to be compared with the simulation results.

Just as Wind Engineering itself, CWE is a broad field. It encompasses topics ranging from turbulence model development to applications such as pedestrian-level wind conditions, natural ventilation of buildings, pollutant dispersion, wind-structure interaction and wind energy. This wide range of topics is reflected in the international journals on Wind Engineering, the proceedings of conferences on (Computational) Wind Engineering and the variety of projects carried out at universities, knowledge institutes and companies all over the world.

This chapter on CWE focuses on CFD in Wind Engineering. It starts with some theoretical information on CFD, including the governing equations, information on their approximate forms used in CWE and on the important parameters for CFD simulations. Next, the historical background of CWE is discussed. Also the main efforts that have been made towards providing best practice guidelines for CWE are reported. The historical background and the best practice guidelines provide the starting point for discussing two important categories of CWE applications: pedestrian-level wind conditions and pollutant dispersion around buildings. The application discussions are not intended to be complete, rather they are meant to provide some important headlines. Finally, the chapter ends with a discussion and conclusions section.

2. Theory

2.1. Governing equations and approximate forms

Navier-Stokes equations. The governing equations are the three laws of conservation: (1) conversation of mass (continuity); (2) conservation of momentum (Newton's second law); and (3) conservation of energy (first law of thermodynamics). While strictly, the term Navier-Stokes (NS) equations only covers Newton's second law, in CFD it is used to refer to the entire set of conservation equations. The instantaneous three-dimensional NS equations for a confined, incompressible, viscous flow of a Newtonian fluid, in Cartesian co-ordinates and in partial differential equation form are:

$$\frac{\partial u_i}{\partial x_i} = 0 \tag{1a}$$

$$\frac{\partial u_i}{\partial t} + u_j \frac{\partial u_i}{\partial x_j} = -\frac{1}{\rho}\frac{\partial p}{\partial x_i} + \frac{\partial}{\partial x_j}\left(2 \nu s_{ij}\right) \tag{1b}$$

$$\frac{\partial \theta}{\partial t} + u_j \frac{\partial \theta}{\partial x_j} = \frac{1}{\rho c_p}\frac{\partial}{\partial x_j}\left(k \frac{\partial \theta}{\partial x_j}\right) \tag{1c}$$

The vectors u_i and x_i are instantaneous velocity and position, p is the instantaneous pressure, θ the instantaneous temperature, t is time, ρ is the density, ν is the molecular kinematic viscosity, c_p the specific heat capacity, k the thermal conductivity and s_{ij} is the strain-rate tensor:

$$s_{ij} = \frac{1}{2}\left(\frac{\partial u_i}{\partial x_j} + \frac{\partial u_j}{\partial x_i}\right) \tag{1d}$$

In case of multi-component flow, an advection-diffusion equation for species concentration, similar to that for temperature, is added:

$$\frac{\partial c}{\partial t} + u_j \frac{\partial c}{\partial x_j} = \frac{\partial}{\partial x_j}\left(D \frac{\partial c}{\partial x_j}\right) \tag{1e}$$

where c is the instantaneous concentration and D the molecular diffusion coefficient or molecular diffusivity. Additional terms can be added to these equations, e.g. the gravitational acceleration term and the buoyancy term in the momentum equations.

As directly solving the NS equations for the high-Reynolds number flows in CWE is currently prohibitively expensive, approximate forms of these equations are solved. Two main categories used in CWE are RANS and LES. In addition, hybrid RANS/LES approaches are sometimes used.

Reynolds-averaged Navier-Stokes equations. The RANS equations are derived by averaging the NS equations (time-averaging if the flow is statistically steady or ensemble-averaging for time-dependent flows). With the RANS equations, only the mean flow is solved while all scales of the turbulence are modeled (i.e. approximated). The averaging process generates additional unknowns and as a result the RANS equations do not form a closed set. Therefore approximations have to be made to achieve closure. These approximations are called turbulence models. Up to now, RANS has been the most commonly used approach in CWE. Therefore, more detailed information about this approach is given below.
The RANS equations are obtained by decomposing the solution variables as they

appear in the instantaneous NS equations (Eqs. 1a-c,e) into a mean (ensemble-averaged or time-averaged) and a fluctuation component. For an instantaneous vector $\vec{a}$ and an instantaneous scalar ξ this means:

$$\vec{a} = \vec{A} + \vec{a}' \quad ; \quad \xi = \Xi + \xi' \tag{2}$$

where $\vec{A}$ and Ξ are the mean and $\vec{a}'$ and ξ' the fluctuation components (around the mean). Replacing the instantaneous variables in Eq. (1a-c,e) by the sum of the mean and the fluctuation components and taking an ensemble-average or time-average yields the RANS equations:

$$\frac{\partial U_i}{\partial x_i} = 0 \tag{3a}$$

$$\frac{\partial U_i}{\partial t} + U_j \frac{\partial U_i}{\partial x_j} = -\frac{1}{\rho}\frac{\partial P}{\partial x_i} + \frac{\partial}{\partial x_j}\left(2\nu S_{ij} - \overline{u_j' u_i'}\right) \tag{3b}$$

$$\frac{\partial \Theta}{\partial t} + U_j \frac{\partial \Theta}{\partial x_j} = \frac{1}{\rho c_p}\frac{\partial}{\partial x_j}\left(k\frac{\partial \Theta}{\partial x_j} - \overline{u_j' \theta'}\right) \tag{3c}$$

$$\frac{\partial C}{\partial t} + U_j \frac{\partial C}{\partial x_j} = \frac{\partial}{\partial x_j}\left(D\frac{\partial C}{\partial x_j} - \overline{u_j' c'}\right) \tag{3d}$$

Here, U_i, P, Θ and C are the mean velocity, pressure, temperature and concentration, u_i', p', θ' and c' are the fluctuation components and S_{ij} is the mean strain-rate tensor:

$$S_{ij} = \frac{1}{2}\left(\frac{\partial U_i}{\partial x_j} + \frac{\partial U_j}{\partial x_i}\right) \tag{4}$$

The horizontal bar in the equations denotes averaging. When comparing the set of equations (Eq. 1) with the instantaneous set (Eqs. 3-4), the similarity between both sets is observed, but also that the averaging process has introduced new terms, which are called the Reynolds stresses (for momentum), turbulent heat fluxes and turbulent mass fluxes. They represent the influence of turbulence on the mean flow, the heat transfer and the mass transfer. The instantaneous NS equations (Eq. 1a-c,e) form a closed set of equations (six equations with six unknowns: u_i, p, θ and c). The RANS equations do not form a closed set due to the presence of the Reynolds stresses and turbulent heat and mass fluxes (more unknowns than equations). It is impossible to derive a closed set of exact equations for the mean flow variables (Ferziger and Peric, 1996). Closure must therefore be obtained by modeling. The modeling approximations for the Reynolds stresses are called turbulence models. Turbulence models are briefly discussed in section 2.2. Note that for including the effect of density differences due to temperature or species concentrations, generally the Boussinesq approximation for buoyancy is applied (Bejan, 2004).

A distinction has to be made between steady RANS and unsteady RANS (URANS). Steady RANS refers to time-averaging of the NS equations and yields statistically steady descriptions of turbulent flow. However, flow in the atmospheric boundary layer (ABL) is inherently unsteady, and therefore, strictly, an unsteady approach is required. URANS refers to ensemble-averaging of the NS equations. Franke et al. (2007) state that, since URANS also requires a high spatial resolution, it is recommended to directly use LES or hybrid URANS/LES. Regardless of spatial resolution, it is important to note that URANS does not simulate the turbulence, but only its statistics. In fact, URANS only resolves the unsteady mean-flow structures, while it models the turbulence. LES on the other hand actually resolves the large scales of the turbulence. URANS can be a good option when the unsteadiness is pronounced and deterministic, such as von Karman vortex shedding in the wake of an obstacle with a low-turbulence approach flow. However, given the relatively high turbulence in (approach-flow) atmospheric boundary layers, LES or hybrid URANS/LES should be preferred over URANS for these applications.

As will be shown in the section on applications, steady RANS is by far most often used in CWE, in spite of its deficiencies. Studies that have employed unsteady RANS (URANS) are scarce. LES on the other hand is increasingly used.

Large Eddy Simulation. In the LES approach, the NS equations are filtered, which consists of removing only the small turbulent eddies (that are smaller than the size of a filter that is often taken as the mesh size). The large-scale motions of the flow are solved, while the small-scale motions are modeled: the filtering process generates additional unknowns that must be modeled in order to obtain closure. This is done with a sub-filter turbulence model. LES generally shows superior performance compared to RANS and URANS, because a large part of the unsteady turbulent flow is actually resolved. However, the required computational resources increase significantly, the inlet boundary condition requires time and space resolved data and a larger amount of output data is generated. It is expected that for particular applications, when accurate reproduction of large-scale turbulent structures and the related heat and mass transport are important, e.g. in pollutant dispersion studies, LES will increasingly replace RANS.

Hybrid approaches. The hybrid URANS/LES approach employs URANS in the near-wall region and LES in the rest of the domain. This approach is based on the fact that near walls, the turbulent eddies are very small and resolving them with LES could become prohibitively expensive. Note that this does not mean that stand-alone LES can not yield good results for wall-bounded flows; in these situations often wall functions are used. A well-known hybrid approach is Detached Eddy Simulation (Spalart et al., 1997), in which LES is combined with the one-equation Spalart-Allmaras turbulence model (Spalart and Allmaras, 1992). The application of hybrid approaches is not straightforward: URANS and LES are fundamentally different approaches with specific grid requirements which have to be matched where the switch between both occurs.

2.2. Turbulence models for RANS and URANS

Two main types of models can be distinguished: first-order closure and second-order closure models. First-order closure uses the Boussinesq eddy-viscosity hypothesis to relate the Reynolds stresses to the velocity gradients in the mean flow. Similarly, the turbulent heat fluxes are related to the mean temperature gradients, and the turbulent mass fluxes to the mean concentration gradients. Second-order closure refers to establishing and solving additional transport equations for the Reynolds stresses and the turbulent heat and mass fluxes.

First-order closure is the simplest approach. The Boussinesq eddy-viscosity hypothesis calculates the Reynolds stresses as the product of a turbulent (eddy) viscosity and the mean strain-rate tensor:

$$-\overline{u_i'u_j'} = 2\nu_t S_{ij} - \frac{2}{3}k\delta_{ij} \tag{5}$$

where ν_t is the turbulent viscosity (also called momentum diffusivity), k is the turbulent kinetic energy and δ_{ij} is the Kronecker delta:

$$k = \frac{1}{2}\overline{u_i'u_i'} \tag{6}$$

$$\delta_{ij} = \begin{cases} 1 & \text{for } i = j \\ 0 & \text{for } i \neq j \end{cases} \tag{7}$$

In first-order closure, the turbulence models need to provide expressions for the turbulent (eddy) viscosity, and are called eddy-viscosity models. A distinction is made between linear and non-linear eddy-viscosity models. Examples are the one-equation Spalart-Allmaras model (Spalart and Allmaras, 1992), the standard k-ε model (Jones and Launder, 1972) and its many modified versions, such as the Renormalization Group (RNG) k-ε model (Yakhot and Orszag, 1986) and the realizable k-ε model (Shih et al., 1995), the standard k-ω model (Wilcox, 1998) and the k-ω shear stress transport (SST) model (Menter, 1997).

Similarly, the standard approximation for the turbulent flux of scalar quantities is the gradient-diffusion assumption, by which the turbulent heat and mass flux can be obtained as:

$$-\overline{u_j'\theta'} = D_{\theta,t} \frac{\partial \Theta}{\partial x_j} \tag{8}$$

$$-\overline{u_j'c'} = D_{c,t} \frac{\partial C}{\partial x_j} \tag{9}$$

where $D_{\theta,t}$ and $D_{c,t}$ are the turbulent heat and mass diffusivities, which are generally related to the momentum diffusivity by the turbulent Prandtl number Pr_t and the turbulent Schmidt number Sc_t, respectively:

$$Pr_t = \frac{\nu_t}{D_{\theta,t}} \tag{10}$$

$$Sc_t = \frac{\nu_t}{D_{c,t}} \tag{11}$$

Neither $D_{\theta,t}$, nor $D_{c,t}$ is a fluid property. Instead, they are a function of the type of flow pattern. The same holds for Pr_t and Sc_t. Nevertheless, often constant values are used for Pr_t and Sc_t in CFD simulations.

Second-order closure is also referred to as second-moment closure or Reynolds Stress modeling (RSM). It consists of establishing and solving additional transport equations for each of the Reynolds stresses and the turbulence dissipation rate. Second-order closure is also possible for the turbulent heat and mass fluxes, but this option is not often used in CWE.

2.3. Computational model, domain and grid

The computational model refers to the geometrical representation of the obstacles (buildings, bridges, trees, ...) in the computational domain (Fig. 1). Three main modeling levels are distinguished (Blocken et al., 2007a): (1) the region of interest, in which the obstacles are modeled explicitly, i.e. with their actual shape, although less important details (e.g. roof and facade details for buildings) can generally be ignored; (2) the intermediate region at a larger distance from the region of interest, in which the obstacles are also explicitly represented, but only with their main shape; (3) the outer region, in which the obstacles are modeled implicitly, i.e. their geometry is not included in the domain but their effect on the flow is modeled in terms of roughness parameters, e.g. by means of roughness modifications to wall functions applied at the bottom of the computational domain. In Figure 1, the region of interest and the intermediate region are located in the central part of the domain, while the outer region is located in the upstream and downstream part.

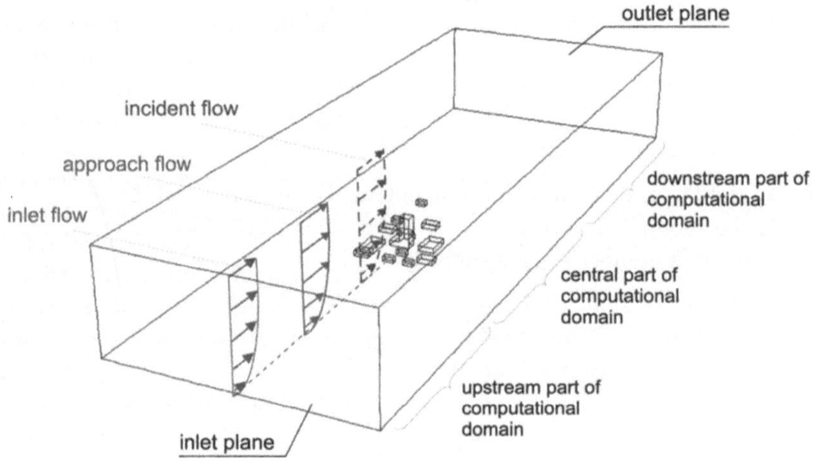

Figure 1. Computational domain with building models, indication of different parts in the domain for roughness modeling and definition of inlet flow, approach flow and incident flow (Blocken et al., 2007a).

The computational domain encloses the region of the explicitly and implicitly modeled obstacles. It defines the location of the inlet boundary, the outlet boundary, the ground boundary, the side boundaries and the top boundary. All boundaries (except the ground boundary) should be located at a sufficiently large distance from the region of interest in order not to significantly influence the calculation results in this region. Several authors have provided guidelines for the distance between the obstacles in the region of interest and the boundaries. These guidelines are generally expressed in terms of minimum distances between the explicitly modeled obstacles in region of interest and the boundaries, or in terms of maximum values of the blockage ratio, i.e. the ratio between the projected frontal area of the explicitly modeled obstacles to the cross-section of the domain. Franke et al. (2004, 2007) and Tominaga et al. (2008a) provided recommendations in terms of both criteria. Generally, it can be stated that the distance between the region of interest and the inlet and side boundaries should be at least $5H_{max}$, where H_{max} is the height of the highest building in the region of interest. The distance from the outlet should be at least $15H_{max}$, and the height of the domain at least $6H_{max}$. The recommended maximum blockage ratio is 3%. In addition, Blocken et al. (2004) mention that, for buildings with an extension in the lateral direction much larger than the height, or vice versa, the ratio of the lateral extension of the computational domain to its height should be similar to the width-to-height ratio of the building. When numerically reproducing wind tunnel experiments on the other hand, the cross-sectional area of the computational domain can be taken equal to that of the wind tunnel. For LES, it should be noted that the size of the domain restricts the size of the largest vortical structures that can be reproduced (Franke et al., 2007, Richards et al., 2007), and that therefore – except when reproducing wind tunnel experiments – a larger domain can be required.

The computational grid is the spatial discretization of the computational domain, which is generally performed with control volumes or finite elements. Generating a high-quality grid is very important to obtain accurate and reliable CFD results. Two main categories can be distinguished: body-fitted (BF) grids and immersed-boundary (IB) grids. BF grids represent the conventional approach in which the grids are built based on the shape of the solid flow boundaries, and are conform to these boundaries (Fig. 1a). In immersed boundary (IB) grids on the other hand, the grid does not conform to the solid boundary defined by the shape of the body in the flow (Mittal and Iaccarino, 2005) (Fig. 1b). Therefore, the boundary conditions on the body surface are not imposed directly, but instead an extra term, called the forcing function, is added to the governing equations, or the discrete numerical scheme is altered near the boundary. An example of BF and IF grids for the same urban area is shown in Fig. 1c-d (Yoshie et al., 2007). An important advantage of IB grids, especially for large computational domains, is the speed of grid generation. Furthermore, IB grids are often structured, while BF grids are generally unstructured and consist of tetrahedral cells which can generate larger discretization errors and can give rise to convergence problems, especially in combination with second-order discretization schemes. BF grids on the other hand allow higher quality boundary cells and more efficient local grid refinement near the boundaries. A main disadvantage of BF grids is that standard automatic or semi-automatic generation of an unstructured grid also allows insufficient control of local grid resolution, grid stretching, control volume skewness and aspect ratio. To remediate these disadvantages of BF grids, van Hooff and Blocken (2010) presented a specific BF grid generation technique to efficiently and simultaneously construct the geometry and the computational grid with full control over grid resolution and quality of the grid cells. This technique consists of a series of extrusion operations, i.e. creating the geometry and the grid based on geometrical translation and rotation operations of pre-meshed 2D cross-sections. It can be applied to simple building geometries as well as very complex urban geometrical configurations. It allows constructing high-resolution and high-quality grids in which tetrahedral cells are avoided (van Hoof and Blocken 2010). Fig. 3 shows an example of the BF grid generated using this technique, to calculate coupled urban wind flow and indoor natural ventilation for a large multifunctional stadium in an urban environment.

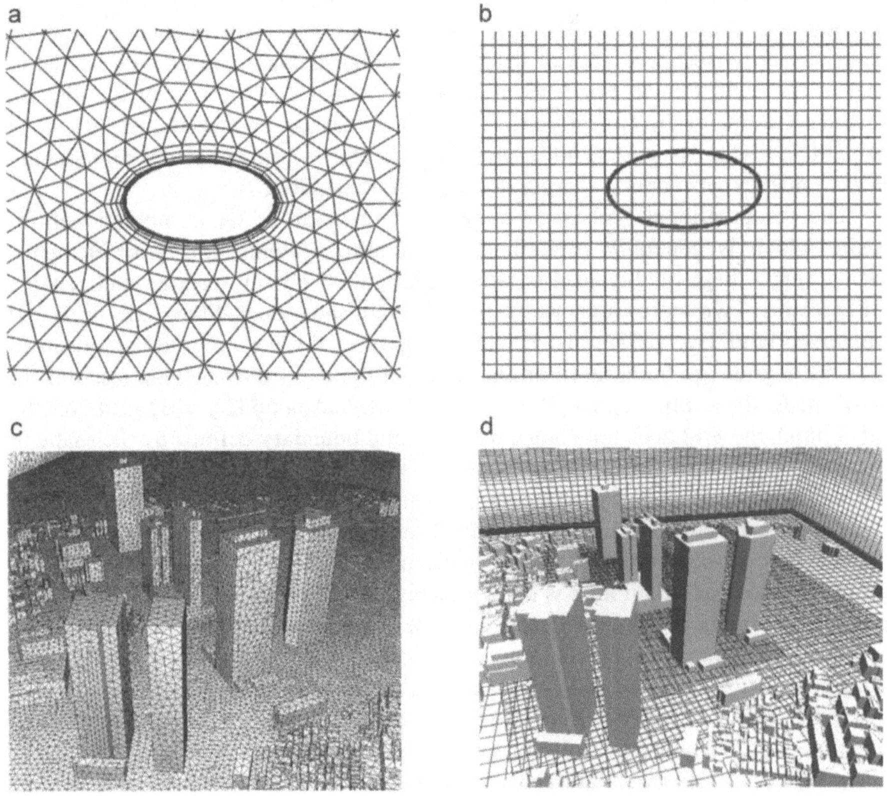

Figure 2. (a-b) Schematic representation of (a) body-fitted grid and (b) immersed boundary grid. (c-d) Body-fitted and immersed boundary grids for the Shinjuku Sub-central area in Tokyo, Japan (Fig. c-d: courtesy of R. Yoshie and Y. Tominaga; (Yoshie et al. 2007)).

Irrespective of the type of grid used, a grid-sensitivity study on at least three systematically refined grids should always be performed. Often, a refinement factor √2 is applied, as a factor 2 in three dimensions ($2^3 = 8$) would lead to refined grids that could be too large compared to the computational resources available. More information on grid generation and grid quality can be found in e.g. Casey and Wintergerste (2000), Scaperdas and Gilham (2004), Bartzis et al. (2004), Franke et al. (2007), Tominaga et al. (2008a) and van Hooff and Blocken (2010).

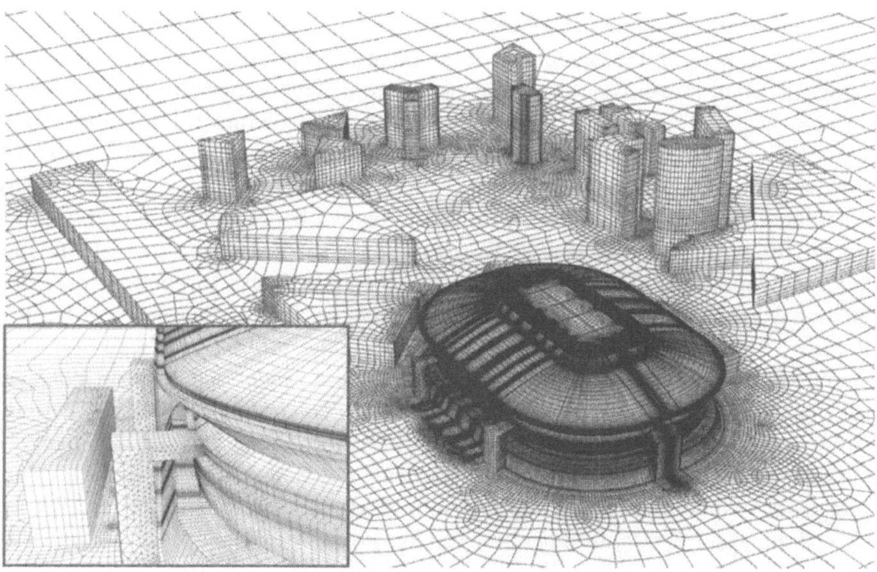

Figure 3. High-resolution body-fitted computational grid (5.6 million cells) for coupled urban wind flow and indoor natural ventilation of a multifunctional stadium (van Hooff and Blocken 2010).

2.4. Boundary conditions

As detailed information on boundary conditions can be found in the available best practice guideline documents (see section 4), only some headlines and the important issue of boundary condition consistency are mentioned here.

The inlet boundary conditions for RANS simulations are the vertical profiles of the mean wind speed (see Fig. 1) and the turbulence quantities (e.g. k and ε for the k-ε turbulence model. Different types of outlet boundary conditions exist, which generally have a limited influence on the results if the outlet is placed at a sufficiently large distance from the region of interest. Symmetry conditions are often used at the sides and the top of the domain, although especially for the top boundary condition, better options exist that take into account the shear exerted by the upper air layers on the air in the domain (Blocken et al., 2007a, Franke et al., 2007, Hargreaves and Wright, 2007). At solid walls, wall functions are often used, e.g. the standard wall functions by Launder and Spalding (1974). Wall functions can be equipped with roughness parameters (the equivalent sand-grain roughness height k_S or the aerodynamic roughness length z_0) to take into account the roughness of the terrain and of the obstacle surfaces.

An important problem in CFD simulations of atmospheric boundary layer flow, which is related to the selection of boundary conditions, are the unintended changes (called streamwise gradients or horizontal inhomogeneity) that occur in the vertical profiles of mean wind speed and turbulence quantities as they travel from the inlet of the computational domain towards the modeled buildings (Fig. 1). This problem is described in detail in (Blocken et al., 2007a). It can dramatically affect the quality of the simulation results, as shown by Blocken et al. (2007b) for pedestrian-level wind conditions. The reason is the inconsistency between the inlet boundary conditions, the wall functions and their roughness parameters, the computational grid and the turbulence model (Blocken et al., 2007a). To solve this problem, Richards and Hoxey (1993) provided inlet profiles and wall boundary conditions that are consistent with the standard k-ε model. Their effort however was focused on z_0-type wall functions, i.e. wall functions in which the aerodynamic roughness length z_0 is present as a roughness parameter. As many – commercial – CFD codes employ k_S-type wall functions, i.e. with the equivalent sand-grain roughness height k_S as a roughness parameter, Blocken et al. (2007a) derived the specific relationships between k_S and z_0, for Fluent 6 and Ansys CFX. It should be noted that the difference between k_S and z_0 is typically a factor 20 or 30. Hargreaves and Wright (2007) provided modifications to the wall functions in CFX to address the same problem. Yang et al. (2009) suggested new and more realistic inlet profiles for the turbulence quantities, to be used in combination with the k_S-z_0 relationships by Blocken et al. (2007a). Finally, Gorlé et al. (2009) demonstrated that further consistency with the inlet profiles by Yang et al. (2009) and the k_S-z_0 relationships by Blocken et al. (2007a) could be achieved by converting turbulence model constants in the standard k-ε model into functions. These efforts focused on steady RANS simulations.

The major difference between boundary conditions for RANS and LES is at the inlet. As mentioned earlier, LES requires time and space resolved inlet boundary conditions. Also with LES, the roughness of the ground boundary is of critical importance.

2.5. Numerical approximations

A wide range of discretization schemes exists. First order schemes should be avoided as they can yield considerable numerical (artificial) diffusion. Higher order schemes should be used for both space and time discretization.

2.6. Time step size and number of time steps

For unsteady simulations, the time step size and the number of time steps need to be chosen. Data sampling should only occur after a sufficient number of travelling times (flow through the domain from inlet to outlet), to remove the effects of the initialization. The time step can be chosen based on the highest frequency or based on the grid cell size and the Courant-Friedrichs-Lewy (CFL) number. The period corresponding to the highest frequency should be divided into at least 10-20 time steps (Menter et al., 2002, Franke et al., 2007). Irrespective of these guidelines, a time-step sensitivity study should be conducted.

2.7. Convergence criteria

Terminating the iterative procedure is based on convergence criteria, often expressed in terms of so-called scaled residuals, the decrease of which is an indication of the obtained level of convergence. It is important to note that many (commercial) CFD codes recommend values for the scaled residuals that are (much) too lenient (e.g. 0.001), which can give rise to large and sometimes very large errors in the CFD results. It is recommended to set the scaled residuals at machine accuracy level, and to monitor convergence of the most important variables at several positions in the region of interest.

2.8. Errors, validation and verification

Five types of errors can be distinguished: (1) physical modeling errors, (2) computer round-off errors, (3) iteration-convergence errors, (4) discretization errors and (5) computer-programming errors. Error assessment can be done by validation and verification studies.

Physical modeling errors are those, which are due to uncertainty in the formulation of the model and to deliberate simplifications of the model (NASA 2004). Examples are employing the RANS equations in combination with the Boussinesq eddy-viscosity hypothesis and/or a given turbulence model, using specific constants in the turbulence model, simplifications of the model geometry, using wall functions, modeling the surface roughness, simplifications in the boundary conditions, etc.

Computer round-off errors are generally not considered significant when compared with other errors (NASA, 2004). If they are suspected to be significant, one can perform a test by running the code at a higher precision.

Iteration-convergence errors are introduced because the iterative procedure has to be stopped at a certain moment in time. As mentioned in section 2.7, setting stringent convergence criteria and monitoring convergence is important.

Discretization errors originate from the representation of the governing equations and the equations of the turbulence model on a mesh that represents a discretized computational domain. For unsteady calculations, also time discretization causes discretization errors. As mentioned earlier, grid-sensitivity and time-step sensitivity studies are imperative. In addition, the discretization error can be estimated using methods described in e.g. (Ferziger and Peric, 1996, Franke and Frank, 2008).

Computer-programming errors are due to mistakes made in writing the computer code. These types of errors can be discovered by systematically performed verification and validation studies and by comparing the results of the code with those of a similar code.

Validation and verification are terms that have a specific meaning in the context of CFD simulations. This meaning can differ from the one that is common in other domains of science. In CWE, validation refers to assessment of physical modeling errors, while verification refers to assessment of the other four types of errors. According to the "*Guide for the Verification and Validation of Computational Fluid Dynamics Simulations*" (AIAA, 1998), validation refers to identifying and quantifying errors by comparing simulation results with experimental data, whereas verification refers to identifying and quantifying the errors in the model implementation and the solution. Verification therefore consists of two aspects: (1) verification of the code and (2) verification of the calculation/solution. The former includes removing computer-programming errors; the latter comprises determining/limiting the iteration-convergence error and the discretization errors. In brief, the difference between validation and verification can be explained as follows: validation refers to the question: "Are the right equations solved?", while verification refers to the question: "Are the equations solved correctly?". It is noted that validation is more an engineering than a mathematical task while for verification, the opposite holds. Ideally, in CFD studies, code verification should be performed first. Next, in a given CFD simulation, solution verification should be applied. Only when the numerical errors have been assessed and limited as much as possible, CFD validation by comparison with high-quality experimental data should be performed. Only performing validation without verification does not allow distinguishing the origin of the discrepancies between CFD results and experimental results, and can be misleading; it is possible that numerical and physical modeling errors (partly) cancel each other for some variables in some part of the computational domain.

3. Historical Background of CWE

CFD simulation of wind flow around buildings started with fundamental studies for isolated buildings, often with a cubical shape, to analyze the velocity and pressure fields (Vasilic-Melling, 1977, Hanson et al., 1986, Paterson and Apelt, 1986, 1989, 1990, Murakami et al., 1987, 1990, 1992, Murakami and Mochida, 1988, 1989, Baskaran and Stathopoulos, 1989, 1992, Stathopoulos and Baskaran, 1990, Murakami, 1990a, 1990b, 1993, Baetke et al., 1990, Mochida et al., 1993). Together with later studies, they laid the foundations for the current best practice guidelines, by focusing on the importance of grid resolution (Murakami and Mochida, 1989, Murakami 1990a, 1990b, Baskaran and Stathopoulos, 1992), the influence of the boundary conditions on the numerical results (Murakami and Mochida, 1989, Paterson and Apelt, 1990, Baetke et al., 1990, Stathopoulos and Baskaran, 1990, Baskaran and Stathopoulos, 1992) and by comparing the performance of various types of turbulence models in steady RANS simulations (Baskaran and Stathopoulos, 1989, Murakami et al., 1992, Murakami, 1993, Mochida et al., 2002). Also comparisons of RANS versus LES were performed (Murakami et al., 1990, 1992, Murakami 1990b, 1993). In the past, especially the deficiencies of using the steady RANS approach with the standard k-ε model for wind flow around buildings were addressed. These include the stagnation point anomaly with overestimation of turbulent kinetic energy near the frontal corner and the resulting underestimation of the size of separation and recirculation regions on the roof and the side faces, and the underestimation of turbulent kinetic energy in the wake resulting in an overestimation of the size of the cavity zone and wake. Various revised linear and non-linear k-ε models and also second-moment closure models were developed and tested, and showed improved performance for several parts of the flow field (Baskaran and Stathopoulos, 1989, Murakami et al., 1992, Murakami, 1993, Wright et al., 2001, Mochida et al., 2002). However, the main limitation of steady RANS modeling remained: its incapability to model the inherently transient features of the flow field such as separation and recirculation downstream of windward edges and vortex shedding in the wake. These large-scale features can be explicitly resolved by LES. While URANS has hardly been used to study wind flow around buildings, early applications of LES for this purpose were already made by Murakami et al. in 1987, and later by Murakami et al. (1990, 1992) and Murakami (1990b). These studies illustrated the superior performance of LES compared to RANS. The studies mentioned above are not all studies that were performed for isolated buildings. But starting from the 1990s, supported by the previous studies and the increased computing performance and availability of CFD codes, fundamental studies gradually shifted their focus to multiple-building configurations, and also application studies were increasingly performed. In addition, the sensitivity of the CFD results to the large number of computational parameters to be set by the user and the possibility of applying CFD in practice led to the development of best practice guidelines, as discussed in the next section.

4. Best Practice Guidelines

As mentioned before, in a typical CFD simulation, the user has to make a large number of choices. He or she has to choose the approximate equations describing the flow (steady RANS, URANS, LES or hybrid URANS/LES), the level of detail in the geometrical representation of the buildings, the size of the computational domain, the boundary conditions, the computational grid, the discretization schemes, the initialization data, the time step size and the iterative convergence criteria. These choices can have a very large impact on the results. Sensitivity studies that systematically investigate the impact of these choices and parameters are therefore important.

Already since the start of the application of CFD for Wind Engineering in the 70-ies and 80-ies, researchers have been testing the influence of these parameters on the results, which has provided a lot of valuable information (e.g. Murakami and Mochida, 1989, Baetke et al., 1990, Stathopoulos and Baskaran, 1990, Cowan et al., 1997, Hall, 1997). However, this information was dispersed over a large number of individual publications in different journals, conference proceedings and reports. In 2000, the ERCOFTAC[1] Special Interest Group on Quality and Trust in Industrial CFD published an extensive set of best practice guidelines for industrial CFD users (Casey and Wintergerste, 2000). The guidelines were focused on RANS simulations in general, but many of these guidelines also apply for CWE. Within the EC project ECORA[2], Menter et al. (2002) published best practice guidelines based on the ERCOFTAC guidelines, but modified and extended specifically for CFD code validation. Within QNET-CFD[3], the Thematic Area on Civil Construction and HVAC (Heating, Ventilating and Air-Conditioning) and the Thematic Area on the Environment presented some best practice advice for the CFD simulations of wind flow and dispersion (Scaperdas and Gilham, 2004, Bartzis et al., 2004).

In 2004, Franke et al. (2004) compiled a set of specific recommendations for the use of CFD in wind engineering from a detailed review of the literature. It was published as a keynote contribution in the final proceedings of the European

[1] ERCOFTAC = European Research Community on Flow, Turbulence and Combustion

[2] ECORA = Evaluation of Computational Fluid Dynamic Methods for Reactor Safety Analysis

[3] QNET-CFD = Network for Quality and Trust in the Industrial Application of CFD

COST[4] Action C14: Impact of Wind and Storm on City Life and Built Environment. Later, Franke et al. (2007) considerably extended this paper into an extensive "Best Practice Guideline for the CFD simulation of flows in the urban environment". Like the ERCOFTAC guidelines, also these guidelines were primarily focused on steady RANS simulations, although also some information on URANS, LES and hybrid URANS/LES was provided.

In Japan, working groups of the Architectural Institute of Japan (AIJ) conducted extensive cross-comparisons between CFD results and high-quality wind tunnel measurements to support the development of guidelines for practical CFD applications. Part of these efforts was reported by Yoshie et al. (2007). In 2008, Tominaga et al. (2008a) published the "AIJ guidelines for practical applications of CFD to pedestrian wind environment around buildings", and Tamura et al. (2008) wrote the "AIJ guide for numerical prediction of wind loads on buildings". While the former document focused on steady RANS simulations, the latter also considered LES, given the importance of time-dependent analysis for wind loading of buildings and structures.

Apart from these general guidelines, also a number of very specific guidelines were published, such as those for modeling equilibrium atmospheric boundary layers in computational domains, i.e. avoiding the occurrence of unintended streamwise gradients, as discussed in section 2.4.

The establishment of these guidelines has been an important step towards more accurate and reliable CFD simulations. Note that, although several of the guideline documents mentioned above have been developed with focus on a particular topic (e.g., pedestrian-level wind conditions), most of the information is also applicable to many other CWE applications.

5. Applications

5.1. Pedestrian-level wind conditions

Importance. High-rise buildings can introduce high wind speed at pedestrian level, which can lead to uncomfortable or even dangerous conditions. Wind discomfort and wind danger can be detrimental to the success of new buildings. Wise (1970) reports about shops that are left untenanted because of the windy environment which discouraged shoppers. Lawson and Penwarden (1975) report the death of two old ladies due to an unfortunate fall caused by high wind speed at the base of a tall building. Today, many urban authorities only grant a building permit for a

[4] COST = European Cooperation in Science and Technology

new high-rise building after a wind comfort study has indicated that the negative consequences for the pedestrian wind environment remain limited.

CFD versus wind tunnel experiments. Wind comfort studies require knowledge of at least the mean wind velocity vector field at pedestrian height (z = 1.75 or 2 m). This information can be obtained by wind tunnel modeling or by CFD. Wind tunnel tests are generally point measurements with Laser Doppler Anemometry (LDA) or Hot Wire Anemometry (HWA). In the past, also area techniques such as sand erosion (Beranek and van Koten, 1979, Beranek, 1982, 1984, Livesey et al. 1990, Richards et al., 2002) and infrared thermography (Yamada et al., 1996, Wu and Stathopoulos, 1997, Sasaki et al., 1997) have been used. They are however considered less suitable to obtain accurate quantitative information. Instead, they can be used as part of a two-step approach: first an area technique is used to qualitatively indicate the most important problem locations, followed by accurate point measurements at these most important locations (Blocken and Carmeliet, 2004a).

One of the main advantages of CFD in pedestrian-level wind comfort studies is avoiding this time-consuming two-step approach by providing whole-flow field data. In spite of its deficiencies, steady RANS modeling with linear k-ε models has become the most popular approach for pedestrian-level wind studies. Two main categories of studies can be distinguished: (1) fundamental studies, which are typically conducted for simple, generic building configurations to obtain insight in the flow behavior, for parametric studies and for CFD validation, and (2) applied studies, which provide knowledge of the wind environmental conditions in specific and often much more complex case studies. Fundamental studies – beyond the case of the isolated building – were performed by several authors including Baskaran and Stathopoulos (1989), Bottema (1993), Baskaran and Kashef (1996), Franke and Frank (2005), Yoshie et al. (2007), Blocken et al. (2007b, 2008a), Blocken and Carmeliet (2008), Tominaga et al. (2008b) and Mochida and Lun (2008). Apart from these fundamental studies, also several CFD studies of pedestrian wind conditions in complex urban environments have been performed (Murakami, 1990a, Gadilhe et al., 1993, Takakura et al., 1993, Stathopoulos and Baskaran, 1996, Baskaran and Kashef, 1996, He and Song, 1999, Ferreira et al., 2002, Richards et al., 2002, Miles and Westbury, 2002, Westbury et al., 2002, Hirsch et al., 2002, Blocken et al., 2004, Yoshie et al., 2007, Blocken and Carmeliet, 2008, Blocken and Persoon, 2009). Almost all these studies were conducted with the steady RANS approach and a version of the k-ε model. An exception is the study by He and Song (1999) who used LES.

Accuracy of CFD. Attempts to provide general statements about the accuracy of steady RANS CFD for pedestrian-level wind studies can easily be compromised by the presence of a combination of physical modeling and numerical errors. Statements on the accuracy of steady RANS with a certain turbulence model should therefore be based on CFD studies that have undergone solution verification, i.e. it should be proven that numerical errors are limited, so clear conclusions about the physical modeling errors can be made. Several studies have adopted this approach in their validation of CFD with wind tunnel measurements and on-site measurements. A general observation from these studies is that the prediction accuracy is a pronounced function of the location in the flow pattern (and therefore of the wind direction). While several validation studies have been performed for multi-building configurations, at least two of those have provided conclusions on the accuracy of steady RANS CFD that can be generalized: the extensive validation study by Yoshie et al. (2007) for four different building and urban configurations and the validation study by Blocken and Carmeliet (2008). These two studies are discussed next.

In the framework of the development of the AIJ guideline for wind environment evaluation, Yoshie et al. (2007) reported validation studies for four different building and urban configurations: (1) an isolated square prism with ratio L:W:H = 1:1:2, (b) an idealized high-rise building surrounded by regularly spaced low-rise buildings, (c) building complexes in the actual urban area of Niigata, Japan, and (d) building complexes in the actual Shinjuku sub-central area in Tokyo, Japan. In all four cases, the simulations were performed with steady RANS, combined with the standard k-ε model or with revised k-ε models, and compared with the results of wind tunnel experiments. Note that the simulations included a grid-sensitivity analysis, careful application of the boundary conditions, higher-order discretization schemes, a complete report of the computational settings and parameters and a detailed comparison with the wind tunnel measurements. This is required in order to support the validity of the conclusions.

The simulations for the isolated building were made with the standard k-ε model and with two revised k-ε models: the Launder-Kato k-ε model (Kato and Launder, 1993) and the RNG k-ε model. Comparison of the standard k-ε model results with the wind tunnel measurements showed that the amplification factor U/U_0 (which is the ratio of the local pedestrian-level wind speed U to the wind speed U_0 that would occur at the same position without buildings) is generally predicted within an accuracy of 10% in the regions where $U/U_0 > 1$. In the wake region behind the building however, the predicted wind speed is generally significantly underestimated, at some locations by a factor 5 and more. The results of the other turbulence models showed a slight improvement in the high wind-speed regions, but worse results in the wake region. The underestimations in the wake region are attributed to the underestimation of turbulent kinetic energy in the wake, due to the fact that steady RANS with turbulence models such as the k-ε model is not capable of reproducing the vortex shedding in the wake of buildings (Yoshie et al., 2007, Tominaga et al., 2008b).

The simulations for the idealized high-rise building surrounded by low-rise buildings were made with the standard k-ε model and the RNG k-ε model. In the high wind-speed regions, the standard k-ε model underestimated the wind tunnel results by about 15%. In the lower wind speed regions, differences up to a factor 4 were found. The results of the RNG k-ε model showed improved performance in the high wind speed regions, but again a deteriorated performance in the lower wind speed regions. Similar conclusions on the different performance in high versus low wind speed regions were found for the CFD study for the actual urban area in Niigata: in high wind speed regions, the predictions are generally within 20% of the measurements, while the wind speed in low wind speed regions is generally significantly underestimated, at some positions with a factor 5 or more. The comparisons for the fourth configuration, the Shinjuku sub-central area, confirmed the findings for the other configurations. While for all four studies, large discrepancies are found in the low wind speed regions, it should be noted that the high wind speed regions are those of interest for pedestrian-level wind studies. In these regions, steady RANS was shown to provide a good to very good accuracy (10-20%).

Blocken and Carmeliet (2008) performed steady RANS CFD simulations with the realizable k-ε model for three configurations of parallel buildings and compared the results with the sand-erosion wind tunnel experiments by Beranek (1982). Their observations were very similar to those by Yoshie et al. (2007): a close to very close agreement between CFD and wind tunnel measurements in the region of high U/U_0 (about 10% accuracy), and significant underestimations in the regions of lower U/U_0. Note that these underestimations were not only found in the wake of the buildings, but also in the low-speed stagnation region upstream of the buildings. Similar to the results by Yoshie et al. (2007), the underestimations can go up to a factor 5 or more. Note that also these simulations were based on grid-sensitivity analysis, careful application of the boundary conditions and higher order discretization schemes. It should be noted that sand-erosion measurement results are generally considered to be less suitable for CFD validation, although in this study a very close agreement – both qualitatively and quantitatively – was found in the high U/U_0 region.

For assessing of the accuracy of CFD for pedestrian-level wind studies, it is important to compare them not only with wind tunnel measurements – where the boundary conditions are generally well-known – but also with on-site measurements. However, CFD pedestrian-level wind studies in complex urban environments including a comparison with on-site measurements are very scarce. To the knowledge of the authors, only two such studies have been published: the study by Yoshie et al. (2007) for the Shinjuku Sub-central area in Tokyo and the study by Blocken and Persoon (2009) for the area around the multifunctional

ArenA stadium in Amsterdam. Although these measurements were quite limited, overall, the comparisons confirmed the main conclusions made earlier, albeit that the discrepancies in the high wind speed regions can exceed 10%.

Practical applicability. In spite of the very limited number of validation studies based on on-site measurements, CFD is gaining increasing acceptance as a tool for pedestrian-level wind studies. This was recently confirmed by the publication of the new Dutch Wind Nuisance Standard, NEN8100 (NEN, 2006, Willemsen and Wisse, 2007) that specifically allows the user to choose between wind tunnel testing and CFD for analyzing the pedestrian wind environment. The standard also demands quality assurance, both for wind tunnel testing and for CFD. CFD solution verification and validation and complete reporting of the followed procedure are essential components of quality assurance. In practical situations in the case of complex urban environments, when measurements are often not available, CFD model validation should be performed for simpler configurations, the flow features of which show resemblance with those expected in the actual complex urban configuration (Oberkampf et al., 2004, Blocken et al., 2004, Franke et al., 2007, Yoshie et al., 2007, Blocken and Carmeliet, 2008, Tominaga et al., 2008a). Blocken and Carmeliet (2008) called this approach sub-configuration validation. For these simpler cases, wind tunnel measurement data are generally available in the literature. Note that steady RANS is the commonly used method, while LES is still considered out of reach for practical pedestrian-level wind studies in actual urban environments (Yoshie et al., 2007). This is mainly attributed to the much larger calculation time. For pedestrian-level wind studies, simulations need to be performed for many (e.g. 12 or 16) wind directions, and this needs to be repeated for configurations with remedial measures implemented (Yoshie et al., 2007). Nevertheless, it is expected that the increase in computing power and speed together with the superior performance of LES will render it increasingly more attractive in the years to come.

5.2. Pollutant dispersion

Importance. Outdoor air pollution is one of the major environmental problems today. It is associated with a broad spectrum of acute and chronic health effects (e.g. Brunekreef and Holgate, 2002). In the built environment, both the outdoor exposure of pedestrians and the indoor exposure of building inhabitants are of concern. Outdoor and indoor air pollution are a main concern of building and air-conditioning engineers that design the ventilation inlets and outlets on building facades or roofs (Drivas and Shair, 1974, ASHRAE, 1999, 2007). Indoor air pollution by outdoor air pollutants can be caused by the re-ingestion of the contaminated exhaust air by the same building or by the intake of exhaust from other sources such as nearby buildings, street traffic, vehicle parking lots and loading docks and emergency generators (Smeaton et al., 1991). The precise prediction of pollutant concentration distributions on and near buildings is important for building design and evaluation. The prediction of such concentrations however is a difficult task, especially in the urban environment. It does not only require the knowledge of air pollution meteorology and dispersion, it also requires knowledge of building aerodynamics because wind and buildings can strongly affect plume behavior.

CFD versus measurements and semi-empirical formulae. Pollutant concentration distributions can be assessed by on-site measurements, wind tunnel measurements, (semi-)empirical formulae and CFD. Several on-site measurement campaigns have been performed (e.g., Barad, 1958, Wilson and Lamb, 1994, Lazure et al., 2002, Stathopoulos et al., 2002, 2004). They are very valuable because they are conducted in the real atmospheric boundary layer and provide information on the real complexity of the phenomenon, but they are also time-consuming, expensive, and not an option in the design stage of a new building or urban area. Many wind tunnel experiments have been conducted in the past to better understand the mechanisms of wind-induced pollutant dispersion (e.g., Halitsky, 1963, Huber and Snyder, 1982, Li and Meroney, 1983, Saathoff et al., 1995, 1998, Leitl et al., 1997, Meroney et al., 1999, Stathopoulos et al., 2002, 2004). The drawbacks of wind tunnel tests are that they can be time-consuming and costly, that they are not applicable for light wind conditions, and that scaling – similarity – can be a difficult issue. Semi-empirical models, such as the Gaussian model (Turner, 1970, Pasquill and Smith, 1983) and the so-called ASHRAE models (Wilson and Lamb, 1994, ASHRAE, 1999, 2003) are relatively simple and easy-to-use, at the expense of limited applicability and less accurate estimates. The Gaussian model, in its original form, is not applicable when there are obstacles between the emission source and the receptor, and the ASHRAE models only evaluate the minimum dilution factor on the plume centerline. CFD can (potentially) avoid a number of these disadvantages.

CFD for dispersion around an isolated cubic building. In the past two decades, a very large number of so-called micro-scale CFD simulations of pollutant dispersion around buildings and in urban areas have been conducted. Micro-scale generally refers to simulations with horizontal length scales smaller than 5 km. These simulations can be divided into generic and applied studies. Many generic studies have focused on very simplified configurations, such as the isolated building (e.g., Li and Meroney, 1983, Leitl et al., 1997, Selvam, 1997, Tominaga et al., 1997, Li and Stathopoulos, 1997, Meroney et al., 1999, Tominaga and Stathopoulos, 2007a, 2007b, 2008, 2009, Blocken et al., 2008b) and the idealized street canyon (e.g., Leitl and Meroney, 1997, Chan et al., 2002, Baik and Kim, 2002, Kim and Baik, 2004, Di Sabatino et al., 2008, Gromke et al., 2008, Buccolieri et al., 2009). These generic studies have proven to be very suitable for verification, validation and sensitivity analysis. The reason is that, even although both situations are strong simplifications of reality, the flow and dispersion processes involved are very complex and contain most of the salient features that are also present in the complex urban environment. Apart from these generic studies, several applied studies have recently been performed. Some of these have included LES and RANS simulations of gas plume spreading in very large computational domains (e.g., Hanna et al., 2006, Patnaik et al., 2007, Löhner et al., 2008). However, the grid resolution (e.g. 2-6 m) in these studies was often much lower than in the generic studies (e.g. 0.1 m). The lower resolution could compromise the prediction accuracy around individual buildings and close to the building surfaces, but it should be noted that this was not the primary intent of these large-scale studies.

As opposed to pedestrian-level wind conditions, where the body of CWE literature is relatively limited, many publications exist on CFD simulation of dispersion around buildings and in urban areas. For the purpose of discussion in this chapter, only one single and simple case is considered: wind-induced dispersion of low-momentum exhaust from a vent in the middle of the roof of an isolated cubic building. Apart from limiting the extent of this chapter, the reasons for this very narrow selection are: (1) in this case the focus is at the scale of the individual building (i.e. on pollution / contamination near potential air intake openings). As mentioned above, the resolution in studies in very large computational domains is often too low (> 2 m) to allow accurate predictions across individual building surfaces; (2) detailed experimental data are available for this case (Li and Meroney, 1983); (3) several researchers independently performed CFD simulation and validation studies for this case (both with steady RANS and LES); (4) the influence of different RANS turbulence models and of the value of the turbulent Schmidt number Sc_t was analyzed; and (5) the results allow a clear identification of the main difficulties in CFD modeling of pollutant dispersion.

Figure 4 illustrates the situation by mean wind-velocity vectors in a vertical plane through the middle of the building. It shows that the emitted pollutant is caught in the recirculation bubble and is advected to the upstream building edge. CFD simulations for this configuration have been performed by e.g. Wang (2006), Tominaga and Stathopoulos (2007a, 2008) and Blocken et al. (2008a). Figure 5a shows the wind tunnel results by Li and Meroney (1983) as contours of the dimensionless concentration coefficient K on the roof. K is defined as CU_HL^2/Q_e, where C is the mass fraction of the tracer gas, U_H the undisturbed wind speed at roof height (m/s), L the size of the cubic building model (m) and Q_e the emission rate of the pollutant (m³/s).

To numerically reproduce these experiments, Wang (2006) used the realizable k-ε model and the Reynolds Stress Model (RSM; Launder et al., 1975) with a linear pressure–strain model and wall-reflection effects (Gibson and Launder 1978, Launder 1989). She employed two values of the turbulent Schmidt number: Sc_t = 0.3 and Sc_t = 0.7 and the QUICK discretization scheme (Leonard, 1979). The results are given in Figures 5b-d. While the upstream advection could not be adequately predicted with the realizable k-ε model, the results by the RSM are much better, although lateral dispersion is considerably underestimated by the simulations.

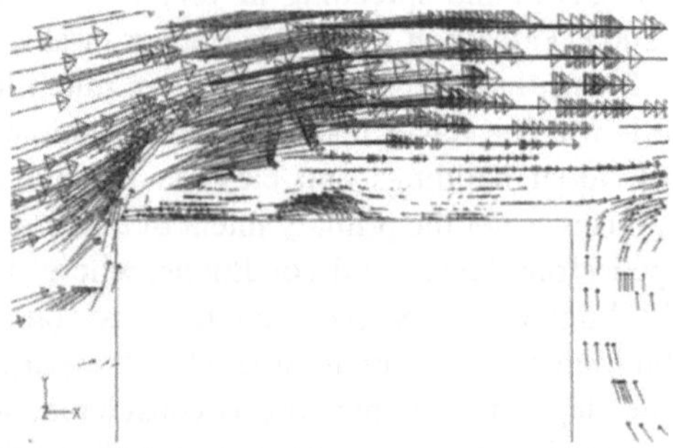

Figure 4. Vertical cross-section through vent in middle of roof. Wind velocity vectors illustrating the wind flow over the building and the escape of pollutants from the vent: the exhaust is trapped in the separation bubble and is advected upstream (results by the present author).

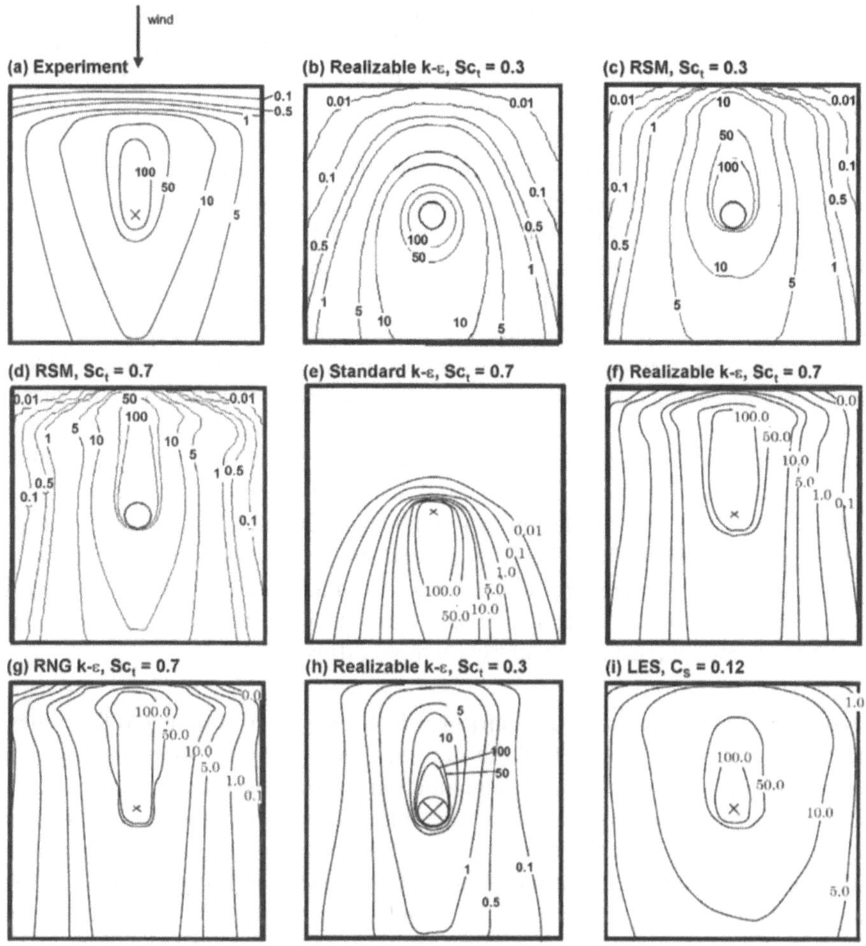

Figure 5. Contours of dimensionless concentration coefficient K on the cubic building roof: (a) wind tunnel measurements (Li and Meroney, 1983); (b-i) CFD results by: (b-d): Wang (2006); (e-g) Tominaga and Stathopoulos (2007a); (h) Blocken et al. (2008b); (i) Tominaga and Stathopoulos (2008). Results are presented for different turbulence models and/or turbulent Schmidt numbers Sc_t. C_S is the Smagorinsky constant.

Tominaga and Stathopoulos (2007a) tested different turbulence models, including the standard k-ε model, the RNG k-ε model and the realizable k-ε model, in combination with Sc_t = 0.3, 0.7 and 1.0, and the QUICK discretization scheme. Some of their results are given in Figure 5e-g. The standard k-ε model could not predict the recirculation on the roof and the realizable k-ε model underestimated it, but better predictions were obtained with the RNG k-ε model, which showed the best agreement with wind tunnel experiments of the separation bubble. Nevertheless, the results still show some underestimation of the lateral dispersion, which is clear by observing the concentrations near the side edges. Blocken et al. (2008b) tested the realizable k-ε model and the RSM, for $Sc_t = 0.2, 0.3, 0.5, 0.7, 1.0$, with second order upwind discretization. Both turbulence models correctly predicted the upstream dispersion, with the best result for $Sc_t = 0.7$, but the lateral underestimation of dispersion found by these authors was significantly more pronounced than that found by Wang (2006) and Tominaga and Stathopoulos (2007a) (Fig. 5h). They attributed this partly to an underestimation of the approach flow turbulent kinetic energy due to unintended streamwise gradients in the approach flow. The differences in the realizable k-ε model results between Wang (2006) (no prediction of recirculation) and Tominaga and Stathopoulos (2007a) and Blocken et al. (2008b) are probably due to artificial diffusion and near-wall treatment limitations associated with the lower grid resolution in Wang's case (2006). Finally, Tominaga and Stathopoulos (2008) applied LES (Fig. 5i) and showed that taking into account the unsteadiness of the separation bubble leads to a much better prediction of lateral dispersion and a much closer agreement with the wind tunnel experiments, although the numerical results are a little bit more diffusive.

Accuracy of CFD and practical applicability. A general conclusion from these and other generic studies is that the accuracy of pollutant dispersion modeling depends highly on the choice of computational parameters by the user. Fig. 5 shows that errors exceeding one order of magnitude can occur at some positions on the roof. Another general conclusion is that RANS simulations in combination with typical turbulent Schmidt numbers of 0.7-0.9 systematically provide too low lateral turbulent diffusion compared to wind tunnel testing. In the past, several authors have attributed this to the fact that steady RANS modeling cannot incorporate the inherently transient behavior of separation and recirculation downstream of windward edges, and of von Karman vortex shedding in the wake, which are particularly important for pollutant dispersion (Leitl et al., 1997, Meroney et al., 1999, Blocken et al., 2008b). In many simulations, Sct has been explicitly used as a tuning factor to compensate for these deficiencies of steady RANS modeling (e.g., Tominaga and Stathopoulos, 2007b, Blocken et al., 2008b). Note that, while Sc_t is generally taken as a constant, it has been shown to be a function of the flow field (e.g., Koeltzsch, 2000). Leitl et al. (1997), using steady RANS on quite dense grids, found deviations between CFD results and wind tunnel measurements of up to a factor 5, and Meroney et al. (1999) found differences up to one order of magnitude. The latter authors made two important remarks: (1) no attempt was made to force-fit agreement between measurements and simulations; and (2) the discrepancies were attributed to the lack of intermittency (unsteadiness) in the simulations. Tominaga and Stathopoulos (2007a, 2008) showed that LES modeling, which takes into account the unsteadiness because it actually solves the large eddies in the flow, can strongly improve pollutant concentration predictions compared to steady RANS. This seems to indicate that LES modeling is a requirement for pollutant dispersion modeling. This is certainly the case when unsteady releases have to be considered, and/or when concentration fluctuations are important. However, the computational demands associated with this are very large. The statement by Yoshie et al. (2007) that LES is still out of reach for practical pedestrian-level wind studies may equally apply to pollutant dispersion modeling. Note that, as mentioned before, even the very extensive LES modeling efforts in complex urban environments, supported by efficient grid generation techniques and parallel computing facilities, were performed with relatively low resolution ($\geq$ 2 m). The application of LES to practical dispersion problems in urban environments however is not only limited by increased model complexity and computational resources, but also by the lack of detailed validation and sensitivity studies. This is important because, as the model formulation increases in complexity, the likelihood of degrading the model's performance due to input data and model parameter uncertainty increases as well (Hanna, 1989). In spite of this, the superior performance of LES is expected to drive the gradual replacement of steady RANS by LES as the modeling approach for dispersion studies.

6. Discussion

The atmospheric boundary-layer wind-flow pattern in an actual urban area is very complex. The flow around a simple isolated cubic building model however contains many of the salient features that are also present in the flow in an actual urban area. Partly because of this reason, CWE model development, verification and validation in the past three decades have for a large part focused on the simplest generic case: the isolated (cubic) building model. Another main reason is that many previous wind tunnel experiments were performed for this simple model and that – as a result – these data were available for CFD validation.

The focus on the simple isolated (cubic) building has allowed researchers to clearly identify some important difficulties, advantages and disadvantages of CFD for various applications. For both applications treated in this chapter, the same main limitation of steady RANS CFD was noted: the inability to accurately reproduce the flow field downstream of the windward facade, i.e. in the separation and wake regions that are inherently transient and characterized by low-velocity recirculations.

The majority of CWE studies in the past three decades employed steady RANS modeling. An increasing number of studies is exploring the use of LES. LES can resolve the inherently transient behavior of separation and recirculation downstream of windward edges, and of von Karman vortex shedding in the wake. As a result, the *potential* accuracy of LES is clearly superior. However, simulations with LES are considerably more complex and more computationally expensive than their steady RANS counterparts. LES requires specific time and space resolved inlet boundary condition data, adjusted grid cell distributions, temporal resolutions, sampling times, etc. There is still a lack of detailed verification, validation and sensitivity studies for LES modeling of atmospheric boundary layer flows. Note that currently a set of extensive best practice documents exist for steady RANS applications in CWE, but that this is not (yet) the case for LES.

In spite of some current deficiencies, CFD has become an indispensible tool in Wind Engineering, in different categories of studies. Examples are:

1. Studies for which accurate wind tunnel experiments are very difficult or not possible (e.g. similarity issues for multiphase flow, flow with thermal effects or too extensive terrains);
2. Studies for which (semi-)empirical models are not applicable or do not provide sufficient accuracy;
3. Studies for which high-resolution or "whole-flow-field" data are requested; and
4. Studies in which the influence of a large number of (small) geometrical design changes need to be analyzed.

Topics belonging to these categories include buoyancy-induced pollutant dispersion (e.g., Baik et al., 2003, Xie et al., 2007, Huang et al., 2008, Cheng et al., 2009), wind-driven rain on buildings (e.g., Choi 1993, 1994, Hangan, 1999, Davidson et al. 2000, Blocken and Carmeliet, 2004b, 2006, Tang and Davidson, 2004, Briggen et al., 2009, Blocken et al., 2010), buoyancy-induced natural ventilation of buildings (e.g., Jiang and Chen, 2003, van Hooff and Blocken, 2010, Norton et al., 2010), urban-scale heat transfer (e.g. Baik et al., 2003, 2009, Huang et al., 2008) and wind conditions for very extensive terrains (e.g. Hanna et al., 2006, Patnaik et al., 2007, Löhner et al., 2008, Tamura, 2008, Baik et al., 2009).

7. Summary and conclusions

The application of CFD in Wind Engineering is part of Computational Wind Engineering (CWE). CFD can be a powerful alternative for and/or addition to measurements, because it can avoid some typical measurement limitations. It can provide detailed information on the relevant flow variables in the whole calculation domain ("whole-flow field data"), under well-controlled conditions and without similarity constraints. However, the accuracy of CFD is an important matter of concern, and verification and validation studies to assess and limit numerical and physical modeling errors are imperative.

CFD simulation of wind flow around buildings started with fundamental studies for isolated buildings, often with a cubical shape. Much has been learned from these pioneering efforts, which included verification and validation studies as well as more general sensitivity studies in which the influence of many computational parameters was investigated. The early sensitivity studies have provided the basis for the current best practice guidelines. Because the user of a CFD code has to make a large number of choices, and because these choices can have a very large impact on the results, best practice guidelines are very important and provide strong support for CWE applications.

Applications in CWE are many, and only two have been briefly addressed in this chapter: pedestrian-level wind conditions around buildings and pollutant dispersion across the roof of an isolated (cubic) building. The two most commonly used approaches in CWE are steady RANS and LES. Past studies seem to indicate that steady RANS can be suitable for the qualitative and quantitative assessment of high wind speed regions around buildings for pedestrian-level wind studies. Previous studies also indicate that steady RANS modeling with the commonly used turbulence models is often deficient for pollutant dispersion studies, while LES shows superior performance. The reason is that it actually resolves the large vortical structures that are important for mass transfer. Choosing between RANS and LES is a trade-off between computational cost and complexity on the one hand and *potential* accuracy on the other. The computational cost for LES is significantly higher than for RANS. LES is also more complex than steady RANS modeling and requires more input data. Because steady RANS and LES are fundamentally different approaches, it is important that specific and extensive sets of best practice guidelines for LES for CWE applications are developed, as has been done for RANS in the past decade. This will allow increasing the accuracy and reliability of LES in CWE.

8. References to Literature

AIAA. 1998. *AIAA Guide for the verification and validation of Computational Fluid Dynamics simulations*. American Institute of Aeronautics and Astronautics Staf. (AIAA G-077-1998).

ASHRAE. (1999). Building air intake and exhaust design. *Applications Handbook*. Chapter 43.

ASHRAE. (2003). Building air intake and exhaust design. *ASHRAE Application Handbook*. Chapter 44.

ASHRAE. (2007). Building air intake and exhaust design. *Applications Handbook*. Chapter 44.

Baetke, F., Werner, H., and Wengle, H. (1990). Numerical simulation of turbulent flow over surface-mounted obstacles with sharp edges and corners. *Journal of Wind Engineering and Industrial Aerodynamics* 35(1-3): 129-147.

Baik, J. J., and Kim, J. J. (2002). On the escape of pollutants from urban street canyons. *Atmospheric Environment* 36(3): 527-536.

Baik, J. J., Kim, J. J., and Fernando, H. J. S. (2003). A CFD model for simulating urban flow and dispersion. *Journal of Applied Meteorology* 42(11): 1636-1648.

Baik, J. J., Park, S. B., and Kim, J. J. (2009). Urban flow and dispersion simulation using a CFD model coupled to a mesoscale model. *Journal of Applied Meteorology and Climatology* 48(8): 1667-1681.

Barad, M. L. (1958). Project prairie grass. A field program in diffusion. *Geophysical Research Paper*, No. 59, vols. I and II, Report AFCRC-TR-58-235. Air Force Cambridge Research Center, Bedford, MA.

Bartzis, J. G., Vlachogiannis, D., and Sfetsos, A. (2004). Thematic area 5: Best practice advice for environmental flows. *The QNET-CFD Network Newsletter* 2(4): 34-39.

Baskaran, A., and Kashef, A. (1996). Investigation of air flow around buildings using computational fluid dynamics techniques. *Engineering Structures* 18(11): 861-873.

Baskaran, A., and Stathopoulos, T. (1989). Computational evaluation of wind effects on buildings. *Building and Environment* 24(4): 325–333.

Baskaran, A., and Stathopoulos, T. (1992). Influence of computational parameters on the evaluation of wind effects on the building envelope. *Building and Environment* 27(1): 39–49.

Bejan, A. (2004). *Convection heat transfer.* 3rd Ed., Wiley.

Beranek, W. J. (1982). *Beperken van windhinder om gebouwen, deel 2*, Stichting Bouwresearch no. 90, Kluwer Technische Boeken BV, Deventer (in Dutch).

Beranek, W. J. (1984). Wind environment around single buildings of rectangular shape. *Heron* 29(1): 4–31.

Beranek, W. J., and Van Koten, H. (1979). *Beperken van windhinder om gebouwen, deel 1*, Stichting Bouwresearch no. 65, Kluwer Technische Boeken BV, Deventer (in Dutch).

Blocken, B., and Carmeliet, J. (2004a). Pedestrian wind environment around buildings: Literature review and practical examples. *Journal of Thermal Envelope and Building Science* 28(2): 107-159.

Blocken, B., and Carmeliet, J. (2004b). A review of wind-driven rain research in building science. *Journal of Wind Engineering and Industrial Aerodynamics* 92(13): 1079-1130.

Blocken, B., Roels, S., and Carmeliet, J. (2004). Modification of pedestrian wind comfort in the Silvertop Tower passages by an automatic control system. *Journal of Wind Engineering and Industrial Aerodynamics* 92(10): 849-873.

Blocken, B., and Carmeliet, J. (2006). The influence of the wind-blocking effect by a building on its wind-driven rain exposure. *Journal of Wind Engineering and Industrial Aerodynamics* 94(2): 101-127.

Blocken, B., Stathopoulos, T., and Carmeliet, J. (2007a). CFD simulation of the atmospheric boundary layer: wall function problems. *Atmospheric Environment* 41(2): 238-252.

Blocken, B., Carmeliet, J., and Stathopoulos, T. (2007b). CFD evaluation of wind speed conditions in passages between parallel buildings—effect of wall-function roughness modifications for the atmospheric boundary layer flow. *Journal of Wind Engineering and Industrial Aerodynamics* 95(9-11): 941-962.

Blocken, B., and Carmeliet, J. (2008). Pedestrian wind conditions at outdoor platforms in a high-rise apartment building: generic sub-configuration validation, wind comfort assessment and uncertainty issues. *Wind and Structures* 11(1): 51-70.

Blocken, B., Moonen, P., Stathopoulos, T., and Carmeliet, J. (2008a). A numerical study on the existence of the Venturi-effect in passages between perpendicular buildings. *Journal of Engineering Mechanics – ASCE* 134(12): 1021-1028.

Blocken, B., Stathopoulos, T., Saathoff, P., and Wang, X. (2008b). Numerical evaluation of pollutant dispersion in the built environment: comparisons between models and experiments. *Journal of Wind Engineering and Industrial Aerodynamics* 96(10-11): 1817-1831.

Blocken, B., and Persoon, J. (2009). Pedestrian wind comfort around a large football stadium in an urban environment: CFD simulation, validation and application of the new Dutch wind nuisance standard. *Journal of Wind Engineering and Industrial Aerodynamics* 97(5-6): 255-270.

Blocken, B., Deszö, G., van Beeck, J., and Carmeliet, J. (2010). Comparison of calculation methods for wind-driven rain deposition on building facades. *Atmospheric Environment* 44(14): 1714-1725.

Bottema, M. (1993). *Wind climate and urban geometry.* PhD thesis, Eindhoven University of Technology, 212p.

Briggen, P. M., Blocken, B., Schellen, H. L. (2009). Wind-driven rain on the facade of a monumental tower: numerical simulation, full-scale validation and sensitivity analysis. *Building and Environment* 44(8): 1675-1690.

Brunekreef, B., and Holgate, S. T., (2002). Air pollution and health. *Lancet,* 360(9341): 1233-1242.

Buccolieri, R., Gromke, C., Di Sabatino, S., and Ruck, B. (2009). Aerodynamic effects of trees on pollutant concentration in street canyons. *Science of the Total Environment* 407(19): 5247-5256.

Casey, M., and Wintergerste, T. (2000). *Best Practice Guidelines*, ERCOFTAC Special Interest Group on Quality and Trust in Industrial CFD, ERCOFTAC, Brussels.

Chan, T. L, Dong, G., Leung, C. W., Cheung, C. S., Hung, T. W. (2002). Validation of a two-dimensional pollutant dispersion model in an isolated street canyon. *Atmospheric Environment* 36(5): 861-872.

Cheng, W. C., Liu, C. H., and Leung, D. Y. C. (2009). On the correlation of air and pollutant exchange for street canyons in combined wind-buoyancy-driven flow. *Atmospheric Environment* 43(24): 3682-3690.

Choi, E. C. C. (1993). Simulation of wind-driven rain around a building. *Journal of Wind Engineering and Industrial Aerodynamics* 46&47: 721-729.

Choi, E. C. C. (1994). Determination of wind-driven-rain intensity on building faces. *Journal of Wind Engineering and Industrial Aerodynamics* 51: 55-69.

Cowan, I. R., Castro, I. P., and Robins, A. G. (1997). Numerical considerations for simulations of flow and dispersion around buildings, *Journal of Wind Engineering and Industrial Aerodynamics* 67 & 68: 535-545.

Davidson, C. I., Tang, W., Finger, S., Etyemezian, V., Striegel, M. F., and Sherwood, S. I. (2000). Soiling patterns on a tall limestone building: changes over 60 years, *Environmental Science and Technology* 34(4): 560–565.

Di Sabatino, S., Buccolieri, R., Pulvirenti, B., and Britter, R. E. (2008). Flow and pollutant dispersion in street canyons using FLUENT and ADMS-Urban. *Environmental Modeling & Assessment* 13(3): 369-381.

Drivas, P. J., and Shair, F.H. (1974). Probing the air flow within the wake downwind of a building by means of a tracer technique. *Atmospheric Environment* 8: 1165-1175.

Ferreira, A. D., Sousa, A. C. M., and Viegas, D. X. (2002). Prediction of building interference effects on pedestrian level comfort. *Journal of Wind Engineering and Industrial Aerodynamics* 90(4-5): 305-319.

Ferziger, J. H., and Peric, M. (1996). *Computational methods for fluid dynamics*. Springer Berlin, 356p.

Franke, J., Hirsch, C., Jensen, A. G., Krüs, H. W., Schatzmann, M., Westbury, P. S., Miles, S. D., Wisse, J. A., and Wright, N. G. (2004). Recommendations on the use of CFD in wind engineering. In: van Beeck, J. P. A. J. (Ed.), *Proceedings of the International Conference on Urban Wind Engineering and Building Aerodynamics*. COST Action C14, Impact of Wind and Storm on City Life Built Environment. Von Karman Institute, Sint-Genesius-Rode, Belgium, 5–7 May 2004.

Franke, J., and Frank, W. (2005). Numerical simulation of the flow across an asymmetric street intersection. In: *Proceedings of the Fourth European and African Conference on Wind Engineering (4EACWE)*, 11–15 July 2005, Prague, Czech Republic.

Franke, J., Hellsten, A., Schlünzen, H., and Carissimo, B. (Eds.), (2007). *Best practice guideline for the CFD simulation of flows in the urban environment*. COST Office Brussels.

Franke J., and Frank, W. (2008). Application of generalized Richardson extrapolation to the computation of the flow across an asymmetric street intersection. *Journal of Wind Engineering and Industrial Aerodynamics* 96(10-11): 1616-1628.

Gadilhe, A., Janvier, L., and Barnaud, G. (1993). Numerical and experimental modelling of the three-dimensional turbulent wind flow through an urban square. *Journal of Wind Engineering and Industrial Aerodynamics* 46-47: 755-763.

Gibson, M. M., and Launder, B. E. (1978). Ground effects on pressure fluctuations in the atmospheric boundary layer. *Journal of Fluid Mechanics* 86: 491–511.

Gorlé, C., van Beeck, J., Rambaud, P., and Van Tendeloo, G. (2009). CFD modelling of small particle dispersion: the influence of the turbulence kinetic energy in the atmospheric boundary layer. *Atmospheric Environment* 43(3): 673-681.

Gromke, C., Buccolieri, R., Di Sabatino, S., and Ruck, B. (2008). Dispersion study in a street canyon with tree planting by means of wind tunnel and numerical investigations - Evaluation of CFD data with experimental data. *Atmospheric Environment* 42(37): 8640-8650.

Halitsky, J. (1963). Gas diffusion near buildings. *ASHRAE Transactions* 69: 464–485.

Hall, R. C. (Ed.) (1997). *Evaluation of modelling uncertainty. CFD modelling of near-field atmospheric dispersion.* Project EMU final report, European Commission Directorate–General XII Science, Research and Development Contract EV5V-CT94- 0531, WS Atkins Consultants Ltd., Surrey.

Hangan, H. (1999). Wind-driven rain studies. A C-FD-E approach. *Journal of Wind Engineering and Industrial Aerodynamics* 81: 323–331.

Hanna, S. R. (1989). Plume dispersion and concentration fluctuations in the atmosphere. Encyclopedia of Environmental Control Technology. In: *Air Pollution Control*, Vol. 2. Gulf Publishing Company, Houston, TX, 547-582.

Hanna, S. R., Brown, M. J., Camelli, F. E., Chan, S. T., Coirier, W. J., Hansen, O. R., Huber, A. H., Kim, S., Reynolds, R. M. (2006). Detailed simulations of atmospheric flow and dispersion in downtown Manhattan: An application of five computational fluid dynamics models. *Bulletin of the American Meteorological Society* 87(12): 1713+.

Hanson, T., Summers D. M., and Wilson, C.B. (1986). Validation of a computer simulation of wind flow over a building model. *Building and Environment* 21: 97–111.

Hargreaves, D.M., and Wright, N.G. (2007). On the use of the k–ε model in commercial CFD software to model the neutral atmospheric boundary layer. *Journal of Wind Engineering and Industrial Aerodynamics* 95(5): 355-369.

He, J., and Song, C.C.S. (1999). Evaluation of pedestrian winds in urban area by numerical approach. *Journal of Wind Engineering and Industrial Aerodynamics* 81, 295–309.

Hirsch, C., Bouffioux, V., and Wilquem, F. (2002). CFD simulation of the impact of new buildings on wind comfort in an urban area. *Workshop Proceedings, Cost Action C14, Impact of Wind and Storm on City Life and Built Environment*, Nantes, France.

Huang, H., Ooka, R., Chen, H., Kato, S., Takahashi, T., and Watanabe, T. (2008). CFD analysis on traffic-induced air pollutant dispersion under non-isothermal condition in a complex urban area in winter. *Journal of Wind Engineering and Industrial Aerodynamics* 96(10-11): 1774-1788.

Huber, A.H., and Snyder, W.H. (1982). Wind tunnel investigation of the effects of a rectangular shaped building on dispersion of effluent from short adjacent stacks. *Atmospheric Environment* 16(12): 2837–2848.

Jiang, Y., and Chen, Q. (2003). Buoyancy-driven single-sided natural ventilation in buildings with large openings. *International Journal of Heat and Mass Transfer* 46(6): 973-988.

Jones, W. P., and Launder, B. E. (1972). The prediction of laminarization with a 2-equation model of turbulence. *International Journal of Heat and Mass Transfer* 15: 301.

Kato, M., and Launder, B. E., (1993). The modelling of turbulent flow around stationary and vibrating square cylinders. In: *9th Symposium on Turbulent Shear Flows*, pp. 10–14.

Kim J. J., and Baik J.J. (2004). A numerical study of the effects of ambient wind direction on flow and dispersion in urban street canyons using the RNG k–ε turbulence model. *Atmospheric Environment* 38(19): 3039-3048.

Koeltzsch, K. (2000). The height dependence of the turbulent Schmidt number within the boundary layer. *Atmospheric Environment* 34: 1147-1151.

Launder, B. E., Reece, G. J., and Rodi, W. (1975). Progress in the development of a Reynolds-stress turbulence closure. *Journal of Fluid Mechanics* 68(3): 537–566.

Launder, B. E. (1989). Second-moment closure and its use in modeling turbulent industrial flows. *International Journal for Numerical Methods in Fluids* 9: 963–985.

Launder, B. E, and Spalding, D.B. (1974). The numerical computation of turbulent flows. Computer Methods in Applied Mechanics and Engineering 3:269-89.

Lawson, T. V., and Penwarden, A.D. (1975). The effects of wind on people in the vicinity of buildings. *4th Int. Conf. Wind Effects on Buildings and Structures*, Heathrow.

Lazure, L., Saathoff, P., and Stathopoulos, T. (2002). Air intake contamination by building exhausts: tracer gas investigation of atmospheric dispersion models in the urban environment. *Journal of the Air Waste Management Association* 52: 160–166.

Leitl, B. M., Kastner-Klein, P., Rau, M., and Meroney, R.N. (1997). Concentration and flow distributions in the vicinity of U-shaped buildings: wind-tunnel and computational data. *Journal of Wind Engineering and Industrial Aerodynamics* 67&68: 745-755.

Leitl, B. M., and Meroney, R. (1997). Car exhaust dispersion in a street canyon. Numerical critique of a wind tunnel experiment. *Journal of Wind Engineering and Industrial Aerodynamics* 67-68: 293-304.

Leonard, B. P. (1979). A stable and accurate convection modelling procedure based on quadratic upstream interpolation. *Computer Methods in Applied Mechanics and Engineering* 19: 59–98.

Li, W., and Meroney, R.N. (1983). Gas dispersion near a cubical model building. Part I. Mean concentration measurements. *Journal of Wind Engineering and Industrial Aerodynamics* 12(1): 15-33.

Li, Y., and Stathopoulos, T. (1997). Numerical evaluation of wind-induced dispersion of pollutants around a building. *Journal of Wind Engineering and Industrial Aerodynamics* 67&68: 757-766.

Livesey, F., Inculet, D., Isyumov, N., and Davenport, A.G. (1990). A Scour technique for evaluation of pedestrian winds, *Journal of Wind Engineering and Industrial Aerodynamics* 36: 779–789.

Löhner, R., Cebral, J. R., Camelli, F. E., Appanaboyina, S., Baum, J. D., Mestreau, E. L., and Soto, O.A. (2008). Adaptive embedded and immersed unstructured grid techniques. *Computer Methods in Applied Mechanics and Engineering* 197(25-28): 2173-2197.

Menter, F. (1997). Eddy viscosity transport equations and their relation to the k–ε model. *Journal of Fluids Engineering* 119: 876–884.

Menter, F., Hemstrom, B., Henrikkson, M., Karlsson, R., Latrobe, A., Martin, A., Muhlbauer, P., Scheuerer, M., Smith, B., Takacs, T., and Willemsen, S. (2002). *CFD Best Practice Guidelines for CFD Code Validation for Reactor-Safety Applications*, Report EVOLECORA-D01, Contract No. FIKS-CT-2001-00154.

Meroney, R. N., Leitl, B. M., Rafailidis, S., and Schatzmann, M. (1999). Wind-tunnel and numerical modeling of flow and dispersion about several building shapes. *Journal of Wind Engineering and Industrial Aerodynamics* 81: 333-345.

Miles, S. D., and Westbury, P. S. (2002). Assessing CFD as a tool for practical wind engineering applications. *Proc. Fifth UK Conf. Wind Engineering*, September.

Mittal, R., and Iaccarino, G. (2005). Immersed boundary methods. *Annual Review of Fluid Mechanics* 37: 239-261.

Murakami, S., Mochida, A., and Hibi, K. (1987). Three-dimensional numerical simulation of airflow around a cubic model by means of large eddy simulation. *Journal of Wind Engineering and Industrial Aerodynamics 25:* 291–305.

Murakami, S., and Mochida, A. (1988). 3-D numerical simulation of airflow around a cubic model by means of the k-ε model. *Journal of Wind Engineering and Industrial Aerodynamics* 31(2-3): 283-303.

Murakami, S., and Mochida, A. (1989). Three-dimensional numerical simulation of turbulent flow around buildings using the k–ε turbulence model. *Building and Environment* 24(1): 51-64.

Murakami, S. (1990a). Computational wind engineering. *Journal of Wind Engineering and Industrial Aerodynamics* 36(1): 517-538.

Murakami, S. (1990b). Numerical simulation of turbulent flowfield around cubic model: current status and applications of k-e model and LES. *Journal of Wind Engineering and Industrial Aerodynamics* 33(1-2): 139-152.

Murakami, S., Mochida, A., and Hayashi, Y. (1990). Examining the k-ε model by means of a wind tunnel test and large-eddy simulation of the turbulence structure around a cube. *Journal of Wind Engineering and Industrial Aerodynamics* 35: 87-100.

Murakami, S., Mochida, A., Hayashi, Y., and Sakamoto, S. (1992). Numerical study on velocity-pressure field and wind forces for bluff bodies by k-ε, ASM and LES. *Journal of Wind Engineering and Industrial Aerodynamics* 44(1-3): 2841-2852

Murakami, S. (1993). Comparison of various turbulence models applied to a bluff body. *Journal of Wind Engineering and Industrial Aerodynamics* 46 & 47: 21-36.

Mochida, A., Murakami, S., Shoji, M., and Ishida, Y. (1993). Numerical Simulation of flow field around Texas Tech Building by Large Eddy Simulation. *Journal of Wind Engineering and Industrial Aerodynamics* 46-47: 455-460.

Mochida, A., Tominaga, Y., Murakami, S., Yoshie, R., Ishihara, T., and Ooka, R. (2002). Comparison of various k–ε models and DSM applied to flow around a high-rise building—report on AIJ cooperative project for CFD prediction of wind environment. *Wind and Structures* 5(2–4): 227–244.

Mochida, A., and Lun, I. Y. F. (2008). Prediction of wind environment and thermal comfort at pedestrian level in urban area. *Journal of Wind Engineering and Industrial Aerodynamics* 96(10-11): 1498-1527.

NASA 2004. www.grc.nasa.gov/WWW/wind/valid/tutorial/glossary.html#verification Retrieved from the internet on March 6, 2004.

NEN. (2006). Wind comfort and wind danger in the built environment, NEN 8100 (in Dutch) Dutch Standard.

Norton, T., Grant, J., Fallon, R., and Sun, D. W., (2010). Assessing the ventilation performance of a naturally ventilated livestock building with different eave opening conditions. Computers and Electronics in Agriculture 71(1): 7-21.

Oberkampf, W. L., Trucano, T. G., and Hirsch, C. (2004). Verification, validation, and predictive capability in computational engineering and physics. *Applied Mechanics Reviews* 57(5): 345 - 384.

Pasquill, F., and Smith, F. B. (1983). *Atmospheric Diffusion*, 3rd Ed. Ellis Horwood Ltd., Chichester, England.

Paterson, D. A., and Apelt C. J. (1986). Computation of wind flows over three-dimensional buildings. *Journal of Wind Engineering and Industrial Aerodynamics* 24: 192–213.

Paterson, D. A., and Apelt, C. J. (1989). Simulation of *wind* flow around three-dimensional buildings. *Building and Environment* 24: 39–50.

Paterson, D. A., and Apelt, C.J. (1990). Simulation of flow past a cube in a turbulent boundary layer. *Journal of Wind Engineering and Industrial Aerodynamics* 35: 149-176.

Patnaik, G., Boris, J. P., Young, T. R., and Grinstein, F. F. (2007). Large scale urban contaminant transport simulations with MILES. *Journal of Fluids Engineering* 129(12): 1524-1532.

Richards, P. J., and Hoxey, R. P. (1993). Appropriate boundary conditions for computational wind engineering models using the k-ε turbulence model. *Journal of Wind Engineering and Industrial Aerodynamics* 46&47: 145-153.

Richards, P. J., Mallison, G. D., McMillan, D., and Li, Y.F. (2002). Pedestrian level wind speeds in downtown Auckland. *Wind and Structures* 5(2-4): 151-164.

Richards, P. J. Hoxey, R. P., Connell, B. D., and Lander, D. P. (2007). Wind-tunnel modelling of the Silsoe Cube, *Journal of Wind Engineering and Industrial Aerodynamics* 95(9-11): 1384-1399.

Saathoff, P. J., Stathopoulos, T., and Dobrescu, M., (1995). Effects of model scale in estimating pollutant dispersion near buildings. *Journal of Wind Engineering and Industrial Aerodynamics* 54&55: 549–559.

Saathoff, P., Stathopoulos, T., and Wu, H. (1998). The influence of freestream turbulence on nearfield dilution of exhaust from building vents. *Journal of Wind Engineering and Industrial Aerodynamics* 77–78: 741–752.

Sasaki, R., Uematsu, Y., Yamada, M., and Saeki, H. (1997). Application of infrared thermography and a knowledge-based system to the evaluation of the *pedestrian*-level *wind* environment around buildings. *Journal of Wind Engineering and Industrial Aerodynamics* 67-68: 873-883.

Scaperdas, A., and Gilham, S. (2004). Thematic Area 4: Best practice advice for civil construction and HVAC, *The QNET-CFD Network Newsletter* 2(4): 28-33.

Selvam, R. P. (1997). Numerical simulation of pollutant dispersion around a building using FEM. *Journal of Wind Engineering and Industrial Aerodynamics* 67&68: 805-814.

Shih, T. H., Liou, W. W., Shabbir, A., and Zhu, J. (1995). A new k–e eddy-viscosity model for high Reynolds number turbulent flows: model development and validation. *Computers & Fluids* 24(3): 227–238.

Smeaton, W. H., Lepage, M. F., and Schuyler, G.D. (1991). Using wind tunnel data and other criteria to judge acceptability of exhaust stacks. *ASHRAE Transactions* 97(2): 583-588.

Spalart, P., and Allmaras, S. (1992). A one-equation turbulence model for aerodynamic flows. *Technical Report AIAA-92-0439*, American Institute of Aeronautics and Astronautics.

Spalart, P., Jou, W. H., Strelets, M., and Allmaras, S. (1997). Comments on the feasibility of LES for wings and on the hybrid RANS/LES approach, *Advances in DNS/LES, 1st AFOSR Int. Conf. on DNS/LES*, Greden Press, 1997.

Stathopoulos, T., and Baskaran, A. (1990). Boundary treatment for the computation of 3D turbulent conditions around buildings. *Journal of Wind Engineering and Industrial Aerodynamics* 35: 177–200.

Stathopoulos, T., and Baskaran, A. (1996). Computer simulation of wind environmental conditions around buildings. *Engineering Structures* 18(11): 876–885.

Stathopoulos, T., Lazure, L., Saathoff, P., and Wei, X. (2002). Dilution of exhaust from a rooftop stack on a cubical building in an urban environment. *Atmospheric Environment* 36: 4577–4591.

Stathopoulos, T., Lazure, L., Saathoff, P., and Gupta, A. (2004). *The effect of stack height, stack location and rooftop structures on air intake contamination: a laboratory and full-scale study*, Report R-392, IRSST, Quebec.

Takakura, S., Suyama, Y., and Aoyama, M. (1993). Numerical simulation of flowfield around buildings in an urban area. *Journal of Wind Engineering and Industrial Aerodynamics* 46-47: 765-771.

Tamura, T., Nozawa, K., and Kondo, K. (2008). AIJ guide for numerical prediction of wind loads on buildings. *Journal of Wind Engineering and Industrial Aerodynamics* 96(10-11): 1974-1984.

Tamura, T. (2008). Towards practical use of LES in wind engineering. *Journal of Wind Engineering and Industrial Aerodynamics* 96(10-11): 1451-1471.

Tang, W., and Davidson, C. I. (2004). Erosion of limestone building surfaces caused by wind-driven rain. 2. Numerical modelling. *Atmospheric Environment* 38(33): 5601-5609.

Tominaga, Y., Murakami, S., and Mochida, A. (1997). CFD prediction of gaseous diffusion around a cubic model using a dynamic mixed SGS model based on composite grid technique. *Journal of Wind Engineering and Industrial Aerodynamics* 67-68: 827-841.

Tominaga, Y., Mochida, A., Yoshie, R., Kataoka, H., Nozu, T., Yoshikawa, M., and Shirasawa, T. (2008a). AIJ guidelines for practical applications of CFD to pedestrian wind environment around buildings. *Journal of Wind Engineering and Industrial Aerodynamics* 96(10-11): 1749-1761.

Tominaga, Y., Mochida, A., Murakami, S., and Sawaki, S. (2008b). Comparison of various revised k–ε models and LES applied to flow around a high-rise building model with 1:1:2 shape placed within the surface boundary layer. *Journal of Wind Engineering and Industrial Aerodynamics* 96(4): 389-411.

Tominaga, Y., and Stathopoulos, T. (2007a). Numerical simulation of dispersion around an isolated cubic building - influence of turbulence models and turbulent Schmidt number. *Proc. 12th Int. Conf. Wind Engineering*, Cairns, Australia.

Tominaga, Y., and Stathopoulos, T. (2007b). Turbulent Schmidt numbers for CFD analysis with various types of flowfield. *Atmospheric Environment* 41(37): 8091-8099.

Tominaga, Y., and Stathopoulos, T. (2008). Numerical simulation of plume dispersion around an isolated cubic buildings: comparisons between RANS and LES computations. *BBAA VI International Colloquium on: Bluff Bodies Aerodynamics & Applications*, Milano, Italy.

Tominaga, Y., and Stathopoulos, T. (2009). Numerical simulation of dispersion around an isolated cubic building: Comparison of various types of k-ε models. *Atmospheric Environment* 43(20): 3200-3210.

Turner, D.B. (1970). *Workbook of atmospheric dispersion estimates*. Environmental Protection Agency, Environmental Health Series Air Pollution, 84.

van Hooff, T., and Blocken, B. (2010). Coupled urban wind flow and indoor natural ventilation modelling on a high-resolution grid: A case study for the Amsterdam ArenA stadium. *Environmental Modelling & Software* 25(1): 51-65.

Vasilic-Melling, D. (1977). *Three dimensional turbulent flow past rectangular bluff bodies*, Ph.D. Thesis, Imperial College of Science and Technology, London.

Wang, X. (2006). *Numerical simulation of wind–induced dispersion of emissions from rooftop stacks*. M.A.Sc Thesis, Department of Building, Civil and Environmental Engineering, Concordia University, Montreal, Canada.

Westbury, P. S., Miles, S. D., and Stathopoulos, T. (2002). CFD application on the evaluation of pedestrian-level winds. *Workshop on Impact of Wind and Storm on City Life and Built Environment, Cost Action C14*, CSTB, June 3–4, Nantes, France.

Wilcox, D. C. (2004). *Turbulence modelling for CFD*, 2nd Ed., DCW Industries, Inc.

Willemsen, E., and Wisse, J.A. (2007). Design for wind comfort in The Netherlands: Procedures, criteria and open research issues. *Journal of Wind Engineering and Industrial Aerodynamics* 95(9-11): 1541-1550.

Wilson, D. J., and Lamb, B. (1994). Dispersion of exhaust gases from roof level stacks and vents on a laboratory building. *Atmospheric Environment* 28(19): 3099–3111.

Wise, A. F. E. (1970). *Wind effects due to groups of buildings*. Royal Society Symposium Architectural Aerodynamics, London.

Wu, H. Q., and Stathopoulos, T. (1997). Application of infrared thermography for pedestrian wind evaluation, *Journal of Engineering Mechanics – ASCE* 123(10): 978–985.

Xie, X. M., Liu, C. H., and Leung, D.Y.C. (2007). Impact of building facades and ground heating on wind flow and pollutant transport in street canyons. *Atmospheric Environment* 41(39): 9030-9049.

Yakhot, V., and Orszag, S.A. (1986). Renormalization group analysis of turbulence. *Journal of Scientific Computing* 1(1): 3–51.

Yamada, M., Uematsu, Y., and Sasaki, R. (1996). A visual technique for the evaluation of the pedestrian-level wind environment around buildings by using infrared thermography. *Journal of Wind Engineering and Industrial Aerodynamics* 65(1-3): 261-271.

Yang, Y., Gu, M., Chen, S., and Jin, X. (2009). New inflow boundary conditions for modelling the neutral equilibrium atmospheric boundary layer in computational wind engineering. *Journal of Wind Engineering and Industrial Aerodynamics* 97(2): *88-95*.

Yoshie, R., Mochida, A., Tominaga, Y., Kataoka, H., Harimoto, K., Nozu, T., and Shirasawa, T. (2007). Cooperative project for CFD prediction of pedestrian wind environment in the Architectural Institute of Japan. *Journal of Wind Engineering and Industrial Aerodynamics* 95(9-11): 1551-1578.

PART II

ON THE DESIGN OF WIND ENERGY STRUCTURES

Aero-Servo-Elastic Design of Wind Turbines: Numerical and Wind Tunnel Modeling Contribution.

Alberto Zasso and Paolo Schito
Dipartimento di Meccanica, Politecnico di Milano,
Via La Masa 1, I-20156 Milano, Italy

Carlo L. Bottasso and Alessandro Croce
Dipartimento di Ingegneria Aerospaziale, Politecnico di Milano,
Via La Masa 34, I-20156 Milano, Italy

1 Introduction

The main purpose of this contribution is to provide a basic understanding of the fundamental interaction mechanism between the wind flow and the wind turbine, responsible for the power generation, as well as for the aerodynamic and inertial loading of the machine. A specific focus will be given at this proposal to the role of the control laws by which the turbine is operated, in determining both the performance as well as the structural loading of the machinery.

This notes are divided in three main parts. In the first the general aerodynamic principles and typologies of wind turbines are presented to provide the basic aerodynamic and control tools for the following evaluation of design criteria and loading of the machinery. This contribution is given by A. Zasso in Chapter 2. Here are presented the basic energetic approach, continuing with the simple aerodynamics of the airfoil and concluding with the performance of the entire blade to explain the basic aerodynamic behavior of a wind turbine. These simple records are important to understand the complex flow-structure interaction and the control of horizontal and vertical axis wind turbines.

In the second part (Chapter 3 and 4) C.L. Bottasso describes the mathematical models and the optimization of a wind turbine. In Chapter 3 are described the mathematical models used for the aero-servo-elastic modeling of wind turbines using comprehensive finite element based multibody procedures. Such models can support a wide range of analysis types on wind turbines, including the estimation of power production, the computation of maximum loads, the verification of stability margins, the assessment of the vibratory characteristics of the machine, the study of its response to gusts, turbulence and to fault and emergency conditions, the design and tuning of control laws, etc. The description of the models and of the algorithms covered in these notes are tailored to the wind turbine simulation software `Cp-Lambda`, developed at the Department of Aerospace Engineering of the Politecnico di Milano. In Chapter 4 are described the procedures for the multi-disciplinary (holistic) design optimization of wind turbines. These are methods and algorithms that, building on the core computational capabilities offered by an aero-servo-elastic simulation software such as `Cp-Lambda`, support and enable the work of the wind turbine designer. The wind turbine design problem in fact has in general multiple objective functions (e.g., maximum production, minimum weight/cost, etc.), and should satisfy a large number of design constraints; the complexity of the problem requires software that can help the designer by proposing feasible solutions that account as much as possible for all couplings among the sub-disciplines and the relevant physical effects. In the second are described the mathematical models used for the aero-servo-elastic modeling of wind turbines using comprehensive finite element based multibody procedures. At the same time are described the multidisciplinary (holistic) design optimization procedures for the wind turbines.

In the third part of this notes (Chapter 5), A. Zasso introduces the role of Wind Tunnel testing in the analysis of the performances of a wind turbine. A brief introduction is given on the scaling laws of the dimensional parameters that must be considered in wind tunnel testing. Three test cases experienced at Politecnico di Milano Wind Tunnel are then presented, addressing the performances of horizontal axis, vertical axis wind turbines and finally a wake interaction problem on a downwind aeroelastic structure.

A conclusion is finally drawn concerning the complementary role of wind tunnel testing as a possible validation tool for the numerical simulation design approach.

2 General aerodynamic principles and typologies of wind turbines

Authored by A. Zasso and P. Schito

Wind turbines are a very well known object since many centuries: everybody has seen at least one windmill in his life and most people remember Don Quixote and his fight against them. Only in the last 100 years electricity has come to use and windmills have been converted in wind turbines and connected to an electric generator. Since the first experiments on farm small wind turbines in Denmark at the beginning of '900, many studies have been conducted in Northern Europe and USA after Second World War, but only after the 1973 oil crisis wind turbines become very interesting as a substitute for oil power.

The principle of power production of wind turbines is the transformation of wind kinetic energy in mechanical energy in a shaft and to electrical energy in a generator. The maximum energy available in the wind is: $P_{Tot} = \frac{1}{2}\dot{m}V_1^2 = \frac{1}{2}\rho A V_1^3$ where $\dot{m}$ is the mass flow, ρ is the density of the air, V_1 is the wind speed and A the worked out area where the wind speed is reduced. Not all wind energy can be extracted from the flow, but there is a maximum stated by the Betz Law, where the maximum power is $P_{max} = \frac{16}{27}P_{Tot}$. To approach at best the Betz limit, the wind turbines have been largely studied, optimizing the aerodynamic efficiency and the structural loads. Wind Turbines can be divided in two main categories: Horizontal Axis Wind Turbines (HAWT) and Vertical Axis Wind Turbines (VAWT). Vertical Axis Wind Turbines are used mainly for micro, mini and medium applications and can be further divided into Savonius (self starting) and Darrieus (non self starting) wind turbines. VAWT have usually a very simple construction, a good resistance to strong winds and a low cost; the drawback is a low efficiency and the difficulty to build big plants. HAWT are the most used solution for big plants, usually they have a propeller with high efficiency, but they need a more advanced technology to produce energy.

2.1 Ideal Wind Turbine 1-D model: Betz Law

The simplest way to examine the wind turbine performances is a 1-D model of an ideal rotor. In this model the rotor is modeled as a permeable disc with no hub and the flow is inviscid, incompressible and only axial. The rotor acts as a drag device to slow down the wind speed from V_1 far upstream in the undisturbed flow to V at the rotor plane and finally to V_2 in the wake as can be seen in figure 1. The thrust T is obtained as a pressure drop Δp over the rotor and is transformed to the shaft power P through the local velocity at the disk section V .

$$T = \Delta p A \tag{1}$$

where $A = \pi R^2$ is the area of the rotor. The flow is stationary, incompressible, frictionless and no external forces act on the fluid, therefore Bernoulli equation is valid from far upstream to just in front of the rotor and from just behind the rotor to far downstream (holding the hypothesis of constant

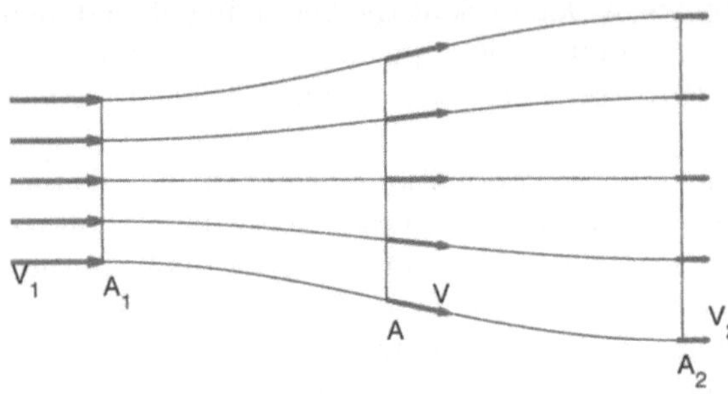

Figure 1. Betz model: streamlines and model assumptions

static pressure p_1 in the far upstream and downstream boundaries):

$$\begin{cases} p_1 + \frac{1}{2}\rho V_1^2 = p + \frac{1}{2}\rho V^2 \\ p - \Delta p + \frac{1}{2}\rho V^2 = p_1 + \frac{1}{2}\rho V_2^2 \end{cases} \tag{2}$$

Combining the two equations it is possible to obtain the pressure drop Δp over the rotor

$$\Delta p = \frac{1}{2}\rho(V_1^2 - V_2^2) \tag{3}$$

Conservation of mass states that:

$$\dot{m} = \rho V A = \rho V_1 A_1 = \rho V_2 A_2 \tag{4}$$

Through a momentum balance on a control volume external to the streamlines of figure 1, it is possible to demonstrate that the thrust T on the rotor can be calculated as

$$T = \dot{m}(V_1 - V_2) = \rho V A(V_1 - V_2) \tag{5}$$

Replacing the thrust T in equation 5 with the pressure drop Δp in equation 1 it is possible to obtain the following peculiar result:

$$V = \frac{1}{2}(V_1 + V_2) \tag{6}$$

showing that the velocity on the rotor plane is the mean of the undisturbed wind speed V_1 and the final speed on the wake V_2 i.e., in an equivalent way, the velocity drop before and after the rotor plane is equal:

$$V_1 - V = V - V_2 \tag{7}$$

The shaft power P can be finally calculated as the total power extracted from the considered flow tube:

$$P = \frac{1}{2}\rho V A(V_1^2 - V_2^2) \tag{8}$$

It is possible to define the axial induction factor a as a function of the undisturbed wind speed V_1 and the rotor wind speed V:

$$a = \frac{V_1 - V}{V_1} \tag{9}$$

Combining equation 6 and equation 9 it is possible to obtain the following relationship between V_1 and V_2:

$$V_2 = (1 - 2a)V_1 \tag{10}$$

Introducing this definition in equations 5 and 8 it is possible to rewrite the power P and the thrust T as:

$$P = 2\rho V_1^3 a(1-a)^2 A \tag{11a}$$
$$T = 2\rho V_1^2 a(1-a)A \tag{11b}$$

The ideal available power P_{Tot} and thrust T_{Tot} associated to a section equal to the rotor area A are then defined as:

$$P_{Tot} = \frac{1}{2}\rho A V_1^3 \tag{12a}$$
$$T_{Tot} = \frac{1}{2}\rho A V_1^2 \tag{12b}$$

With this definition it is possible to give a non-dimensional expression of the power and the thrust defining the power coefficient C_P and the thrust coefficient C_T:

$$C_P = \frac{P}{P_{Tot}} \tag{13a}$$
$$C_T = \frac{T}{T_{Tot}} \tag{13b}$$

Using the power and the drag defined in equations 11 it is possible to write the C_P and C_T as a function of the axial induction factor a:

$$C_P = 4a(1-a)^2 \tag{14a}$$

$$C_T = 4a(1-a) \tag{14b}$$

From these equations it is possible to differentiate C_P and C_T with respect to a:

$$\frac{dC_P}{da} = 4(1-a)(1-3a) \tag{15a}$$

$$\frac{dC_T}{da} = 4(1-2a) \tag{15b}$$

and calculate the maximum power coefficient $C_{Pmax} = \frac{16}{27} = 0.59$ for an axial induction factor $a = \frac{1}{3}$. This theoretical maximum for an ideal wind turbine is known as the Betz Limit, stating that a maximum 59% of the total power associated to the flow kinetic energy can be extracted by the wind turbine. In figure 2 are presented the power and thrust coefficients as functions of the axial induction factor, showing that the maximum thrust is reached for a higher induction factor value than the maximum power limit. From the continuity equation 4 it is possible to calculate the ratio between the areas A_1 and A_2 as a function of a:

$$\frac{A_1}{A_2} = 1 - 2a \tag{16}$$

It is possible finally to demonstrate and experiments show that the Betz assumptions are valid only for values of a lower than 0.4 and that the wind turbines can reach those high axial induction factors a just at low wind speeds.

2.2 Blade aerodynamics and wing profiles

Almost all wind turbines, both horizontal and vertical axis, use generally aerodynamic profiles to convert the wind pressure into torque. For both HAWT and VAWT the flow reaching the profiles is mainly bi-dimensional while the third direction (along the blade) is usually very small and can be at a first assumption neglected. Speaking of wind turbines it is straight forward the opportunity of giving some mentions about the wing profiles describing the blade geometry.

Wing Profiles

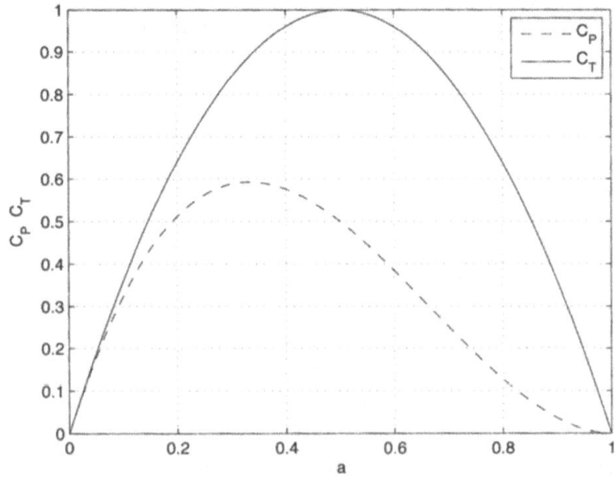

Figure 2. Power and Drag Coefficients C_P and C_T as a function of the axial induction factor a

Aerofoils are composed by a round leading edge where also the stagnation point is located and a sharp trailing edge as shown in figure 3. The line connecting the leading edge and the trailing edge is called chord, c. The camber of the wing profile is described as the displacement of the mean line from the chord, the thickness of the airfoil is given as a displacement from the profile mean line. One of the most used wing profiles categories is known as NACA classification. The forces on the airfoil can be decomposed

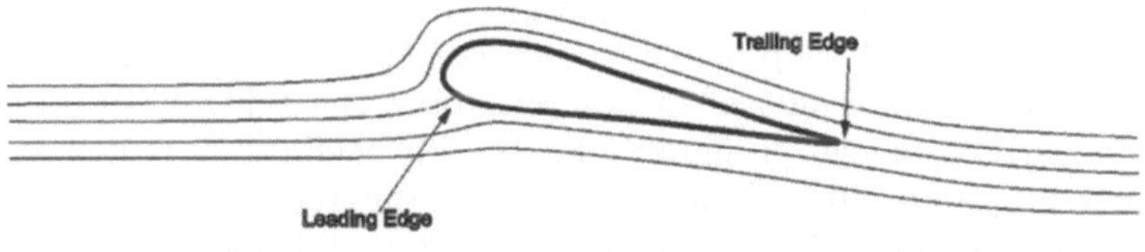

Figure 3. Airfoil design and streamlines

into two components: the one in the direction of the undisturbed mean flow V_∞ is called "drag", D, the force in the perpendicular direction is called "lift", L. On the wing profile acts also a moment M around the reference point selected for the reduction of the forces: this point is usually located in the chord line at $c/4$ from the leading edge. The moment is positive when it tends to turn the airfoil nose up as shown in figure 4. Lift, drag and

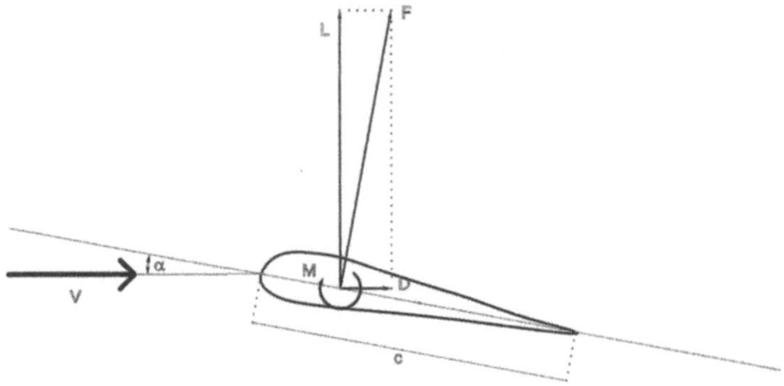

Figure 4. Airfoil reference for the lift and drag forces

moment for an airfoil are usually given in non dimensional form as lift, drag and moment coefficients C_L, C_D and C_M:

$$C_L(\alpha) = \frac{L}{\frac{1}{2}\rho V_\infty^2 c} \tag{17a}$$

$$C_D(\alpha) = \frac{D}{\frac{1}{2}\rho V_\infty^2 c} \tag{17b}$$

$$C_M(\alpha) = \frac{M}{\frac{1}{2}\rho V_\infty^2 c^2} \tag{17c}$$

where ρ is the fluid density. The aerodynamic behavior depends on the angle of attack α that is the angle between the chord c and the undisturbed wind direction V_∞. It is possible to measure the forces acting on the profile by changing the angle of attack obtaining the typical curves $C_L(\alpha)$, $C_D(\alpha)$, $C_M(\alpha)$ as shown in figure 5 for the lift coefficient. It is possible to note that the two profiles of the example have different aerodynamic performances, due to geometrical differences in camber and thickness distribution along the chord. The major aerodynamic characteristic for an airfoil is the aerodynamic efficiency and the stall behavior: the aerodynamic efficiency is defined as the lift over drag ratio ($e = C_L/C_D$). The stall is the rapid decrease in aerodynamic efficiency at high incidence angles; it can be seen in figure 5 that for airfoil B the stall occurs at about 8 of angle of attack and the lift coefficient remains constant, while for airfoil A it is possible to obtain higher lift coefficients for higher angles of attack, but after stall

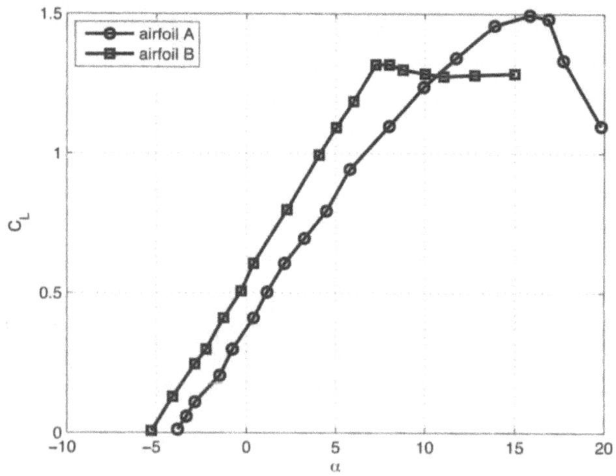

Figure 5. $C_L(\alpha)$ curves for two different airfoils.

there is a rapid decrease in lift. The flow condition where the $C_L(\alpha)$ curve is linear can be seen in figure 6 on the left: the flow is attached to the airfoil. The image on the right represents a typical flow field of a stalled airfoil: it is possible to se the separated zone on the rear part of the profile.

Figure 6. Streamlines for the same airfoil at two different angles of attack: $\alpha = 5\,\mathrm{deg}$ attached airflow on the left, $\alpha = 15\,\mathrm{deg}$ separated flow on the right. Cfr. Hansen (2008)

Blade aerodynamics Generally speaking, the aerodynamic interaction between the air flow and the wind turbines is provided by the blades, that are a tridimensional extrusion of airfoils. The flow field around the blades is not constant for both horizontal and vertical axis wind turbines: usually on HAWT the flow conditions changes along the length of the blade, while on VAWT the flow field changes in time during the rotation of the turbine as will be more in detail described later on. Therefore the blade must grant satisfying performances for a quite wide range of flow conditions and, at the same time, support structural loads and resist to fatigue.

The air speed incident on the wind turbines blades has a different and a more complex formulation than what has been presented for 2-D airfoils. Considering the section of a blade of a wind turbine as an airfoil it is possible to analyze the single performance of each section: in this case it is clear that the incident wind velocity must be defined in a different way than what stated in the previous paragraph since the blade itself has a velocity associated to the shaft rotation. The relative flow incident on the airfoil is therefore defined as the vectorial sum of the undisturbed wind speed V_∞ and the local speed of the rotating blade section $V_t = R\omega$ (changed of sign), being R the distance from the section to the center of rotation of the rotor and $\omega = \dot{\theta}$ the rotational speed of the wind turbine (where θ is the angular position of the blade). This consideration is valid for both vertical and horizontal axis wind turbines. With a simplified approach it is possible to say that for HAWT the relative incident air speed is calculated as

$$\overrightarrow{V}_{r,HAWT}(R) = \overrightarrow{V}_\infty - \overrightarrow{R} \times \overrightarrow{\omega} \tag{18}$$

as a function of the distance from the rotor hub R, while for VAWT the incident air speed is calculated as

$$\overrightarrow{V}_{r,VAWT}(\theta) = \overrightarrow{V}_\infty + \overrightarrow{R}(\theta) \times \overrightarrow{\omega} \tag{19}$$

It is clear that for horizontal axis wind turbines the relative flow speed $\overrightarrow{V}_{r,HAWT}(R)$ increases with the distance from the rotor hub (changing also significantly the incidence angle), while for vertical axis machines the incident air speed $\overrightarrow{V}_{r,VAWT}(\theta)$ changes in both, magnitude and direction with the angular position of the blade, and has a periodic cycle.

2.3 Horizontal axis Wind Turbines: generalities

The description of the aerodynamic flow around a wind turbine is quite complex for a horizontal axis machine. For a first analysis in figure 7 is presented the typical configuration of a horizontal axis wind turbine. Analyzing a single blade of a modern HAWT, it is possible to observe that it is the result of compromises and optimization studies from both the aerodynamic and structural point of view. The aerodynamics of the blade must provide the necessary torque to the shaft connected to the electrical generator in order to produce the maximum available power. It is also necessary on the other hand, that, from the structural point of view, the structure must be

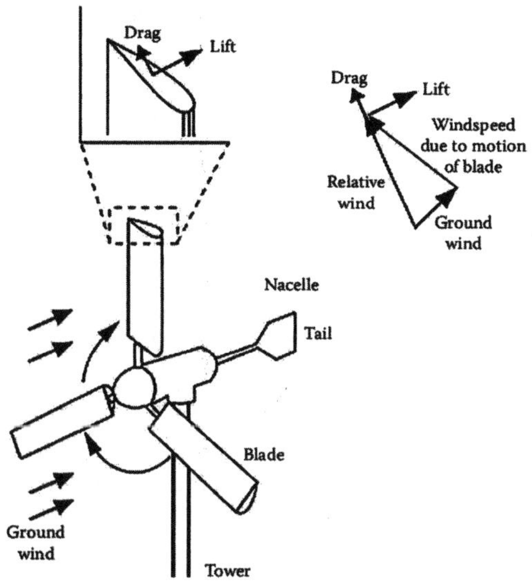

Figure 7. Horizontal Axis Wind Turbine: general aerodynamic considerations. Cfr. Vaughn (2009)

safe and survive to the most extreme atmospheric conditions for a long time. This considerations lead to an optimization of the blade: this already optimized blade scheme will be presented in this section. The classical approach to the numerical evaluation of the overall aerodynamics of a horizontal axis wind turbine is the Blade Element Momentum Method (BEM), also known as strip theory. With this approach it is possible to analyze the blade considering the contribution of the different stripes of length dr as shown in figure 8. As quoted in equation 18 the relative air velocity V_R magnitude and direction incident on the single blade element, changes with the distance R from the rotor hub. It is important that each profile works at his best performance in order to maximize the overall efficiency of the wind turbine. For this reason the blades of a HAWT present a twist angle θ i.e. the angle between the chord and the rotor plane. The local angle of attack α is then defined in equation 20 as the difference between the relative velocity angle ϕ and the twist angle θ as shown in figure 9.

$$\alpha = \phi - \theta \tag{20}$$

The aerodynamic force R generated by the airfoil can be decomposed, as indicated in figure 4, in drag and lift components by projecting R along

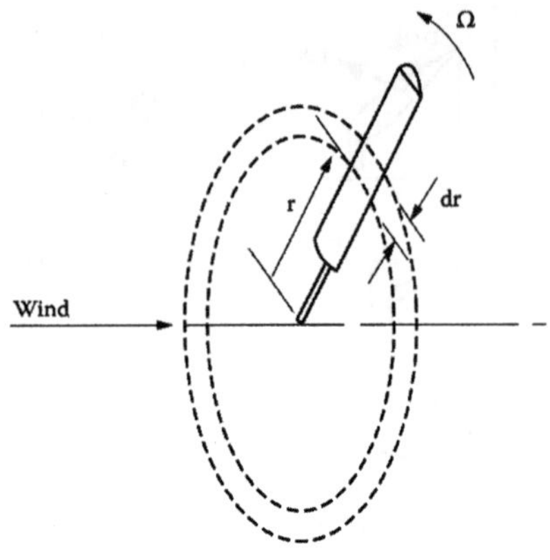

Figure 8. Diagram of a blade element. Cfr. Vaughn (2009)

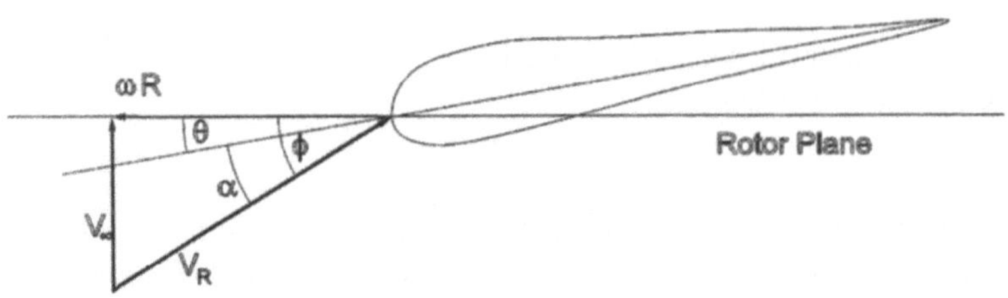

Figure 9. Velocities and angles at the rotor plane

the relative air velocity direction (and its orthogonal direction), but can be also decomposed by projecting R along the rotor plane (and its orthogonal direction) as indicated in figure 10. In this case it is obtained the driving force T i.e. the only component of the aerodynamic force responsible for the power generation, being the axial force A just responsible of the structural load on the wind turbine supporting pylon and of the bending moment on the basement. With a simple trigonometric calculation it is possible to calculate the driving and the axial force as functions of the relative velocity angle ϕ:

$$T = L \sin\phi - D \cos\phi \tag{21a}$$
$$A = L \cos\phi + D \sin\phi \tag{21b}$$

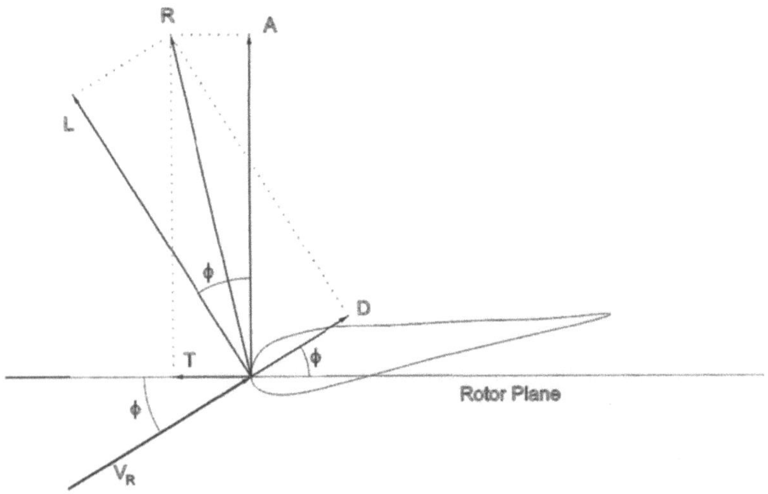

Figure 10. Force R decomposition on the velocity direction in lift L and drag D or in the rotor plane in driving T and axial A forces

The negative contribution of the drag force D to the driving force T shown in 21a clearly highlights the fundamental importance of a high efficiency of the wing profile in order to enhance the wind turbine global performances. In figure 11 is reported the twist and the chord distribution on a horizontal axis wind turbine blade: it is possible to see that at the root the blade has a very high twist and a very long chord. The high twist is due to the very low contribute to the relative wind V_r given by the rotational component ΩR as indicated in equation 18: if no twist is applied to the blade the local airfoil incidence angle α is too high and the section works in separated flow with the inefficient performances of the stalled conditions. It is possible to note also that the chord of the airfoil is decreasing with the distance from the rotor hub: this is a both structural and aerodynamic choice: a bigger section supports better the solicitations that occur at the root of the blade and also generates bigger forces than smaller sections with the same incoming wind conditions. Another issue for both structural and aerodynamic design of the root chord is the choice of a thicker profile: this ensures a structure that supports better the torque and the bending moment while from the aerodynamic point of view the profile thickness guarantees a higher stall angle and a smoother loss of performance beyond the stall. In figure 11 it is possible to see the change in the angles of attack and of thickness of the

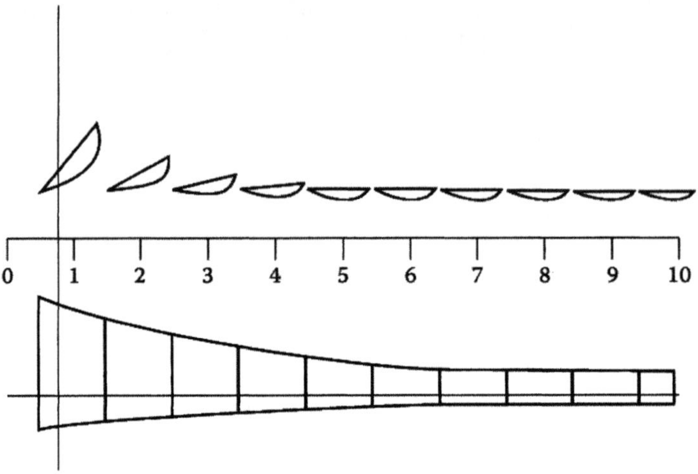

Figure 11. Twist and Planform for a Carter 25 wind turbine blade. The blade is divided in ten sections. The station is at the midpoint of the section. Cfr. Vaughn (2009)

single airfoils with the distance from the rotor hub. The parameter that affects in major part the twist and the thickness is the non-dimensional value of the tip speed ratio λ, defined as the ratio of the speed of the tip of the blade ωR and undisturbed wind speed V_∞:

$$\lambda = \frac{\omega R}{V_\infty} \tag{22}$$

where λ gives the influence of the rotor speed on the relative wind angle locally incoming on the blade. Usually tip speed ratio is below the value of 10, since higher values will not extract optimally the power from the wind and will increase the loads on the structure at risk of structural failure. Combining the considerations on aerodynamic efficiency, power generation and structural analysis it is possible to produce the graph reproduced in figure 12 where the power generated by each station is reproduced in different undisturbed wind speed V_∞. It is possible to see that the power generated from tip stations is increasing with the wind speed, while for root stations the power does not increases as much.

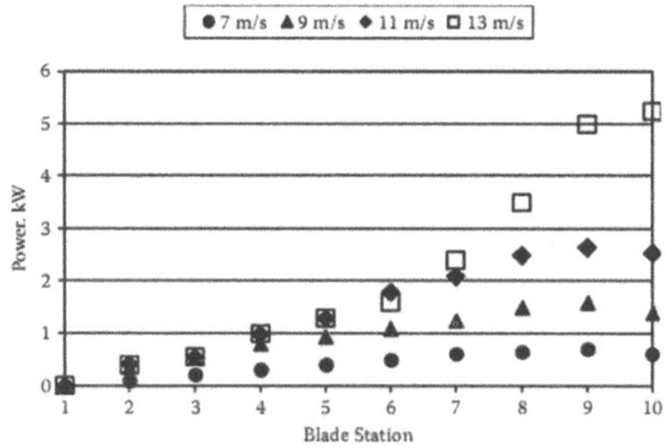

Figure 12. Prediction of power output for each station of a Carter 25 Wind Turbine blade for four wind speeds (tip speed ratio = 6.1). Cfr. Vaughn (2009)

2.4 Vertical axis Wind Turbines: generalities

The aerodynamics of VAWT is much more complicated than that of HAWT since the rotational axis is orthogonal to the incoming flow. The main advantage of such a configuration is the insensibility to the direction of the wind and therefore the control is simpler because no alignment device is needed. On the other hand there are many aerodynamic problems related to the periodic regime that characterizes the machine even in steady state operating condition. In order to better understand the aerodynamic behaviour of a VAWT it is possible to study a bidimensional cross section of the turbine (see figure 13). The cross section of the turbine is then divided in two sides: a) the upstream side where the flow enters the rotor and is almost unperturbed; b) the downstream side where the flow exits the rotor and the wake effects are very important because of the interaction between the rotating blades and the wakes generated in the upstream side. Let's now consider the simplified aerodynamics of a single blade (figure 14), assuming the local incoming wind speed V_i as uniform and always parallel to the y axis. A moving reference frame $(\vec{n}, \vec{t})$ is then considered as located in the aerodynamic center of the blade with $\vec{n}$ pointing to the shaft of the rotor and $\vec{t}$ along the chord. Finally the position of the blade is identified through the azimuth angle θ. The relative velocity V_R, conventionally projected along the local reference frame $(\cdot\vec{n}, \vec{t})$ of the blade, is obtained through the vector sum of the translational speed of the profile (changed of

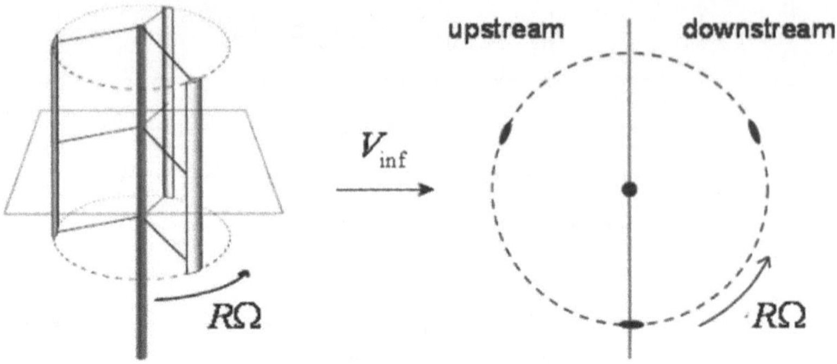

Figure 13. Bidimensional cross section of a VAWT

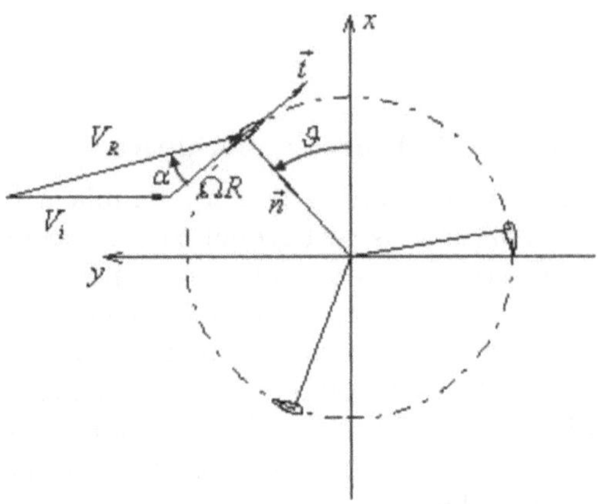

Figure 14. VAWT: aerodynamics of a single blade

sign) ΩR and the local incoming wind speed V_i:

$$V_t = V_i \cos\theta + \Omega R \tag{23a}$$
$$V_n = V_i \sin\theta \tag{23b}$$

The projections of the relative speed can be then used to calculate the angle of attack α and the relative speed V_r:

$$\alpha = \arctan(\frac{V_n}{V_t}) \tag{24a}$$

$$V_r = \sqrt{V_n^2 + V_t^2} \tag{24b}$$

Since V_t and V_n are functions of the azimuth angle θ, the angle of attack α and the relative speed V_r change during the revolution as shown for three blades in figure (15) considering V_i constant. The variation of the angle of

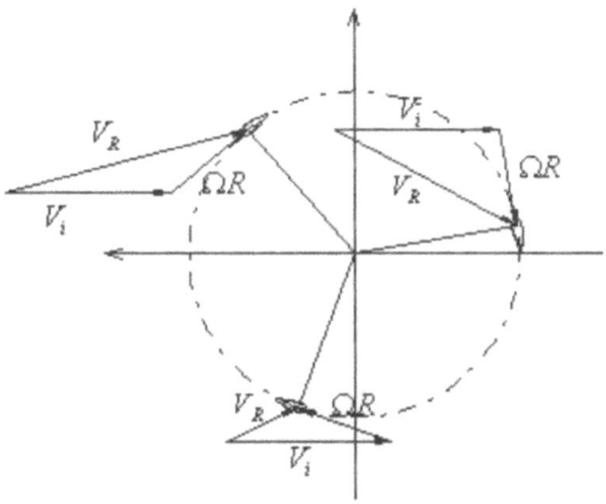

Figure 15. Speed triangles of three blades with V_i uniform

attack and the relative speed has a double effect on the aerodynamic forces because the aerodynamic coefficients are functions of the angle of attack and the Reynolds number. In fact the real Reynolds number Re_L felt by the blade depends on the relative speed V_r:

$$Re_L = \frac{\rho V_r C}{\mu} \tag{25}$$

Let's now introduce the tip speed ratio λ that is the ratio between the translational speed ΩR of the airfoil and the undisturbed flow speed V_{inf}:

$$\lambda = \frac{\Omega R}{V_{inf}} \tag{26}$$

The tip speed ratio is a non dimensional parameter describing the shape of the speed triangle. In fact, if λ is high, the relative speed V_r is mainly composed by the translational speed of the blade and therefore the angle of attack remains low. On the other hand, if λ is low, the relative speed V_r is mainly composed by the wind speed and therefore the angle of attack reaches large values. This effect is shown in figure 16 where the speed triangles for $\lambda = 1$ and $\lambda = 3$ are compared. Therefore, if the tip speed

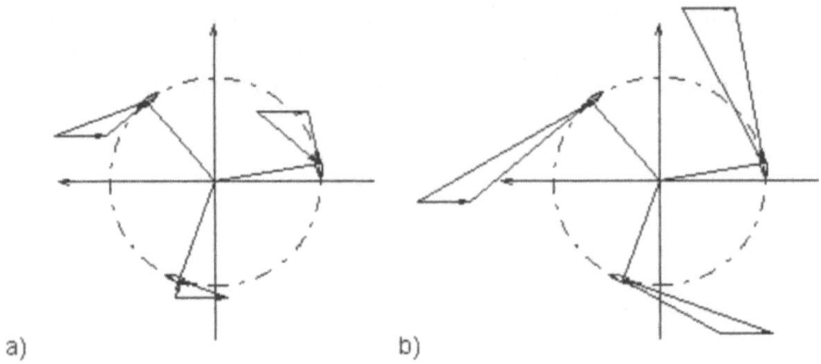

Figure 16. Speed triangles of three blades with V_i uniform for $\lambda = 1$ (a) and $\lambda = 3$ (b)

ratio is low enough, the angle of attack on the blade of a VAWT can exceed the stall angle (figure 17). Since the angle of attack changes rapidly and can overcome the stall limit, the aerodynamic coefficients can be subjected to the dynamic stall phenomenon resulting in an hysteretic cycle of the lift, drag and pitching moment characteristics. Therefore the dynamic stall is an essential phenomenon to be accounted for in order to correctly simulate the wake effects on the downstream side of the rotor. Usually there is an optimal value of the tip speed ratio maximizing the generated power. Let's now consider the aerodynamic forces on a single blade (figure 18). The instantaneous torque T contribution due to a single blade is the product of the aerodynamic forces projections on the tangential direction times the radius of the rotor:

$$T = L \sin\alpha \cdot R - D \cos\alpha \cdot R = \frac{1}{2}\rho c (C_L(\alpha, Re_L)\sin\alpha - C_D(\alpha, Re_L)\cos\alpha) V_r^2 R \tag{27}$$

The lift gives a positive contribution to the power generation while the drag contribution is always negative, reducing therefore the energetic efficiency of the turbine. Finally there is another parameter affecting the aerodynamics

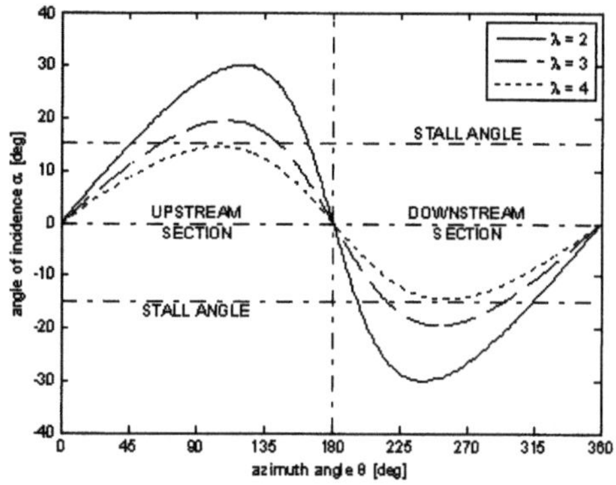

Figure 17. Angle of attack as function of the azimuth angle and the tip speed ratio considering V_i uniform

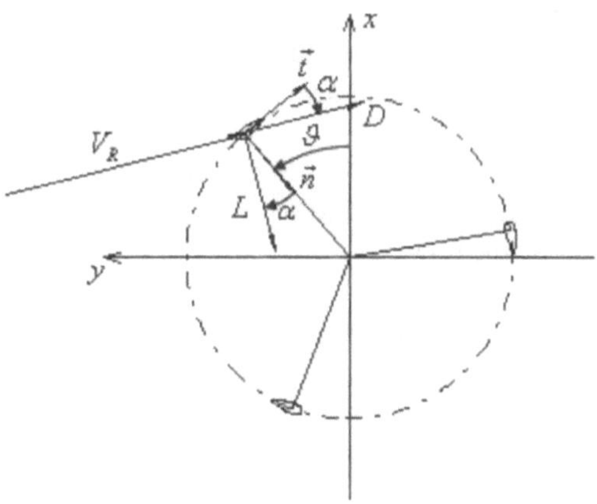

Figure 18. Aerodynamic forces on a single blade

of the blade: the chord/radius ratio c/R. The classic aerodynamic coefficients of an airfoil are valid only for uniform straight flow but the blade feels a "non-straight rotating flow". In fact the local angle of attack changes along the surface of the blade and this effect becomes more important for increasing values of the chord/radius ratio as shown in figure 19. In this condition the classic experimental and numerical aerodynamic coefficients don't describe correctly the interaction between fluid and structure.

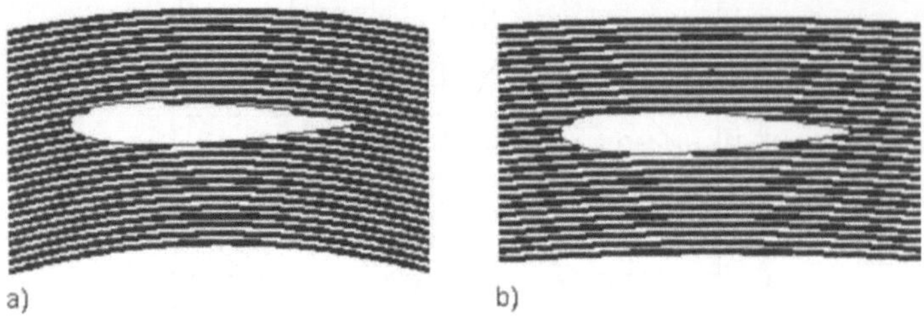

Figure 19. Effect of chord/radius ratio for $c/R = 0.33$ (a) and $c/R = 0.1$ (b)

Summarizing, a reliable analytical model of a VAWT is therefore very complicated for the following reasons:

- the correct knowledge of the aerodynamic coefficients of the airfoil profile for different Reynolds numbers and angles of attack;
- the correct simulation of the hysteretic effects connected with the dynamic stall;
- the characterization of the wake effects on the downstream side;
- the three-dimensional effects due to finite aspect ratio;
- the effects of rotating flow.

2.5 Typical control laws and aerodynamic implications

The control system of a wind turbine must assure the correct operation of the turbine in terms of:

- keep the rotational speed within a certain range;
- yaw the turbine axis;
- keep the power output within a certain range;
- start and stop the turbine.

The control can be performed following three main strategies: stall regulation, pitch regulation and yaw control.

Stall Regulation This is the most simple machine, the blades are fixed and cannot be pitched. The rotational speed of the rotor is kept almost constant (generally the rotation speed control is implicitly given by the constant frequency of the electrical network in which the generator is connected) and thus the angle of attack increases with the wind speed. Increasing the wind speed the blade can locally stall, causing an increase in drag coefficient and a decrease in lift coefficient, decreasing the load on the blade. The power decrease depends on the geometrical characteristics of the blade (pitch angle, twist, chord and aerofoil distributions). If the power is not limited sufficiently it is necessary to unbolt the blades to change the pitch settings or to change the design of the blade itself. When the rotor angular speed ω becomes dangerous for the wind turbine structure a safety system is designed to slow down the rotation of the blades: it is common on stall regulated wind turbines that the outer part of the blades are activated by centrifugal force to turn 90 deg acting as an aerodynamic brake (see figure 21). In figure 20 it is possible to see the power generation curve for a stall regulated wind turbine: there is a cut-in velocity for the generation of energy, to reach then a maximum to be maintained up to the cut-out wind speed. The maximum power generation is constant for a range of wind speeds using the stall to control the speed of the rotor.

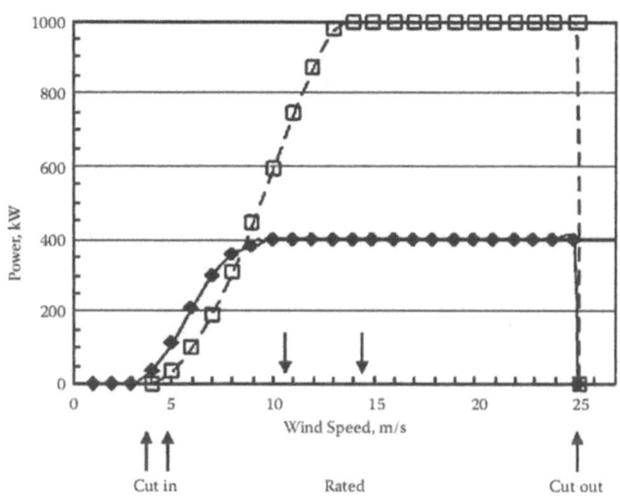

Figure 20. Typical power curve for a horizontal axis wind turbine running with a stall or pitch controlled machine. The curves are reported for a rotor with two different generator sizes. Cfr. Vaughn (2009)

Pitch Regulation (Constant Rotational Speed) Pitch regulated machines permit to actively pitch the entire blade, and thus to change the angle of attack on the entire length. A pitched blade can also act as an aerodynamic brake. Regulating the pitch angle of the blade it is possible to control

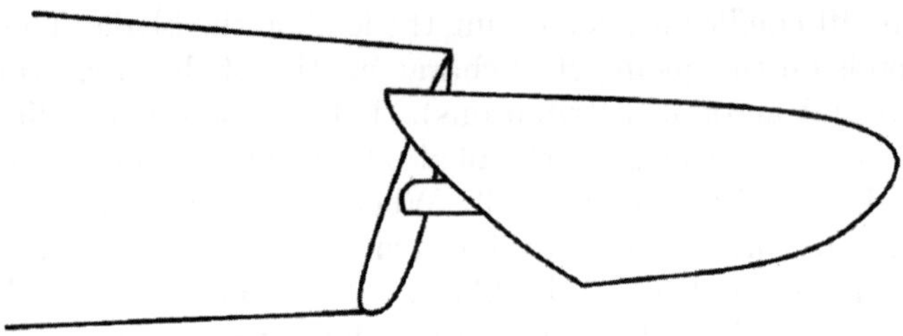

Figure 21. Turnable tip used as aerodynamic brake and activated by the centrifugal force. Cfr. Hansen (2008)

the power output. The power can be reduced by decreasing the local angle of attack α in equation 20 by increasing θ. Alternatively it is possible to reduce the power output increasing the angle of attack and forcing the blades to stall: this is called active stall. In figure 20 it is possible to see the typical power generation curve for a pitch regulated wind turbine that is analogous to what said for stall regulated machines. In this case the first part of the curve is obtained using the most efficient pitch regulation, the second part of the curve, at almost constant power, is obtained decreasing the relative angle of incidence (i.e. increasing the blade pitch), reducing the aerodynamic efficiency of the blades until the cut out wind speed is reached. In figure 22 are quoted the pitch angle curves for an active stall controlled and a pitch regulated wind turbine: it can be seen that until the generation of the maximum power the pitch control law is similar, and the angle does not changes very much. The differences start when generating the maximum power: the active stall regulated wind turbine maintains low pitch angles to have stalled blades, while the pitch regulated machine increases the pitch to generate lower aerodynamic forces.

Yaw Control It is possible to control the power output of a wind turbine by varying the yaw of the rotor. In normal pitch or stall regulated machines it is always present a yaw drive aligning the rotor hub with the incoming wind speed. For yaw controlled machines the rotor is turned out of the wind in high wind speeds to limit the airflow through the rotor and thus the power extraction.

Variable Speed In case the electrical power generation system allows as an additional degree of freedom the control of the turbine rotational speed, writing the power generation as a function of the tip speed ratio and of

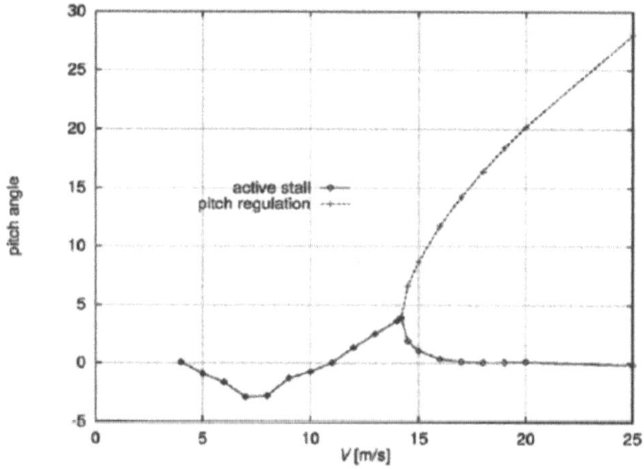

Figure 22. Optimum values for the pitch (left side single curve) and pitch regulation to avoid exceeding the nominal power (bifurcated right side curve). These curves are referred to the "squared line" power characteristic quoted in figure 20. . Cfr. Hansen (2008)

the pitch angle and applying variable speed on a pitch regulated rotor, it is possible to run a turbine at a optimum point occurring at a optimum twist θ_{opt} and a optimum tip speed ratio λ_{opt} as a function of the incoming wind speed. In figure 23 are indicated the power curves, parameterized on the incoming wind speed, as functions of the turbine rotational speed, so that as functions of the tip speed ratio. It is possible to calculate that, for this pitch angle, the optimum tip speed ratio is $\lambda = 9.8$. A variable speed wind turbine runs more efficiently than a stall or pitch regulated machine that runs at fixed rotational speed: this corresponds to a variable tip speed ratio giving a maximum power efficiency for wind speeds around 7 m/s. The line that intersects all optimum rotational speed for each wind speed gives the optimum regime for the wind turbine. Due to noise emission, the tip speed is limited to 70 m/s and therefore the optimum tip speed ratio can be obtained only for lower wind speeds. In the upper range of operating wind speeds, up to the cut off limit, the turbine will be then controlled through pitch control strategies. The drawback of this system is a more complex generator architecture, where alternating current (AC) is first transformed into direct current (DC) and then back to alternating current with the correct frequency using a so-called ACDC/DCAC device.

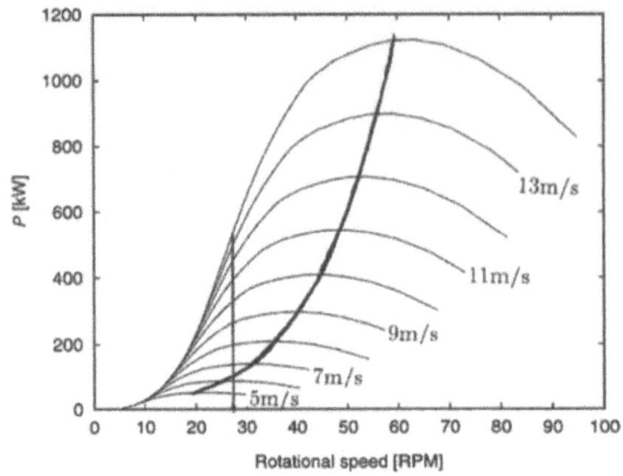

Figure 23. Power generated at constant speed (vertical thin line) versus variable speed (thick line) for a NTK 500/41 wind turbine. Cfr. Hansen (2008)

3 Aero-Servo-Elastic Modeling of Wind Turbines using Finite Element Based Multibody Procedures

Authored by C.L. Bottasso and A. Croce

The simulation of wind turbines requires the ability to model flexible structural mechanisms made of several components in relative motion with respect to one another, connected by various kinds of mechanical linkages, assembled in topological configurations of arbitrary complexity. Such structural elements are typically interacting with multiple coupled fields, including aero and hydro dynamic ones, and with a variety of actuators (e.g., pitch and yaw actuators, the electrical generator, brakes, etc.), possibly experiencing contacts and impact events among the bodies of the system (e.g., within the gear-box, or at the flap stop in a teetering rotor, etc.), operating in closed-loop under the action of control systems.

These complex multi-field models find applicability in a multitude of tasks related to the design, verification and simulation of wind energy systems, as for example the evaluation of loads in critical components, including those made of composite materials as the blades, the analysis of vibratory levels, the estimation of damping and flutter margins, the synthesis of control laws, etc. Clearly, there is an ever increasing demand from industry regarding the fidelity and reliability of such simulations.

In recent years, multibody dynamics has emerged as the technology of choice for addressing these simulation needs. In fact, the success of modern multibody analysis systems can be traced to their ability to model mechanisms by assembling generalized elements from within a library, much like in classical finite element methods (FEM). Each element provides a basic functional building block, such as for example a rigid body, a hinge, an electrical or hydraulic actuator, an interacting field embedded in an external application, etc. By assembling and connecting such generalized elements together, one has a powerful way of describing complex systems with the desired level of modeling fidelity.

Figures 24 through 26 exemplify possible simple topological sketches of wind turbine models. The goal of these examples is just to give an idea to the reader of how it is possible to build more or less detailed models of wind turbines by using just a few elements and joint types; it is clear that the need for capturing certain effects or estimating loads in some components of the machine can require models of greater complexity than the ones shown here for illustrative purposes. Notice further that the figures are meant to represent topological assemblies of joints and bodies, i.e. they indicate connections among the various parts, and hence are not to scale.

Figure 24 shows the topological representations of a three-bladed direct-drive machine at left, and a two-bladed teetering one at right. The blades, as the tower, are modeled with beam elements with span-varying structural properties. The pitch and the yaw bearings are modeled with revolute joints, possibly including frictional effects, connected to actuators. Another revolute joint connected with an electrical torque actuator models the generator and includes a non-linear damper modeling mechanical losses. Clearly, the simple arrangement shown in the figure does not model the actual load path in the nacelle-generator-shaft group, but nonetheless it is often sufficient for conducting load analyses on the rotor or tower; if a greater detail in the nacelle-generator-shaft group is necessary, this can be achieved with a proper assembly of elements. At the tower base, elastic foundations are represented with equivalent springs and dampers in revolute joints. The two-bladed teetering model (Figure 24 at right) has an additional revolute joint modeling the teetering hinge; by simply defining the angle between the hinge axis and the pitch axis once can model the effects of a δ_3 pitch-flap coupling.

Figure 25 depicts a hypothetical possible model of a wind turbine drive-train, comprising a sequence of flexible shafts, flexible couplers and rigid bodies. The inclusion of contact elements in the revolute joints of the gearbox enables a simple modeling of backlash in the drive-train.

Finally, Figure 26 shows the topological sketch of a wind turbine with an

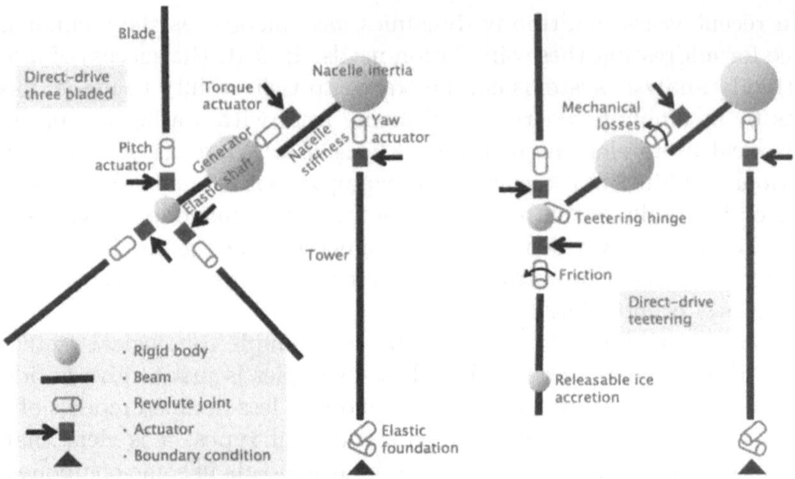

Figure 24. Topological view of wind turbine multibody models. Left: three-bladed direct-drive turbine. Right: two-bladed teetering.

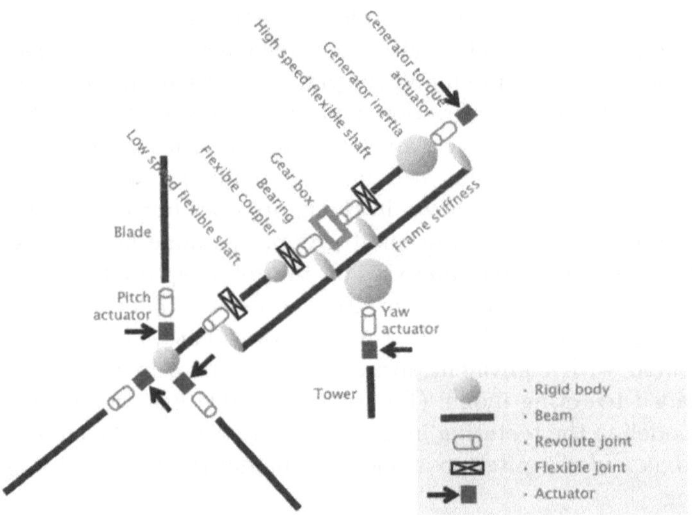

Figure 25. Topological view of drive-train multibody model.

up-tilting rotor, a passive control system that allows the rotor to tilt away from the wind when the aerodynamic thrust exceeds a design value. This tilting motion is aided with a preloaded torsional spring and has two stops which bound the rotation. Two additional hydraulic deceleration cylinders are used to smooth the dynamic transitions. In the multibody model, the up-hinge system is modeled with a revolute joint with a torsional preloaded spring and a backlash contact model, while the deceleration cylinders are modeled with prismatic joints with non-linear dampers.

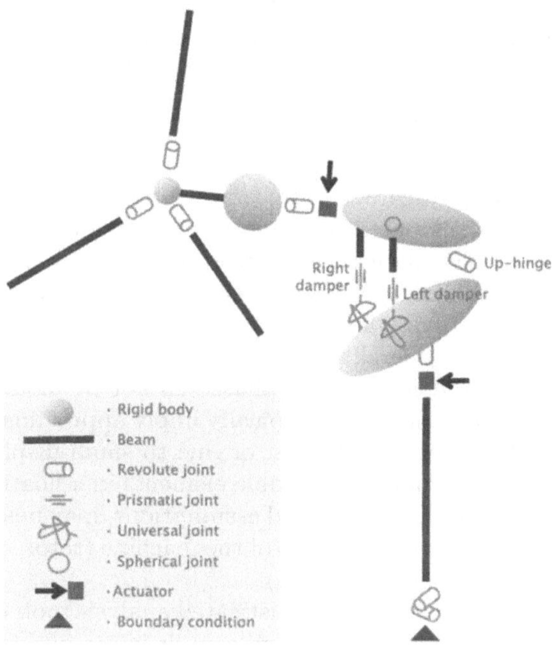

Figure 26. Topological view of wind turbine with up-tilting rotor.

In all these illustrative cases, the wind turbine models are simply obtained by assembling components from the library of elements implemented in a general purpose code, which has no a priori built-in knowledge of the topology of the system being modeled, in marked contrast with non-multibody-based formulations. Notice that this modeling system also opens the way to a hierarchical approach, where more or less refined models of the same system are built based on required accuracy, the need to resolve or not certain scales in the solution, the limitation of computational cost or the maximization of execution speed, or various possible trade-offs.

Key to this modeling approach is the capability of handling constraints; this, in turn, affects the choice of coordinates, the writing of the governing equations and their integration in time, the coupling of the multiple interacting fields, and a number of related algorithmic details. Several ad-hoc formulations have been developed in the multibody dynamics literature for niche applications: for example, methods have been developed for maximum execution speed in the analysis of systems in tree-like topologies. While some of these niche formulations find applicability in the wind energy field, in this work we focus exclusively on methods that can be used for modeling complex flexible systems with arbitrary topological configurations.

Historically, the modeling needs of wind turbine manufacturers were answered by dedicated software programs, as for example `GH Bladed` (Bossanyi, 2008), `Flex` (Øye, 1996) and `FAST` (Jonkman et al., 2005). Such specialized codes are built around given specific configurations of the wind turbine, and typically employ few degrees of freedom; furthemore, the modeling of the elastic components is based on some form of modal reduction. Although such approaches have seen a widespread adoption by industry, have effectively enabled the design of the current generation of wind turbines, and have often compared very favorably with respect to experimental data, it is clear that they lack generality. In fact, topological configurations not contemplated during the code design phase can not be modeled; furthermore, the modal reduction approaches typically imply approximations (e.g., on the stiffening effects of rotating beams, or due to small displacement/rotation assumptions when formulating flexible elements in a floating frame of reference), and, due to linearizations and assumptions, may miss subtle couplings among the various sub-components of the machine (rotor, drive-train, tower, floating platform, moore lines, etc.).

More recently, the need for sophisticated analysis tools capable of capturing all relevant physical processes and couplings has pushed the wind energy field towards the more general paradigm of multibody dynamics, grown out of demanding applications in other engineering fields, as for example the rotorcraft and automotive areas. General purpose multibody codes which support wind turbine modeling applications include, among others, `Adams` (Adams, 2010), `HAWC2` (HAWC2, 2010), `Mecano` (Mecano, 2010), `Simpack` (SIMPACK, 2010).

The following sections describe a finite element multibody formulation that is at the heart of the wind turbine modeling software `Cp-Lambda` (Code for Performance, Loads and Aeroelasticity by Multi-Body Dynamic Analysis). `Cp-Lambda` is based on a geometrically exact formulation, where the kinematics are handled exactly without approximations even in the presence of arbitrarily large displacements and rotations; this can improve on the current modeling abilities provided by legacy codes, for example in the case of slender blades undergoing large deflections, or for the modeling of off-shore machines.

To better motivate the choice of algorithms and methods employed in `Cp-Lambda`, and at the same give a quick view of the main methods formulated within the broader field of multibody dynamics, alternative approaches that have been proposed in the literature will also be briefly reviewed.

3.1 Equations of Motion and Solution Techniques

Choice of Coordinates. As for many other aspects of multibody dynamics, even the very basic choice of the coordinates used for formulating the equations of motion can be made in several very different ways (Geradin and Cardona, 2001). Clearly, each way has its own specific features, which make it more or less suited to a given application niche.

A classical approach used mainly for analytical developments, is to adopt a minimal set of coordinates (Geradin and Cardona, 2001). This approach leads to a minimum number of unknowns and of equations which, however, are typically very complicated and highly non-linear. This approach is difficult to apply to arbitrary topologies, and it is unsuitable for flexible systems.

Another approach uses relative, recursive Lagrangian coordinates. Within this category, the Denavit-Hartenberg (Geradin and Cardona, 2001) method has been developed for describing with a minimum number of coordinates the relative motion of a body connected to another one by a joint, which leaves some degrees of freedom free and constraints the others. Clearly, such an approach is particularly efficient for tree topologies, and in fact it was used in the formulation of the `SymDyn` software (Stol et al., 2005), a rigid body code for wind turbine modeling where the flexibility of blades and tower is rendered by means of equivalent hinges. For systems with arbitrary topologies, and in the presence of closed loops, the method becomes much more complicated, and can not avoid the introduction of loop-closure constraints. Furthermore, the resulting equations are usually very complex and highly non-linear, and hence expensive to evaluate. For flexible systems modeled using modal-based elasticity, this approach has been used for expressing the motion of a floating frame of reference, whose purpose is to describe the gross rigid motion of the body; in addition to this motion, the elastic body is assumed to undergo a small deformation about the floating frame (Shabana, 1998). This approach leads again to complex expressions, most notably of the kinetic energy, and does not readily account for the geometric stiffening effect, which is important in certain applications, as for example those involving rotating blades.

A formulation that is more amenable to problems with arbitrarily complex configurations is the one based on Cartesian coordinates. In this case, each body is regarded as free in space and described with its own set of coordinates. Next, constraints are introduced to represent the presence of mechanical joints among the various bodies of the system. This way, the multibody system is described by a redundant set of unknowns, composed by the collection of the sets of each body, often with the addition of Lagrange multipliers for the enforcement of the constraints (see further on in this document). This set of unknowns is redundant, as opposed to the previous choices that were minimal or quasi-minimal. On the other hand the equations of motion are much simpler, because they are obtained by assembly of the equations of the individual bodies and of the constraints, with a procedure similar to the one routinely used in finite element analysis. Furthermore, since each body or joint is typically connected with a small number of adjacent bodies or joints, the equations of motion have a high degree of sparsity, and present a banded nature, possibly after a suitable reordering of equations and unknowns; these aspects are crucial for the efficiency of the solution when analyzing large scale problems.

Another important characteristic of a formulation expressed in terms of Cartesian coordinates, is that it is readily amenable to the analysis of flexible systems using the finite element method. In fact, each elastic body in a multibody system may be described in terms of finite element unknowns, i.e. positions and possibly rotations, as in the case of beams, at mesh nodes. Specific nodes in the mesh may then be used for connecting the elastic body with other bodies or joints in the system, again using a standard assembly process. Clearly, the finite element formulation of elastic bodies must be capable of describing in exact terms the kinematics of the body motion. Such formulations are termed geometrically exact, and are based on definitions of the (invariant) strain energy V in terms of strain measures $\boldsymbol{\epsilon}$ that are unaffected by arbitrarily large rigid body motions, i.e. $V(\boldsymbol{\epsilon}) = V(\boldsymbol{R}\boldsymbol{\epsilon})$ if $\boldsymbol{R}$ is an arbitrary rotation (Borri et al., 2000; Geradin and Cardona, 2001; Bauchau et al., 2003). Contrary to the floating frame approach, these formulations correctly account for geometrically non-linear effects due to the kinematics being exact (finite rotation approach), and lead to particularly simple expressions of the inertial forces.

Lagrange Equations and the Index 3 Form. The Lagrangian of a multibody system $\mathcal{M}$ can be written as

$$L^* = \sum_i L_i(\dot{\boldsymbol{q}}, \boldsymbol{q}) + \boldsymbol{\lambda} \cdot \boldsymbol{c} + \frac{1}{2} \rho \, \boldsymbol{c} \cdot \boldsymbol{c}, \tag{28}$$

where $L_i(\dot{\boldsymbol{q}}, \boldsymbol{q})$ is the Lagrangian of the ith body in the system, and $\boldsymbol{q}$ are generalized coordinates. The system is subjected to constraints

$$\boldsymbol{c} = 0, \tag{29}$$

which are enforced in `Cp-Lambda` through two modifying terms of the Lagrangian: the first, $\boldsymbol{\lambda} \cdot \boldsymbol{c}$, uses Lagrange multipliers $\boldsymbol{\lambda}$, while the second, $1/2 \, \rho \, \boldsymbol{c} \cdot \boldsymbol{c}$ is a penalty-like term with penalty factor ρ. Clearly, both terms are null when the solution satisfies the constraints.

For $\boldsymbol{\lambda} = 0$ one obtains a pure penalty formulation. To enforce the presence of the constraints, it is necessary to choose ρ as a large number since $\boldsymbol{c} \to 0$ only if $\rho \to \infty$; this however leads to ill-conditioning of the problem, so that this practice is not recommended for the large scale complex applications that are the focus of the present work.

For $\rho = 0$ one obtains a purely Lagrangian formulation, and this increases the number of unknowns with respect to the penalty approach, since one has now to solve for both the coordinates $\boldsymbol{q}$ and the multipliers $\boldsymbol{\lambda}$. The multipliers, however, rigorously account for the presence of the constraints and their reactions on the bodies of the system.

The formulation retaining both terms, Equation (28), is termed the augmented Lagrangian approach. Since constraints are enforced by Lagrange multipliers, ρ does not need to be chosen as a large number, and hence one avoids the resulting ill-conditioning. Nonetheless, the penalty-like term proves to be useful because it allows one to factorize the system Jacobian (as required for the solution of the equations using a Newton-like method) without pivoting (Bauchau et al., 2009), and this is crucial for efficiency when dealing with large scale problems by retaining the bandedness of the iteration matrix.

Imposing the stationarity of L^*, one obtains the equations of Lagrange:

$$\boldsymbol{M}\ddot{\boldsymbol{q}} + \boldsymbol{A}^T \boldsymbol{\mu} = \boldsymbol{g}, \tag{30a}$$

$$\boldsymbol{c} = 0, \tag{30b}$$

where $\boldsymbol{\mu}$ are augmented Lagrange multipliers, $\boldsymbol{\mu} = \boldsymbol{\lambda} + \rho \boldsymbol{c}$, and $\boldsymbol{A}^T \boldsymbol{\mu}$ are the constraint reactions. Clearly, at convergence when $\boldsymbol{c} = 0$ then $\boldsymbol{\mu} = \boldsymbol{\lambda}$, and the penalty-like term has no effect on the computed solution. For holonomic constraints, i.e. when $\boldsymbol{c} = \boldsymbol{c}(\boldsymbol{q})$, then $\boldsymbol{A} = \partial \boldsymbol{c}/\partial \boldsymbol{q}$ and $\boldsymbol{A}$ is termed the constraint Jacobian.

Equations (30) are differential algebraic equations (DAEs), since they have among their unknowns the algebraic variables $\boldsymbol{\lambda}$. These equations can be turned into ordinary differential equations (ODEs) by taking three analytical differentiations of Equation (30b); this has the effect of introducing terms in $\dot{\boldsymbol{\lambda}}$ and therefore eliminates the algebraic nature of the problem. For this reason, Equations (30) are said to be in index 3 form (Hairer and Wanner, 1996).

Transformation into ODEs and DAE Index Reduction Approaches. It was recognized early in the literature that the solution of high index DAEs can cause severe numerical difficulties. For example, the analysis of Petzold and Lötstedt (1986) demonstrated that the condition number of the iteration matrix for the index 3 form of Equations (30) is $\mathcal{O}(h^{-3})$, where h is the time step size. Furthermore, it was shown that errors grow in the Lagrange multipliers as $\mathcal{O}(h^{-3})$, and at lower but still unfavorable rates in the displacement and velocity fields. Therefore, for DAEs one obtains a rather surprising behavior: the solution, which should converge to the true one as the time step size vanishes, on the contrary becomes polluted by the unavoidable errors that are due to the use of finite precision arithmetic and of finite convergence tolerances for arresting Newton iterations. Hence, contrary to intuition, mesh refinement makes the problem harder to solve. This unusual behavior manifests itself when using small time steps, which on the other hand are often necessary for achieving the desired level of accuracy or for resolving fast solution components present in the solution, for example when analyzing contact/impact phenomena.

The multibody dynamics literature abounds with methods for turning problem (30) into an ODE one or for reducing the index from 3 to 2 or 1. A comprehensive treatment of the methods that have been proposed goes beyond the scope of this work. The interested reader may consult the review offered in Laulusa and Bauchau (2008) and Bauchau and Laulusa (2008), which also contain a rich list of relevant references.

Very synthetically, a first family of methods formulates the problem in ODE terms by using a minimum set of unknowns. This can be accomplished at the level of the coordinates, or at the level of the velocities.

Considering the first of these two options, if the system is described in terms of n (redundant) coordinates $\boldsymbol{q}$ and has m constraints, it is conceptually possible to split the coordinate vector into an independent set of $n-m$ coordinates $\boldsymbol{q}_i$ and a dependent set $\boldsymbol{q}_d$ of size m, whereby the dependent variables can be expressed in terms of the independent ones: $\boldsymbol{q}_d = \boldsymbol{q}_d(\boldsymbol{q}_i)$.

This leads to a formulation of the equations of motion without Lagrange multipliers, and hence in ODE form.

Similarly, working at the level of velocities, the so called Maggi-like methods seek to write the generalized velocities $\dot{\boldsymbol{q}}$ in terms of a minimal set of $n - m$ generalized speeds $\boldsymbol{e}$

$$\dot{\boldsymbol{q}} = \boldsymbol{B}\boldsymbol{e} + \boldsymbol{b}, \tag{31}$$

while accounting for the fact that generalized velocities are constrained as

$$\boldsymbol{A}\dot{\boldsymbol{q}} + \boldsymbol{a} = 0. \tag{32}$$

This is true by definition in the case of non-holonomic constraints, or it is obtained by differentiating holonomic constraints once with respect to time, i.e. $\dot{\boldsymbol{c}} = 0$. It may be proven that $\boldsymbol{B}$ spans the null space of $\boldsymbol{A}$, i.e. $\boldsymbol{A}\boldsymbol{B} = 0$. Hence, premultiplying (30a) by $\boldsymbol{B}^T$ effectively eliminates the Lagrange multipliers, leaving an ODE in the generalized speeds $\boldsymbol{e}$ and accelerations $\dot{\boldsymbol{e}}$.

Both methods suffer from several drawbacks, and this makes them unsuitable for the FEM applications that are the focus of the present work, although they have found their niches of applicability in other sectors of multibody dynamics. First of all, both when working at the level of coordinates and at the level of velocities, the choice of the minimal set is not unique. This means that one has to devise an automated way of picking a set, typically through some form of optimality. Furthermore, the definition of the set is often local, and this means that the set might become ill-defined in certain configurations of the system; therefore, the set must be continuously monitored and updated throughout the simulation. While this is certainly possible, it significantly complicates the implementation. Furthermore, while equations in the redundant form (30) are typically highly sparse and present a banded pattern, equations in minimal form are usually dense or with limited sparsity; unfortunately, the loss of bandedness incurs in very significant computational costs for large scale FEM-based models. Hence, for large problems, the potential advantage of having lowered the number of unknowns by moving from a redundant set with multipliers to a minimal set, is offset by the increased complexity of the equations. Finally, in the case of Maggi's methods, holonomic constraints are accounted for through their time derivatives $\dot{\boldsymbol{c}}$, and this will cause the position-level constraints to drift away from the manifold, i.e. in general the numerical solution $\boldsymbol{q}_h$ will be such that $\boldsymbol{c}(\boldsymbol{q}_h) \neq 0$ (although this drift is often limited and sufficiently small for being acceptable in practical situations).

A second class of methods is based on the lowering of the index of problem (30), with the bulk of formulations focusing on the index 1 form obtained by appending to Equation (30a) the second derivative of the constraints, i.e. $\ddot{c} = 0$, which yields

$$\boldsymbol{M}\ddot{\boldsymbol{q}} + \boldsymbol{A}^T\boldsymbol{\mu} = \boldsymbol{g}, \tag{33a}$$

$$\boldsymbol{A}\ddot{\boldsymbol{q}} + \dot{\boldsymbol{A}}\dot{\boldsymbol{q}} + \dot{\boldsymbol{a}} = 0. \tag{33b}$$

The Lagrange multipliers can be readily eliminated from system (33) either by straightforward substitution (compute $\ddot{\boldsymbol{q}}$ from (33a), insert into (33b), solve in terms of $\boldsymbol{\lambda}$ and replace $\boldsymbol{\lambda}$ into (33a)), or by using a Moore-Penrose generalized inverse (Laulusa and Bauchau, 2008). Yet another alternative is to compute the null space $\boldsymbol{B}$ of $\boldsymbol{A}$, and premultiply Equation (33a) by $\boldsymbol{B}^T$ to eliminate the constraint reactions by exploiting the fact that $\boldsymbol{AB} = 0$.

Even this second class of methods is affected by several problems. First, here again the manipulations of the equations that are necessary for turning the system in index 1 form and for subsequently solving it, irremediably destroy the banded sparsity of the matrices and render the equations highly involved and complicated. Furthermore, the constraints are now enforced at the acceleration level, instead of the position one. It can be shown that the acceleration-level constraint is unstable, so that in this case the solution will experience a much larger drift from the constraint manifold than in the case of the velocity-level enforcement. Since this is in most cases unacceptable, various techniques have been developed for improving the satisfaction of the constraints. Here again details go beyond the scope of the present work, but in synthesis it can be said that most approaches are based on one of the following ideas: the first is to modify the acceleration constraint so as to make it stable by the addition of the velocity and position level constraints as damping-like and stiffness-like terms, respectively; the second is based on the projection of the solution back onto the constraint manifold (Bauchau and Laulusa, 2008). The former can not guarantee the exact satisfaction of the constraint conditions but only aims at the stabilization of the drift effect, while the latter one can indeed compute solutions compatible with the constraints at machine accuracy or assigned levels of tolerance.

As a notable exception, Betsch and Leyendecker (2006) have proposed a formulation based on the null space approach. In this case the null space is computed starting from a discrete (as opposed to the continuous, as usually done) version of the equations of motion, as obtained by applying a conserving time discretization scheme. Furthermore, by exploiting the simpler nature of the discrete equations, the contribution to the null space of each individual constraint condition can be computed explicitly and analytically up-front, and this results in a major simplification of the numerical procedure.

Direct Solution of Index 3 DAEs and the Importance of Scaling. As previously explained, much of the literature dealing with various ways to solve problem (30), does not address the pollution problem directly, but rather tries to avoid the problem by turning the problem into an ODE one or by lowering the index. It was argued that, unfortunately, this strategy incurs in other problems since constraints, being imposed at the velocity or acceleration level rather than at the displacement one, typically drift away from the constraint manifold. This in turn calls for additional corrective actions, in the form of constraint stabilization and/or projection back onto the manifold. An alternative approach is to re-write the governing equations so as to include both the position-level and velocity-level constraints (Gear et al., 1985; Borri et al., 2006), but this comes at the cost of additional problem unknowns.

What is even more important for the applications of interest here, is that all these approaches are unsuitable for FEM-based applications, due to the loss of banded structure of the problem and to the complexity and numerical cost of the manipulations of the equations. From this point of view, it is clear that the redundant but highly sparse, algebraic formulation expressed by Equations (30) is highly preferable, as recognized already by several authors (Orlandea et al., 1977; Cardona, 1989). Furthermore, since the index 3 form enforces the constraints at the position level, the "inherited" velocity and acceleration-level constraints, although not explicitly accounted for, will be approximatively satisfied since they are obtained by differentiation of the enforced ones; as noted above, the opposite is not true for the lower index formulations, where drift effects are present since "inherited" constraints are obtained by integration and not differentiation. Therefore, by adopting an index 3 formulation one avoids all complications associated with constraint stabilization and projection, while working with relatively simple equations with a banded structure.

Yet, the problem of numerical pollution has to be solved in order to arrive to fully effective computer implementations. Stiff integrators denoted by high frequency numerical damping help in this regard (Orlandea et al., 1977) and are also useful for other reasons, as noted below. But it is only recently that the pollution problem has been tackled directly (Bottasso et al., 2007, 2008; Bauchau et al., 2009), leading to a remarkably simple solution of the problem, based on an early hint reported but not analyzed in Cardona (1989). The idea is to scale the problem so as to eliminate the unfavorable behavior with respect to the time step length h. To this effect, one defines a non-dimensional time $\tau = t/h$ and works with derivatives with respect to τ instead of t, which are indicated here as $\mathrm{d}(\cdot)/\mathrm{d}\tau = (\cdot)'$. This way, the time step in non-dimensional time becomes of unit length. Of course, this scaling of time can be defined at each time step based on the current value of h, so that the procedure can be applied with no difficulty in the case of variable time step sizes.

As for most problems in mechanics, a further improvement of the overall conditioning of a numerical process can be obtained by using unknowns that are well scaled, i.e. all roughly of $\mathcal{O}(1)$. To this effect, one may use, in addition to the scaling of time, also non-dimensional coordinates $\bar{\boldsymbol{q}}$ instead of the dimensional ones, $\boldsymbol{q}$. Similarly, non-dimensional functions of the non-dimensional coordinates are noted in the following as $\bar{(\cdot)}$.

This way, a non-dimensional augmented Lagrangian may be written as

$$\bar{L}^* = \sum_i \bar{L}_i(\bar{\boldsymbol{q}}', \bar{\boldsymbol{q}}) + h^2 \boldsymbol{\lambda} \cdot \bar{\boldsymbol{c}} + \frac{1}{2} \rho h^2 \, \bar{\boldsymbol{c}} \cdot \bar{\boldsymbol{c}}. \tag{34}$$

Notice that, since the Lagrangians $L_i(\dot{\boldsymbol{q}}, \boldsymbol{q})$ are quadratic in the velocities, the introduction of the non-dimensional time has caused the appearance of a term h^2 in the two augmenting terms, which must now be carefully analyzed.

For the first of the two, the idea is to define $h^2\boldsymbol{\lambda}$ as the scaled Lagrange multiplier of the new problem, i.e. we set $\bar{\boldsymbol{\lambda}} = h^2\boldsymbol{\lambda}$ and we solve for $\bar{\boldsymbol{\lambda}}$ instead of $\boldsymbol{\lambda}$. Once a solution has been computed, $\boldsymbol{\lambda}$ can be straightforwardly recovered as $\boldsymbol{\lambda} = \bar{\boldsymbol{\lambda}}/h^2$, so that constraint reactions can be evaluated.

For the second of the two, it is necessary to recognize that, for the effect of the penalty-like term not to vanish with $h \to 0$, one must set $\rho h^2 = r$, where r is now the penalty-like coefficient of the new scaled formulation.

With these choices, by imposing the stationarity of the non-dimensional augmented Lagrangian $\bar{L}^*$, one gets the scaled equations of motion of multibody systems in index 3 form:

$$\bar{\boldsymbol{M}}\boldsymbol{q}'' + \bar{\boldsymbol{A}}^T \bar{\boldsymbol{\mu}} = h^2 \bar{\boldsymbol{g}}, \tag{35a}$$

$$\bar{\boldsymbol{c}} = 0, \tag{35b}$$

where $\bar{\boldsymbol{\mu}}$ are scaled augmented Lagrange multipliers, $\bar{\boldsymbol{\mu}} = \bar{\boldsymbol{\lambda}} + r\bar{\boldsymbol{c}}$. It may be proven (Bauchau et al., 2009), that the scaled problem (35) is now completely insensitive to the pollution that affects (30). In fact, the error propagation rates in the solution fields and the condition number are all $\mathcal{O}(h^0)$, which is what one would expect from the solution of an ODE. Hence, by scaling alone one may eliminate the problems associated with small time step sizes of multibody DAEs, and make them not harder to treat than standard ODEs.

It should be noted that this result is achieved at the sole price of working in non-dimensional time with an augmented Lagrangian formulation, and this means that one has to recover the Lagrange multipliers by scaling them back, a trivial scalar operation of negligible cost. It should be further remarked that, typically, further improvements in the conditioning and robustness of the problem may be obtained by working with well scaled (non-dimensional) variables and equations, as for most numerical problems in mechanics, although this will not cure by itself the unfavorable behavior with respect to h of the original problem; as previously noted, this can only be eliminated by introducing the non-dimensional time and h^2-scaled Lagrange multipliers.

Numerical Integration Techniques. The numerical integration of the equations governing high index DAEs has received considerable attention in the literature. The solution of Equations (30) (or (35)) presents features that are in common with non-linear transient finite element applications. Specifically, here we consider the case where the equations typically model the low-medium frequency range of aero-servo-elastic wind turbine systems, and where therefore implicit schemes are superior to explicit ones. Notice that we do not address the case of processes with very fast dynamic scales, where explicit schemes would be more effective, as for example the simulation of gear meshing in drive trains or other contact-dominated problems. Furthermore, it is a well known fact that the higher frequency content of FEM models does not accurately represent the behavior of the true system, and it is in fact an artefact of the discretization process. When the response of these higher frequencies is excited, noise of a numerical origin affects the solution, a problem that may be exacerbated in the presence of non-linearities to the point of leading to the blow-up of the computation. Therefore, stiff time integration schemes are typically used, i.e. integrators that act as low pass filters for the lower accurate modes and that damp out the higher unphysical ones (Hairer and Wanner, 1996; Geradin and Cardona, 2001), typical examples being the dissipative members of the Newmark family of schemes (and the related modified-α method), or stiffly accurate Runge-Kutta schemes.

The idea of designing integration schemes for which non-linear proofs of stability are possible, has also been pursued with success. Dot multiplying Equations (30a) by $\dot{\boldsymbol{q}}$, it may be easily proven that for scleronomic (time indipendent, i.e. $\partial \boldsymbol{c}/\partial t = 0$) constraints, one has

$$\dot{E} = P_e, \tag{36}$$

which states that the time rate of change of the total mechanical energy E of the system is equal to the power generated by the external forces, P_e. This implies that constraint reaction forces do not generate nor absorb power, i.e.

$$\dot{\boldsymbol{q}} \cdot \boldsymbol{A\lambda} = 0. \tag{37}$$

Energy methods are then based on the idea of allowing for the proof of these two facts, Equations (36) and (37), at the discrete solution level. The procedure is based in essence on two steps: 1) for each unconstrained body model in the multibody system, one devises a temporal scheme for which it is possible to prove that the discrete rate of change of energy within a time step is equal to the algorithmic power generated by the external forces; 2) for each joint model in the multibody system, one devises an algorithmic discrete version of the constraint reactions for which it is possible to prove that no power is generated or absorbed within a time step. Once such discretizations have been defined for all body and joint models in the system, the assembly of an arbitrary number of such models into a multibody system will imply the existence of a discrete version of Equation (36), i.e.

$$\frac{E_{i+1} - E_i}{h} = P_{e_h}, \tag{38}$$

which leads to the notion of unconditional stability of the integrator in the non-linear regime (see Geradin and Cardona, 2001; Bauchau et al., 2003; Betsch and Leyendecker, 2006 and references therein).

Clearly, energy preserving schemes are unable to damp out the high frequency unphysical modes in the system, by their very design. Hence, a generalization of this concept has been devised that is based on the proof of an energy decay (rather than conservation) statement within each time step; this way one retains non-linear stability since energy is bounded from above, while at the same time achieving the desired goal of removing the unresolved modes from the solution. `Cp-Lambda` implements such an an energy decaying scheme with tunable high frequency damping, as described in Bauchau et al. (2003).

3.2 Element Models

Body Models. A general-purpose multibody code for the modeling of wind turbines must include a library of body models, the simplest being a rigid body that can be used for modeling components whose flexibility can be neglected or for introducing localized masses.

The inclusion of flexible elements in a multibody formulation is a very ample subject, which offers a wide range of possibilities both in terms of mathematical models of the body and of associated algorithms. Here only a very short summary of some important aspects of this topic is offered.

Beam models have attracted a great deal of attention in the literature, with the more sophisticated formulations being devoted to the modeling of rotor blades; clearly, beam models also find applicability in the modeling of the tower, shaft and other slender structural members of wind turbines.

The problem of blade modeling is particularly challenging, since it must be possible to represent shearing deformation effects, the offset of the center of mass and of the shear center from the beam reference line, and all the elastic couplings that can arise from the use of tailored composite materials. Although simple beam models have been used extensively in the past, the very long and slender blades of new large machines are calling for an increased attention in the modeling of such effects. To provide for accurate modeling at affordable computational costs, the three-dimensional elasticity problem is split into two sub-problems (Giavotto et al., 1983). The first problem is a linear, two-dimensional problem defined over the beam cross-section that provides the sectional elastic constants. The problem is solved by using a linear 2-D finite element approach, where a mesh is used for describing with all necessary details the geometric and material characteristics of the blade cross section. The second problem is a classical non-linear, one-dimensional problem defined along the beam reference line that predicts the non-linear response of the structure when subjected to time dependent loads; this problem is handled in `Cp-Lambda` using a geometrically exact formulation. At the post-processing stage, recovery relations provided by the 2-D analysis step can be used for computing the three-dimensional displacement, strain and stress fields in the beam in terms of the generalized one-dimensional strain measures computed using the geometrically exact model. The splitting of the three-dimensional problem into two- and one-dimensional parts results in very significant savings in computing time with respect to a standard three-dimensional finite element analysis. An example of the use of this technology is given later on in this document, with reference to the integrated multi-disciplinary design optimization of wind turbines.

In a wind turbine there are often complex structural components that can be modeled in linear terms, as for example the nacelle system with its structural supporting frame. An effective way of including the effects of linear non-beam-like structural members is to use a modal based approach. This way a few of the lower modes are extracted from possibly very detailed FEM models of the structure. Next, the modal representation of the structure is included in the multibody model using a component mode synthesis approach, whereby modal amplitudes are used for describing the linear elastic response of the body in a suitably defined moving frame, while boundary degrees of freedom are retained as additional unknowns in the model so as to allow its coupling to the rest of the multibody system (Bauchau and Rodriguez, 2003).

Joint Models. Multibody systems are characterized by the presence of joints that impose constraints on the relative motion of the bodies of the model. Most joints used for practical applications can be modeled in terms of the so called lower pairs (Geradin and Cardona, 2001): the revolute, prismatic, screw, cylindrical, planar and spherical joints.

The kinematics of lower pair joints can be described in terms of Cartesian frames. On the (rigid or flexible) body A, we consider a frame with origin at a point on the body whose position vector is $\boldsymbol{u}^A$, denoted by a triad of unit vectors $\mathcal{A} = (\mathbf{e}_1^A, \mathbf{e}_2^A, \mathbf{e}_3^A)$. Similarly, on body B, a frame has origin in $\boldsymbol{u}^B$ and a triad $\mathcal{B} = (\mathbf{e}_1^B, \mathbf{e}_2^B, \mathbf{e}_3^B)$. The relative displacement between the two bodies in the direction aligned with the unit vector $\mathbf{e}_i^A$ is noted d_i, while θ_i is the relative rotation about the same vector. Table 1 defines the six lower pairs in terms of the relative displacement and/or rotation components that can be either free or constrained to a null value.

	d_1	d_2	d_3	θ_1	θ_2	θ_3
Revolute	C	C	C	C	C	F
Prismatic	C	C	F	C	C	C
Screw	C	C	$p\,\theta_3$	C	C	F
Cylindrical	C	C	F	C	C	F
Planar	F	F	C	C	C	F
Spherical	C	C	C	F	F	F

Table 1. Definition of the six lower pair joints. F: free, C: constrained. For the screw joint, p is the screw pitch.

All lower pair constraints can be expressed by one of the following two equations:

$$\mathbf{e}_i^A \cdot (\boldsymbol{u}^A - \boldsymbol{u}^B) - d_i = 0, \tag{39a}$$

$$\cos\theta_i\,(\mathbf{e}_j^A \cdot \mathbf{e}_k^B) - \sin\theta_i\,(\mathbf{e}_k^A \cdot \mathbf{e}_k^B) = 0. \tag{39b}$$

The first equation represents a constraint on the relative displacement by setting $d_i = 0$; on the other hand, by regarding d_i as an additional variable, the same equation serves the purpose of defining the unknown relative displacement in the joint along that direction. Similarly, the second equation either constrains the relative rotation if $\theta_i = 0$, or defines the unknown relative rotation θ_i if this is regarded as a free variable.

The explicit definition of the relative displacements and rotations in a joint as additional unknown variables represents an important detail of the implementation. In fact it allows for the introduction of spring, damper and backlash elements in the joints. Furthermore, the time histories of joint relative motions can be driven according to suitably specified time functions, or can be used as inputs or outputs of control elements. When such additional variables are defined, equations(30) take the following form:

$$\boldsymbol{M}\ddot{\boldsymbol{q}} + \boldsymbol{A}^T\boldsymbol{\mu} = \boldsymbol{g}, \tag{40a}$$

$$\boldsymbol{c}(\boldsymbol{q}) = 0, \tag{40b}$$

$$\boldsymbol{d}(\boldsymbol{\nu}, \boldsymbol{q}) = 0, \tag{40c}$$

where the relative displacements/rotations are denoted by the vector of algebraic variables $\boldsymbol{\nu}$, while (40c) group together all defining equations of the kind (39) in the model. The new constraints and algebraic variables are of index 1, since one single derivative of (40c) is necessary for introducing terms in $\dot{\boldsymbol{\nu}}$, so that system (40) is an index 1-3 DAE.

Multidisciplinary Models. Aerodynamic effects in wind turbines are accounted for by using a variety of approaches, which range from CFD, to free wake models, to more specialized approaches. For example, the models routinely applied in transient simulations of wind turbines typically make a combined use of lifting lines based on two-dimensional strip theory, each lifting line being attached to a beam and moving accordingly, coupled with inflow models that describe the flow field induced by the wake over the rotor disk; the more sophisticated of such models are denoted by their own set of states that are solved together with the multibody DAEs. There aerodynamic models are capable of providing sufficiently accurate estimates of the aerodynamic effects, at least in certain operating conditions, at computational costs that are compatible with their routine use in an industrial environment, something that CFD is not yet capable of delivering. These models are further improved by the use of a variety of sub-models that account for blade tip losses, radial and unsteady flow, dynamic stall and other effects.

In `Cp-Lambda`, lifting lines can be associated with beam elements and their geometric description is given in terms of three-dimensional twisted curves; for generality of the implementation, these aerodynamic reference curves are distinct from the structural reference ones they are associated with. The lifting lines are based on classical two-dimensional blade element theory, and account for the aerodynamic center offset, twist, sweep, and unsteady corrections. At a number of span-wise stations along each lifting line, the aerodynamic characteristics of the aerofoil used at that location are given using look-up tables, which store for a given number of angles of attack and Reynolds numbers the values of the sectional lift, drag and moment coefficients. Lifting lines are used here to model the aerodynamic characteristics of the blades, but also of the tower and of the nacelle. An inflow element can be associated with the blade lifting lines so as to model the rotor inflow effects; the code implements the Peters-He dynamic inflow wake model (Peters and He, 1995) and a classical blade-element momentum (BEM) model (Burton et al., 2001; Manwell et al., 2002; Hansen, 2008) based on the annular stream-tube theory with wake swirl. Tip and hub loss models are also considered. Figure 27 shows the aerodynamic model of the rotor: at different aerodynamic stations, the effective wind value is computed as a superimposition of different effects: a) the steady state mean wind with the turbulent (and/or gust) perturbation, b) the wind shear effect, c) the tower shadow correction and d) the inflow model. This effective wind, combined with the motion of the airstation computed from the one of its associated structural beam, generates the relative wind acting at each airstation.

The wind blowing on the rotor is modeled as the sum of a steady state mean wind and a perturbation wind, accounting for turbulence and/or gusts. The deterministic component of the wind field implements the transients specified by IEC 61400 (IEC 61400, 2005–2006, Parts 1 and 2), the exponential and logarithmic wind shear models, and the tower shadow effects, which include the potential flow model for a conical tower, the downwind empirical model based on Powles (1983), or an interpolation of these two models. The stochastic component of the wind field is computed according to the Mann or Kaimal turbulence models. The turbulent wind is precomputed before the beginning of the simulation for an assigned duration of time and for a user-specified two-dimensional grid of points (see Figure 27). During the simulation, the current position of each airstation is mapped to this grid, and the current value of the wind is interpolated in space and time from the saved data.

The recent literature reports several attempts at developing and validating hydrodynamic models for application in off-shore wind engineering (Jonkman et al., 2007; Passon et al., 2007). Standard hydrodynamic models routinely used by industry, and also implemented in `Cp-Lambda`, include the Jonswap/Pierson-Moskowitz spectrum for irregular waves with upper frequency limit, the computation of wave particle kinematics based on linear Airy theory, Wheeler stretching, the simulation of regular waves using stream function theory and that of irregular waves via digital filtering of pseudo-random white noise, and the modeling of current velocities as a linear combination of near-surface, sub-surface and near-shore currents. The resulting hydrodynamic loads are evaluated according to Morrisons equa-

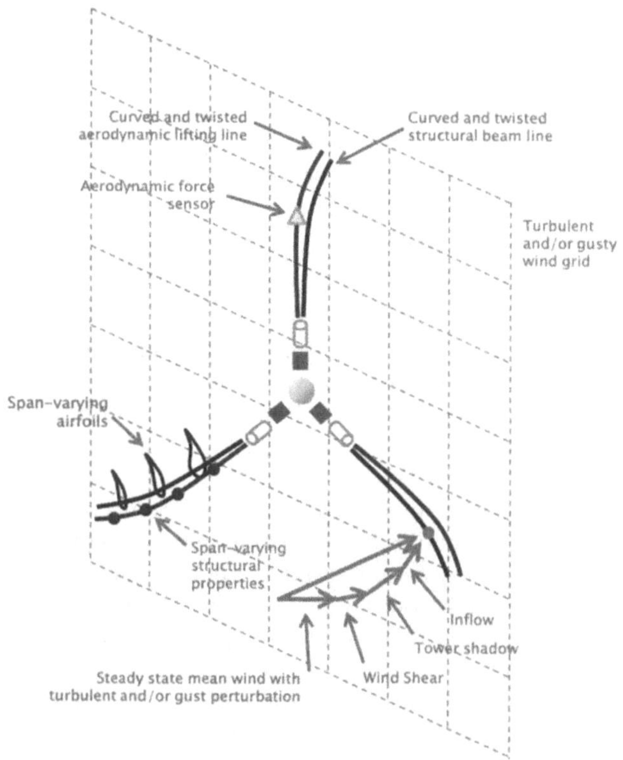

Figure 27. Aerodynamic model of the rotor.

tion, and are transferred to structural beam elements through lifting lines, as in the aerodynamic case.

Contact models account for interactions among bodies. Two main families of approaches are used: methods that explicitly model the deformation processes which take place in the contact zone, typically based on FEM formulations using detailed meshes of the contacting parts, and methods where such processes are rendered in a simplified, equivalent form. `Cp-Lambda` implements the latter class of methods, which are based on the combination of a kinematic model with a contact force model. The kinematic model, given a mathematical description of the geometry of candidate parts, determines whether and where the parts are in contact and the speed of their relative motion, while the contact model, based on the information provided by the kinematic model, provides the interaction forces between the contacting bodies, including friction. Contact models in wind turbines find applicability in the modeling of stops in joints, e.g. the flap stop in a teetering rotor, the modeling of gear meshing or free-play effects in a drive-train, etc.

Other coupled models frequently used in applications include electromechanical models, hydraulic models, sensor models, control elements, etc.

3.3 Solution Procedures

A general-purpose multibody system for wind turbine analysis implements a number of solution procedures, some of which are reviewed next.

The static analysis solves the governing equations obtained by setting all time derivatives to zero, under the action of given static loads, and may be computed by Newton-like methods, often in conjunction with continuation techniques that allow one to incrementally load the structure, considerably easing the convergence process.

A useful generalization of the static analysis concept is to include also the case of assemblies of bodies in the system undergoing rigid body motions about a given point at a constant assigned angular rate. The resulting inertial forces, as well as the aerodynamic forces generated by the same motion, can be regarded as steady configuration-dependent loads, and the equilibrium configuration of the system can be easily computed by a standard static analysis. This procedure finds applicability, for example, for rapidly computing the curves of power C_P and thrust C_T coefficient vs. tip-speed-ratio (TSR) and blade pitch β, as shown in Figure 28. Such curves express not only the fundamental aerodynamic characteristics of a machine, but are also crucial for the design of its regulation policy for varying wind speed, using techniques such those described in Bottasso et al. (2009).

In fact, when a wind turbine operates in constant (in time) wind conditions, the machine settles on a periodic orbit (trimmed operating condition). The periodicity of the response is caused by the non-uniformity of the spatial distribution of the wind over the rotor disk; this is due to the vertical wind shear (Earth boundary layer), possible lateral shear, and tower shadow effect, and to the fact that the wind direction is in general not parallel to the rotor axis, due to the rotor up-tilt and the possible presence of lateral (operation in yawed conditions) and vertical wind components (e.g., due to the terrain orography in proximity of the wind turbine). The vertical wind shear, tower shadow and rotor up-tilt are always present, while the other effects may or may not exist depending on the operating conditions of the machine.

It is a common practice to compute $C_P - \lambda - \beta$ and $C_T - \lambda - \beta$ oper-

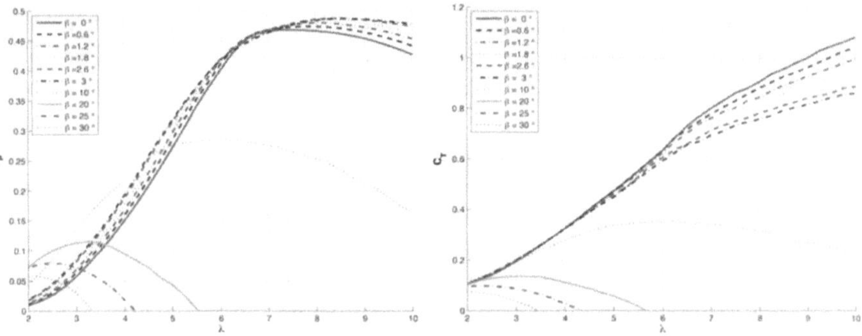

Figure 28. Power coefficient C_P (left) and thrust coefficient C_T (right) vs. TSR λ for varying collective blade pitch settings β, for a hypothetical 3 MW wind turbine.

ating curves of a wind turbine using an *isolated rigid rotor*, by considering axial flow conditions and a uniform distribution of the wind over the rotor, since this yields a steady rather than periodic response. Clearly, better estimates could be obtained by computing such quantities using a complete flexible wind turbine model, rather than the isolated rigid rotor, that includes flexible blades and tower, rotor up-tilt, vertical wind shear layer and tower shadow effect. Using this model, which captures the principal causes for a periodic response of the machine, transient simulations could be run for given values of the wind speed, blade pitch setting and generator torque. Once the solution settles on a periodic orbit, the zeroth harmonic of all quantities of interest could be computed, including values of the power coefficient and TSR.

Instead of conducting transient simulations, an alternative simpler and faster approach is to perform static solutions for varying wind speeds, rotor speeds and blade pitch settings. During such simulations, the deflected wind turbine configuration is computed under the action of the steady aerodynamic and inertial loads due to a steady rotation of the rotor at a constant angular speed. The simulation also include all other steady loads, such as gravity, wind loads on the tower and nacelle, and blade-tower aerodynamic interference loads. In practice, such simulation amounts to a snap-shot of the wind turbine at a given rotor azimuthal position, where accelerations have been neglected except for the inertial loads causes by a constant rotor speed. Each one of these simulations is extremely fast to compute, since it amounts to a simple (although non-linear) static solution. By computing the solution at different values of the rotor azimuthal angle, one obtains a quasi-static picture of the periodic response of the rotor around a full revolution. Averaging over the rotor revolution, one obtains quasi-static estimates of the necessary quantities (C_P, C_T, etc.). Notice that such estimates are obtained using a complete flexible model of the wind turbine that includes all principal causes for a period response, without necessitating of the use of a regulator. Clearly, by repeating such computations for varying values of the TSR and blade pitch, one can obtain the complete $C_P - \lambda - \beta$ and $C_T - \lambda - \beta$ curves.

Once the static deflected configuration has been computed, one may consider the dynamic behavior of small amplitude perturbations about the equilibrium configuration. This is obtained by first linearizing the dynamic equations of motion, and then extracting the eigenvalues and eigenvectors of the resulting linear system. Due to the presence of gyroscopic effects when rotating parts are present, the eigenpairs are, in general, complex. This way, one can readily compile Campbell diagrams, such as the one of Figure 29, that reports the eigenfrequencies of the machine for varying RPM, highlighting the stiffening effects due to the spinning of the rotor and showing the relative placement of the natural frequencies with the fundamental excitation harmonics.

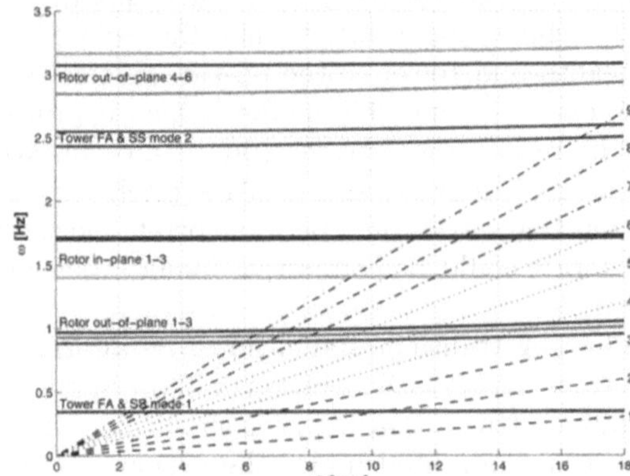

Figure 29. Campbell diagram of a hypothetical 3 MW wind turbine, showing the structural natural frequencies of the machine and fundamental excitation harmonics for varying rotor RPM.

Stability can be assessed by using methods that account for the fact that wind turbines are denoted by periodic coefficients, for example using techniques such as those of Bertogali et al. (1999); Bittanti and Colaneri

(2000).

The dynamic analysis solves the non-linear equations of motion for the complete multibody system. The initial conditions are taken to be at rest, or those corresponding to a previously determined static or dynamic equilibrium configuration. The equations of motion of the system are integrated, starting from the given initial conditions, under the action of externally applied loads, given driving inputs, and the action of closed-loop control systems, which on board wind turbines typically include a supervision module in charge of selecting the operating condition (start up, power production, emergency shut down, etc.) and a pitch-torque controller. Complex multibody systems often involve rapidly varying responses, and this may render the use of a constant time step a computationally inefficient strategy, so that time step size adaptivity is commonly adopted for increased efficiency.

Given the complexity and number of the analyses that need to be performed during the design of a wind turbine, `Cp-Lambda` implements automated procedures that support a number of standard operations, such the computation of Campbell diagrams, the automated analysis of all IEC 61400 DLCs (IEC 61400, 2005–2006, Parts 1 and 2), the determination of trimmed periodic conditions, the generation of C_P vs. tip-speed-ratio (TSR) curves, the tracing of power curves, the determination of fatigue equivalent loads using rain-flow analysis, etc.

3.4 Corollary Supporting Technologies

Parameter Estimation. The problem of system identification is concerned with finding a model that best matches some experimental observations. More precisely, given a system $\mathcal{S}$ (the plant), the problem of *model characterization* is concerned with defining an appropriate model structure, i.e. a suitable class of models $\mathcal{M}(\boldsymbol{p})$ parameterized in terms of free quantities $\boldsymbol{p}$. Then, given class $\mathcal{M}(\boldsymbol{p})$, the problem of *parameter estimation* is to find the parameters $\boldsymbol{p}$ such that the model outputs $\boldsymbol{y}$ best match in some given sense the measured quantities $\boldsymbol{z}$. The problem is of a stochastic nature, since the observations are corrupted by measurement noise, the plant maybe excited by process noise, and the model is affected by unknown modeling errors.

Clearly, effective parameter estimation techniques can have a profound and positive impact on multibody simulation technology, by closing the loop between virtual prototyping and testing of the actual hardware; in other words, parameter estimation allows one to tune a model with a mathematically sound approach, rather than doing manual adjustments to the model parameters. However, given the level of complexity of modern multibody models, which are multi-field and non-linear, the problem of parameter estimation is particularly challenging.

System identification for wind turbines is still in its infancy and much needs to be done before mature techniques that work in practice can be routinely used by industry. No attempt at covering this broad field is made in these notes. We simply mention that typically the problem is broken down in the identification of the properties of the system sub-components (e.g. blades, tower, actuators, drive-train or its individual elements such as shafts and gear-boxes, etc.) using specific experimental observations of each sub-system behavior. Such experiments, which may include the measurement of natural frequencies, mode shapes, structural deflections, temporal responses under known inputs, etc., can be typically conducted in the controlled environment of the lab with good accuracy and relatively low noise and uncertainty levels. A part from structural measurements, one may consider among the sub-system identification steps the identification of sectional aerodynamic models, such as dynamic stall models and the measurement of airfoil characteristics in the wind tunnel.

Next, with properly identified sub-system models, it is conceptually possible to conduct further identification steps on the whole system operating in closed-loop, with the goal of identifying those model parameters that can not be estimated in the first step, although very few examples of such an approach are present in the literature. Possible parameters that could be identified at this stage include coupling interactional terms, such as those arising in rotor empirical interference models with tower and nacelle, parameters appearing in inflow models, or other complex aerodynamic effects. When such experiments are conducted in the field, the expected uncertainly levels are much higher, one of the reasons being that the wind incident on the rotor is typically measured only at one or very few locations so that the wind spatial distribution is unknown or only very sketchily known. Hence, one should probably not try to identify from runs of the full system those parameters that can be more accurately estimated using specific measurements of the sub-component behavior.

[CLB: need some references here]

Model Reduction. As previously described, the response of a multibody system $\mathcal{M}$ to a time history of control inputs $\boldsymbol{u}(t)$ can be obtained by solving the governing Equations (30) starting from given initial conditions. Accordingly, one may compute the response of the system outputs $\boldsymbol{y}(t)$. Consider a second model $\hat{\mathcal{M}}$, described by some system of governing ODEs or DAEs expressed in terms of its own states $\hat{\boldsymbol{x}} \in \mathbb{R}^{\hat{n}}$. When subjected to the same time history of control inputs $\boldsymbol{u}(t)$, $\hat{\mathcal{M}}$ produces a time history of its own associated outputs $\hat{\boldsymbol{y}}(t)$. Model $\hat{\mathcal{M}}$ is said to be a reduced model of $\mathcal{M}$ if $\hat{n} \ll n$ and

$$\hat{\boldsymbol{y}}(t) = \boldsymbol{y}(t) + \boldsymbol{e}(t), \tag{41}$$

where $\boldsymbol{e}(t)$ is the output error, small in an appropriate norm. Equation (41) states that the reduced model $\hat{\mathcal{M}}$ and the original one $\mathcal{M}$ produce output responses that differ by the output error $\boldsymbol{e}(t)$ when subjected to the same input signal $\boldsymbol{u}(t)$. Clearly, the output error measures the fidelity of $\hat{\mathcal{M}}$ to $\mathcal{M}$. Typically, $\hat{\mathcal{M}}$ will give a good approximation of $\mathcal{M}$ at the slower scales, while the error $\boldsymbol{e}(t)$ will be mostly due to unmodeled or unresolved faster solution components.

The ability to reliably generate accurate reduced models of wind turbines from complex multibody systems can find applicability in several areas, as for example the synthesis of model-based control laws, the evaluation of the stability of the system, the use in real-time applications with high-frequency rates and/or limited computational resources, etc. (Brüls et al., 2007; Lehner et al., 2007; Fehr et al., 2010). The area of model reduction is being actively researched. Since the treatment of model reduction methods in these notes would be incompatible with space limitations, here only a very simple case is illustrated. The example chosen here has no aim at generality, and simply shows a very basic reduced model, routinely used for the modeling of a collective pitch machine with a very limited number of degrees of freedom, and the relationship of such a model to a more refined one; specifically, it is briefly shown that one can extract from a refined model information, for example in the form of modal mass or stiffness or aerodynamic power and thrust coefficients, and then use such information within a simpler reduced model.

A collective-pitch non-linear wind turbine reduced model including drive-train shaft dynamics, elastic tower fore-aft motion, blade pitch actuator dynamics and electrical generator dynamics can be written as:

$$(I_R + I_G)\dot{\Omega} + T_l(\Omega) + T_{el_e} - T_a(\Omega, \beta_e, V_w - \dot{d}, V) = 0, \tag{42a}$$

$$M_T\ddot{d} + C_T\dot{d} + K_T d - F_a(\Omega, \beta_e, V_w - \dot{d}, V) = 0, \tag{42b}$$

$$\ddot{\beta}_e + 2\xi\omega\dot{\beta}_e + \omega^2(\beta_e - \beta_c) = 0, \tag{42c}$$

$$\dot{T}_{el_e} + \frac{1}{\tau}(T_{el_e} - T_{el_c}) = 0. \tag{42d}$$

The first equation, (42a), describes the drive-train dynamics; Ω is the rotor angular speed, d is the tower top fore-aft displacement and β_e is the effective blade pitch angle. Moreover, I_R is the sum of the moments of inertia about the rotation axis of the rotor hub and of the three rotor blades, while

I_G is the moment of inertia of the rotating part of the electric generator. The torques acting upon the drive-train include the mechanical losses on the shaft bearings T_l, the effective electrical reaction torque T_{el_e} and the aerodynamic torque T_a. The mechanical loss T_l is modeled by means of a speed-torque look-up table. The second equation, (42b), models the fore-aft tower dynamics. Here, M_T, C_T and K_T are, respectively, the tower equivalent modal mass, structural damping and bending stiffness. These quantities can be obtained by modal reduction of a detailed finite element model of the tower. Finally, F_a indicates the aerodynamic force produced by the rotor. The third equation, (42c), is a second order model of the blade pitch actuator, where ω is the undamped natural frequency, ξ the damping factor, and β_c the blade pitch control input. The model also includes upper and lower limits on the pitch and the pitch rate. The fourth and last equation, (42d), is a first order model of the electrical generator that includes a time delay τ, while T_{el_c} is the commanded electrical torque input.

The rotor aerodynamic force and torque are computed as

$$T_a = \frac{1}{2}\rho\pi R^3 \frac{C_P(\lambda, \beta_e, V)}{\lambda}(V_w - \dot{d})^2, \tag{43a}$$

$$F_a = \frac{1}{2}\rho\pi R^2 C_F(\lambda, \beta_e, V)(V_w - \dot{d})^2, \tag{43b}$$

where ρ is the air density, C_P and C_F the power and force coefficients, respectively, and λ is the TSR, defined as $\lambda = \Omega R/(V_w - \dot{d})$. Finally, $V_w = V + V_t$ is the turbulent upstream wind speed obtained as the sum of the mean wind V and the turbulent wind V_t. For this reduced model, the mean wind V is computed by spatially averaging over the rotor disk the wind speed profile given by the power law (IEC 61400, 2005–2006, Parts 1 and 2). Similarly, the longitudinal turbulent wind V_t is defined, at each time step, as the spatial average over the rotor disk of the wind turbulence model centered at the hub.

Although the model is rather simple, its accuracy can be substantially enhanced by a proper modeling of the crucial aerodynamic coefficients C_P and C_F. In fact, these coefficients can be computed off-line using a fine-scale aero-servo-elastic model, and then stored in tabulated form. This way, the reduced model inherits the aerodynamic modeling of the fine scale one, while keeping a very simple implementation and extremely low computational cost. To estimate the aerodynamic coefficients, several simulations are run with the fine scale model, each one for given constant values of blade pitch, mean wind speed and electric torque. The resulting power and force coefficients are then computed by averaging the periodic responses over a rotor revolution; similarly, the averaged TSR is computed. Finally, the power and force coefficients are stored in a look-up table to be used by the reduced model, the entries of the table being λ, the blade pitch β_e and the mean wind speed V.

A model such as the one expressed by Equations (42) can be effectively used for designing model-based control laws, as for example the collective-pitch/torque linear quadratic regulator (LQR) described in Bottasso et al. (2009).

4 Design and Optimization of Wind Turbines – a Multi-Disciplinary Approach

Authored by C.L. Bottasso and A. Croce

The optimization of the configuration of a wind turbine is a complex multi-disciplinary problem. Many considerations of various nature must be made and taken into account in order to arrive to a design that achieves the right trade-offs between performance and cost, while accounting for a variety of other constraints that make that specific design solution viable from all relevant points of view. Clearly, tools that can effectively support such complex design efforts in a integrated, holistic, manner and with rapid turn-around times can be useful for sizing a new machine, for improving a tentative configuration, or for studying modifications to existing models.

The optimization of wind turbines has been the subject of a number of investigations in the recent literature. For example, ECNBOT (2010); Lee et al. (2007); Maalawi and Badr (2003) and Mendez and Greiner (2006) describe procedures for the determination of the optimal aerodynamic shape of rotor blades. Typically such approaches use aerodynamic wind turbine models based on variants of the blade element momentum theory; Xudong et al. (2009) adopts a more sophisticated aero-elastic model to account for the structural dynamics response of the machine, although even in this case the optimization is limited to the aerodynamic shape of the blade and does not account for the structural sizing aspect of the problem. The purely structural sizing problem given an aerodynamic shape has been studied by Jureczko et al. (2005), and specialized FEM-based software for the detailed structural analysis of rotor blades is described by Laird (2008).

On the other hand, the integrated multi-disciplinary optimization of wind turbine rotors addresses a much more complex problem, which considers the aerodynamic shape optimization, the evaluation of all relevant load conditions which in turn requires the definition of appropriate control laws, the optimal sizing of the structural members under the effects of the loads, considering the mutual couplings between the various sub-disciplines and simultaneously accounting for the presence of a number of design constraints of various nature. It appears that there are very few holistic design tools with such characteristics, and few papers have been devoted to the subject (see Fuglsang and Madsen, 1999; Fuglsang, 2008). Most notably, the two codes `RotorOpt` (Fuglsang, 2008; RotorOpt, 2007) and `FOCUS5` (Duineveld, 2008) implement integrated design environments; on the other hand, Jonkman (2008) describes a suite of tools that cover all primary modeling needs required to perform a full design cycle, although they do not appear to have been yet cast within a unified optimization framework.

The present section describes a suite of procedures for the integrated multi-disciplinary constrained optimization of wind turbines, and that includes aerodynamics, load calculation, synthesis of control laws and structural sizing. The approach described here, although similar in spirit and motivation to the few other published holistic design tools, has two principal distinguishing features.

First, finite element based multibody models are used that can capture the relevant aero-servo-elastic characteristics of the system to a high level of fidelity. The use of sophisticated models of the machine since the very inception of the design, implies that one can immediately account for some important aspects (e.g., the placement of the rotor natural frequencies, the effects on structural blade sizing induced by constraining the maximum tip deflection, and several others), which therefore can express their effects and couplings with all other design requirements; with simpler models some of these interactions are lost or not properly accounted for, so that the obtained design must be modified a posteriori to include the overlooked effects. The approach described here is based on a two-level modeling system. The first level model is a parametric global model of a wind turbine, implemented within the comprehensive aero-servo-elastic non-linear finite-element-based multibody dynamics solver `Cp-Lambda`. The model enables the evaluation of a variety of contributors to the merit function and constraints, by running simulations that include eigen-analysis and transient design load cases according to IEC 61400 (2005–2006, Parts 1 and 2), as necessary. The second level model is a finite element parametric cross sectional model of the blade, that enables the calculation of section-wise stresses and strains under the loads computed on the first model, to support the determination of a minimum weight blade configuration satisfying all associated required constraints. The two-level modeling is closed by synthesizing on the detailed blade model beam-equivalent structural and inertial characteristics that are then used in the definition of the multibody model.

Second, many of the complex considerations that are made by the designer so as to ensure a viable solution are formulated as constraints to the optimization problem. Although this has been in part done in previous works on the subject, in the present effort we have tried to include as many of the crucial design constraints as possible. In fact, many constraints have profound couplings with the functioning of the machine and hence have complex effects on its design. For example, the inclusion of a noise constraint even through a simple limit on the tip speed, alters the regulation strategy of the machine in between regions II and III (Bottasso et al., 2009), and this has an effect on the power curve and hence on the annual energy production; this effect is mitigated by changing the rotor solidity, that in turn alters the loads and might interact with other geometric constraints, for example on the maximum chord length for ensuring the transportability of the blade on-board trucks. Another constraints with complex effects on the design is the relative placement of the rotor natural frequencies with respect to the predominant harmonic excitations, which should be done correctly to avoid the insurgence of resonant conditions within the whole operating envelope of the machine. Clearly, this can only be done while simultaneously ensuring a blade that, under the maximum experienced loads, does not exceed the maximum allowable stresses and strains, does not have excessive deflections and is also of minimum possible weight. For capturing the effects of such constraints in a correct way, they need to be consistently enforced during the design optimization, which is one of the highlights of the present approach.

Space constraints preclude a detailed description of the proposed procedures in this work; the interest reader may find a more complete discussion and further examples in Bottasso et al. (2010).

4.1 Multi-Disciplinary Optimization Algorithm

The multi-disciplinary optimization of a wind turbine is here considered as a multi-objective design problem where one seeks a compromise between the maximization of the Annual Energy Production (AEP) and the weight of the machine. We implicitly assume that weight is well correlated with cost; however we do not use cost here because reliable cost models are not available in the public domain.

The problem is challenging for a number of reasons. First, the problem is subjected to a number of inequality constraints that translate various additional requirements (see below for details); many of these constraints are active at the optimal solution. Second, the evaluation of the merit function contributors and of the constraints can be computationally expensive, especially if using refined models of the machine and a large set of loading conditions. Although powerful methods are available for efficiently dealing with costly functional evaluations, for example using response surfaces or neural networks, these are often not well suited for highly constrained problems as the present one.

The procedure described here was devised for allowing the execution of a complete design loop on a standard desktop computer in a reasonably fast way, typically in one-two days. To satisfy this requirement, the multi-objective design is not formulated as a Pareto optimal problem, but rather using a combined cost defined as AEP divided by total weight. Furthermore, to minimize the number of expensive functional evaluations, a special two-stage procedure was devised, as described next. It is clear that the use of parallel processing or more powerful computational resources might allow for different approaches than the one described here, or for the same approach to be executed in shorter time.

The two-stage process is formulated as follows.

At first, we compute an optimal solution that yields the maximum AEP with the minimum possible weight of the rotor blade. As shown later on, this problem leads to an iteration that alternates between a purely aerodynamic optimization of the blade twist and chord distributions for maximum AEP given a structural configuration, and a purely structural optimization for minimum blade weight given an aerodynamic design. The former problem also determines the weight of the other design-parameter-dependent components of the machine, i.e. drive-train and generator. This analysis is conducted for assigned values of some macro configuration parameters, typically the rotor radius, the maximum chord, and/or a blade taper measure. The design above is repeated for different values of the macro-parameters. This generates a family of different AEP optimal designs, each one achieving the least possible blade weight for that AEP.

Next, the family of optimal design solutions is interpolated with respect to the independent macro-parameters, and the design that achieves the best ratio of AEP and total weight is found by computing the maximum of the interpolated combined cost (see later on for details).

The first step above involves essentially uncoupled aerodynamic and structural optimizations, which therefore can be performed at reasonable computational costs. The coupling between aerodynamic and structural solutions is brought about at the second step of the procedure, where AEP and total weight are combined together (notice that total weight is used in this stage, where the non-blade related weight components are estimated using weight models, so the weight optimization is not limited to the sole rotor blade). This step however still implies reasonable costs, since the combination is performed when optimizing with respect to the macro-parameters, which are usually just very few in number.

Stage 1: Maximum AEP for Minimum Blade Weight. The problem of finding the configuration that yields the maximum possible AEP with the minimum possible weight of the rotor blade is formulated as the following constrained optimization:

$$\textbf{Function } (\boldsymbol{p}_a^*, \boldsymbol{p}_s^*, P_y^*, w^*) = \texttt{MaxAEPMinBladeWeight}(\boldsymbol{p}_a, \boldsymbol{p}_s, D): \tag{44a}$$

$$P_y^* = \min_{\boldsymbol{p}_a} \mathrm{AEP}(\boldsymbol{p}_a, \boldsymbol{p}_s, D) \quad (\text{and } \boldsymbol{p}_a^* = \arg\min_{\boldsymbol{p}_a} \mathrm{AEP}), \tag{44b}$$

$$\text{s.t.:} \quad \boldsymbol{g}_a(\boldsymbol{p}_a) \leq \boldsymbol{0}, \tag{44c}$$

$$v_{\text{tip}} \leq v_{\text{tip}_{\max}}, \tag{44d}$$

$$w_b^* = \min_{\boldsymbol{p}_s} W_b(\boldsymbol{p}_s, D) \quad (\text{and } \boldsymbol{p}_s^* = \arg\min_{\boldsymbol{p}_s} W_b), \tag{44e}$$

$$\text{s.t.:} \quad \boldsymbol{g}_s(\boldsymbol{p}_s) \leq \boldsymbol{0}, \tag{44f}$$

$$\boldsymbol{\omega}(\boldsymbol{p}_s, D) \in [\boldsymbol{\omega}_L, \boldsymbol{\omega}_U], \tag{44g}$$

$$\boldsymbol{E} = \texttt{LoadEnvelope}(\boldsymbol{p}_a, \boldsymbol{p}_s, D), \tag{44h}$$

$$\boldsymbol{\sigma}(\boldsymbol{p}_s, \boldsymbol{E}, D) \leq \boldsymbol{\sigma}_{\text{adm}}, \tag{44i}$$

$$\boldsymbol{\epsilon}(\boldsymbol{p}_s, \boldsymbol{E}, D) \leq \boldsymbol{\epsilon}_{\text{adm}}, \tag{44j}$$

$$\boldsymbol{\delta}_{\text{tip}_{\max}}(\boldsymbol{p}_s, \boldsymbol{E}, D) \leq \boldsymbol{\delta}_{\text{tip}_{\text{adm}}}. \tag{44k}$$

Here and in the following functions are used (as in Equation (44a)) to highlight the input and output data of the various algorithms, which is useful for clarifying how the proposed procedures work.

In problem (44), $\boldsymbol{p}_a$ and $\boldsymbol{p}_s$ are aerodynamic and structural, respectively, unknown parameters to be optimized, while D is a list of given data:

$$D = \{P_r, [V_{\text{in}}, V_{\text{out}}], R, H, \text{AF}, C, v_{\text{tip}_{\max}}, L_{\text{DLC}}, \ldots\}. \tag{45}$$

Among the many possible elements in the list D, we mention here specifically the rated power P_r, the wind speed range $[V_{\text{in}}, V_{\text{out}}]$ between the cut-in V_{in} and cut-out V_{out} speeds, the rotor radius R, the tower height H, the list $\text{AF} = \{\ldots, \text{AF}_i, \ldots\}$ containing the airfoil types used along the blade span, the wind turbine class C (IEC 61400, 2005–2006, Part 1), the maximum allowed tip speed to limit noise emissions $v_{\text{tip}_{\max}}$ (IEC 61400, 2005–2006, Part 11), and the list $L_{\text{DLC}} = \{\ldots, \text{DLC}\, i.j, \ldots\}$ containing all Design Load Conditions (DLCs) (IEC 61400, 2005–2006, Parts 1 and 2) that one wants to consider in the optimization of the machine.

Problem (44) seeks a minimum for the AEP function. The optimal AEP is noted P_y^*, and the aerodynamic parameters describing the corresponding wind turbine configuration are noted $\boldsymbol{p}_a^*$ (see Equation (44b)). The problem is subjected to three sets of constraints. The first is given by Equations (44c), that in general are used to expressed desired conditions on the unknown aerodynamic parameters (often in the form of simple lower and upper bounds, for example to limit the maximum blade chord so as to satisfy transportability constraints). The second constraint ensures that the blade tip velocity does not exceed a given limit, to contain noise emissions. The third constraint is the optimal blade weight problem expressed by Equations (44e–44g); hence problem (44) finds the machine that maximizes the AEP subjected to the constraint, among others, of having a rotor blade of minimum weight.

The optimal blade weight w_b^* minimizes the blade weight function W_b, which is computed on the detailed structural blade model; the associated optimal structural parameters are noted $\boldsymbol{p}_s^*$ (see Equation (44e)). The total weight w of the machine is computed as the sum of the drive-train and generator weight w_{dt+g}, plus the weight of the B blades, i.e.

$$w = w_{dt+g} + Bw_b^*. \tag{46}$$

It is assumed that the weight of the other components of the machine are almost constant with respect to the design parameters, at least within a certain machine category, and hence can be neglected in an optimization, since their gradients are essentially null. The weight of drive-train and generator are given by a weight model in terms of the rated power, maximum rotor speed and maximum torque, i.e.

$$w_{dt+g} = W_{dt+g}(P_r, \Omega_{\max}, T_{\max}). \tag{47}$$

The rated power is an assigned input to the problem; the maximum rotor speed is either given by the tip speed constraint or by the rated rotor speed, and the maximum torque can be computed once the regulation policy is known.

Even the optimal blade weight problem (44e) is subjected to a number of constraints, which are detailed next. Equations (44f) express bounds or other more complex desired conditions on the unknown structural parameters (e.g., constraints on the maximum relative position between sectional center of gravity and pitch axis, or limits on ply taper rates to account for the fact that the increase/decrease of the number of plies in a composite laminate must satisfy certain ply per length constraints imposed by manufacturing and technological considerations).

Inequality (44g) constrains some natural frequencies $\boldsymbol{\omega}$ of the structure to lie within the admissible bounds $[\boldsymbol{\omega}_L, \boldsymbol{\omega}_U]$, to avoid resonant conditions in the operating envelope of the machine. For example, a typical condition is the requirement for the first blade flap natural frequency $\omega_{1\text{ flap}}$, which is the lowest blade eigenfrequency for a conventional configuration, to be larger than the three-per-rev frequency at the rated rotor speed $\omega_{3P}(\Omega_r)$, i.e.

$$\omega_{1\text{ flap}} \geq s_{1f}\,\omega_{3P}(\Omega_r), \tag{48}$$

where s_{1f} is an appropriate multiplicative factor.

Next, Equations (44h) define the load envelope $\boldsymbol{E}$, i.e. the extreme loading conditions at all points of interest obtained by computing all DLCs in list $L_{\text{DLC}} \in D$.

At a number of cross sections along the blade span, the maximum stress components are noted $\boldsymbol{\sigma}(\boldsymbol{p}_s, \boldsymbol{E}, D)$; the notation highlights the fact that these quantities depend on the structural configuration $\boldsymbol{p}_s$, load envelope $\boldsymbol{E}$ (which in turn depends on $\boldsymbol{p}_a$, $\boldsymbol{p}_s$ and D, cfr. Equation (44h)) and given data D. Inequality (44i) constrains the maximum stresses to be lower than a given admissible upper limit $\boldsymbol{\sigma}_{\text{adm}}$; according to IEC 61400 (2005–2006, Parts 1 and 2), safety factors are included in the DLC loads, and therefore are automatically accounted for in the load envelope. Similarly, inequality (44j) constrains the maximum strains $\boldsymbol{\epsilon}$.

Finally, inequality (44k) constrains the maximum blade tip deflection $\boldsymbol{\delta}_{\text{tip}_{\max}}$ measured throughout all DLC simulations in list L_{DLC}, i.e.

$$\boldsymbol{\delta}_{\text{tip}_{\max}} = \max_{L_{\text{DLC}}} \max_{t} \boldsymbol{\delta}_{\text{tip}}(t, \text{DLC}). \tag{49}$$

The definition of the load envelope used in this work includes also the loads that cause the maximum tip deflections; similarly to what done for sectional stresses and strains, we use the notation $\boldsymbol{\delta}_{\text{tip}_{\max}}(\boldsymbol{p}_s, \boldsymbol{E}, D)$ to highlight this fact.

Problem (44) is a *nested* constrained optimization problem, i.e. an optimization problem that has among its constraints a second constrained optimization problem. The direct solution of problem (44) may require a significant computational effort. A very considerable simplification of the problem may be obtained by realizing that the AEP of a machine depends to a large extent on its aerodynamic design, and considerably less so on its structural one (i.e. it strongly depends on airfoils, chord and twist distributions along the blade span, rotor radius, tower height, etc., and much less so on the thickness of the external shell of the blade, the location and sizing of the spars, etc.).

Under this hypothesis, the nested problem (44) can be solved by two consecutive constrained optimizations: the first maximizes the AEP and finds the corresponding optimal aerodynamic parameters assuming frozen structural ones, while the second minimizes the blade weight and finds the optimal structural parameters by using the optimal aerodynamic ones obtained through the first optimization. If one suspects a coupling between structural parameters $\boldsymbol{p}_s$ and AEP, the procedure can be iterated until convergence. The resulting simplified *sequential* constrained optimization algorithm can be expressed as follows

$$\textbf{Function } (\boldsymbol{p}_a^*, \boldsymbol{p}_s^*, P_y^*, w^*) = \texttt{SequentialMaxAEPMinBladeWeight}(\boldsymbol{p}_a, \boldsymbol{p}_s, D): \tag{50a}$$

$$\textbf{do} \tag{50b}$$

$$(\boldsymbol{p}_a^*, P_y^*) = \texttt{MaxAEP}(\boldsymbol{p}_a, \boldsymbol{p}_s, D), \tag{50c}$$

$$(\boldsymbol{p}_s^*, w^*) = \texttt{MinBladeWeight}(\boldsymbol{p}_a^*, \boldsymbol{p}_s, D), \tag{50d}$$

$$\Delta p_a = \|\boldsymbol{p}_a^* - \boldsymbol{p}_a\|, \quad \Delta p_s = \|\boldsymbol{p}_s^* - \boldsymbol{p}_s\|, \tag{50e}$$

$$\boldsymbol{p}_a = \boldsymbol{p}_a^*, \quad \boldsymbol{p}_s = \boldsymbol{p}_s^*, \tag{50f}$$

$$\textbf{while } (\Delta p_a \geq \text{tol}_{p_a} \textbf{ and } \Delta p_s \geq \text{tol}_{p_s}). \tag{50g}$$

The first of the two optimizations in (50) is the maximum AEP problem that, from (44), writes

$$\textbf{Function } (\boldsymbol{p}_a^*, P_y^*) = \texttt{MaxAEP}(\boldsymbol{p}_a, \boldsymbol{p}_s, D): \tag{51a}$$

$$P_y^* = \min_{\boldsymbol{p}_a} \text{AEP}(\boldsymbol{p}_a, \boldsymbol{p}_s, D) \quad (\text{and } \boldsymbol{p}_a^* = \arg\min_{\boldsymbol{p}_a} \text{AEP}), \tag{51b}$$

$$\text{s.t.:} \quad \boldsymbol{g}_a(\boldsymbol{p}_a) \leq \boldsymbol{0}, \tag{51c}$$

$$v_{\text{tip}} \leq v_{\text{tip}_{\max}}. \tag{51d}$$

The second optimization as expressed in problem (44) (see Equations (44e–44g)) can also by itself imply very considerable computational costs, since it requires a computation of the load envelope $\boldsymbol{E}$ for each change in the structural design variables $\boldsymbol{p}_s$. A very considerable simplification is obtained by using an iterative approach where the load envelope is considered as frozen at each step:

$$\textbf{Function } (\boldsymbol{p}_s^*, w^*) = \texttt{MinBladeWeight}(\boldsymbol{p}_a, \boldsymbol{p}_s, D): \tag{52a}$$

$$\boldsymbol{E} = \texttt{LoadEnvelope}(\boldsymbol{p}_a, \boldsymbol{p}_s, D), \tag{52b}$$

$$\textbf{do} \tag{52c}$$

$$(\boldsymbol{p}_s^*, w^*) = \texttt{MinBladeWeightFrozenLoads}(\boldsymbol{p}_a, \boldsymbol{p}_s, D, \boldsymbol{E}), \tag{52d}$$

$$\boldsymbol{E}' = \texttt{LoadEnvelope}(\boldsymbol{p}_a, \boldsymbol{p}_s^*, D), \tag{52e}$$

$$\Delta p_s = \|\boldsymbol{p}_s^* - \boldsymbol{p}_s\|, \quad \Delta E = \|\boldsymbol{E}' - \boldsymbol{E}\|, \tag{52f}$$

$$\boldsymbol{p}_s = \boldsymbol{p}_s^*, \quad \boldsymbol{E} = \boldsymbol{E}', \tag{52g}$$

$$\textbf{while } (\Delta p_s \geq \text{tol}_{p_s} \textbf{ and } \Delta E \geq \text{tol}_E), \tag{52h}$$

where a minimum weight blade structure for a given load envelope is computed as

$$\textbf{Function } (\boldsymbol{p}_s^*, w^*) = \texttt{MinWeightBladeFrozenLoads}(\boldsymbol{p}_a, \boldsymbol{p}_s, D, \boldsymbol{E}): \quad (53a)$$

$$\boldsymbol{p}_s^* = \min_{\boldsymbol{p}_s} W(\boldsymbol{p}_s, D) \quad (\text{and } w^* = \arg\min_{\boldsymbol{p}_s} W), \quad (53b)$$

$$\text{s.t.:} \quad \boldsymbol{g}_s(\boldsymbol{p}_s) \leq \boldsymbol{0}, \quad (53c)$$

$$\boldsymbol{\omega}(\boldsymbol{p}_s, D) \in [\boldsymbol{\omega}_L, \boldsymbol{\omega}_U], \quad (53d)$$

$$\boldsymbol{\sigma}(\boldsymbol{p}_s, \boldsymbol{E}, D) \leq \boldsymbol{\sigma}_{\text{adm}}, \quad (53e)$$

$$\boldsymbol{\epsilon}(\boldsymbol{p}_s, \boldsymbol{E}, D) \leq \boldsymbol{\epsilon}_{\text{adm}}, \quad (53f)$$

$$\boldsymbol{\delta}_{\text{tip}_{\text{max}}}(\boldsymbol{p}_s, \boldsymbol{E}, D) \leq \boldsymbol{\delta}_{\text{tip}_{\text{adm}}}. \quad (53g)$$

Within the minimum weight optimization at frozen load envelope, stresses, strains and maximum deflections are computed for varying structural parameters, but keeping the loads fixed. The use of frozen loads is based on the hypothesis that the load envelope $\boldsymbol{E}$ changes slowly with respect to changes in the structural design parameters $\boldsymbol{p}_s$, which is a reasonable assumption in this case. Note that the effects of the design on the load envelope is just temporarily frozen to reduce the computational cost, and it is recovered by the iteration in (52).

The constrained optimization problems (51) and (53) can be solved in a variety of ways. When refining an already good design solution, which hence provides an initial guess close to the optimal one, gradient-based methods can be used effectively; for the solution of both problems we have used the implementation of the sequential quadratic programming (SQP) method available in the `fmincon` routine of the `MATLAB`® software. For the solution of problem (51), we have also used the commercial optimization software `Noesis Optimus`®, that offers a choice of different global and local optimization methods and supporting functional approximation techniques that allow for a more thorough search of the design space.

Stage 2: Combined AEP and Total Weight Cost with Expansion of Design Parameters. After having computed an optimal solution using algorithm (50), it is in general useful to free one or more of the assumed parameters in list D. For example, one might be interested in optimizing the solution with respect to the rotor radius R; this might have a complex repercussion on the solution, since a different radius will not only modify the area swept by the rotor, but will also imply different loads and, for a given maximum tip speed, a different rated rotor speed and hence a different region II½ in the power curve (Bottasso et al., 2009), and many other effects, such as a change in the weight not only of the blade but also of the drive-train and generator. Clearly, there are several parameters other than the radius that an analyst might want to study for identifying trade-offs and evaluating design sensitivities, such as the solidity, maximum blade chord, etc.

The optimization of the ratio of AEP and weight with respect to parameter $D_i \in D$ can be formally written as

$$\max_{D_i} \frac{P_y^*}{w^*}, \tag{54a}$$

$$\text{s.t.:} \quad D_i \in [D_{i_L}, D_{i_U}], \tag{54b}$$

$$(\boldsymbol{p}_a^*, \boldsymbol{p}_s^*, P_y^*, w^*) = \texttt{SequentialMaxAEPMinBladeWeight}(\boldsymbol{p}_a, \boldsymbol{p}_s, D), \tag{54c}$$

where D_{i_L} and D_{i_U} are lower and upper, respectively, allowable bounds on D_i. Often, optimizations with one or two unknown variables as the one in problem (54), can be solved by simply solving (54c) for different values in the range $[D_{i_L}, D_{i_U}]$. Interpolation of these family of optimal design points, for example using cubic splines, gives a way to readily and inexpensively compute the maximum of the merit function.

4.2 Models and Their Implementations

The optimization procedures described in Section 4.1 require the ability to define parametric aero-servo-elastic models of a wind turbine, and to perform a variety of simulations on the model for each instantiation of its parameters. Such simulations include the computation of a number of given DLCs, the generation of the charts of power, torque and thrust versus tip-speed-ratio, the computation of power curves in turbulent wind, the evaluation of the natural frequencies of the structure and the associated Campbell diagram, etc.

In this work, aero-servo-elastic models of wind turbines are implemented with the software `Cp-Lambda`, described earlier on in this work.

The aerodynamic parameterization is simply obtained by parameterizing the twist and chord span-wise distributions of the lifting lines of the model. For efficiency, the multibody formulation uses beam models for describing blades and tower. On the other hand, the structural parameterization used by the optimization problem is based on detailed structural models of the blade cross sections at a number of span-wise locations (see Figure 30); from the parametric detailed sectional models, equivalent cross sectional stiffness and inertial data are generated using the approach of Giavotto et al. (1983), which leads to the characterization of the sectional beam data necessary for the definition of the multibody model. The sectional models are also used for the calculation of the maximum stresses and strains experienced at each design cross section for the considered DLCs, stresses and strains that appear among the optimization constraints.

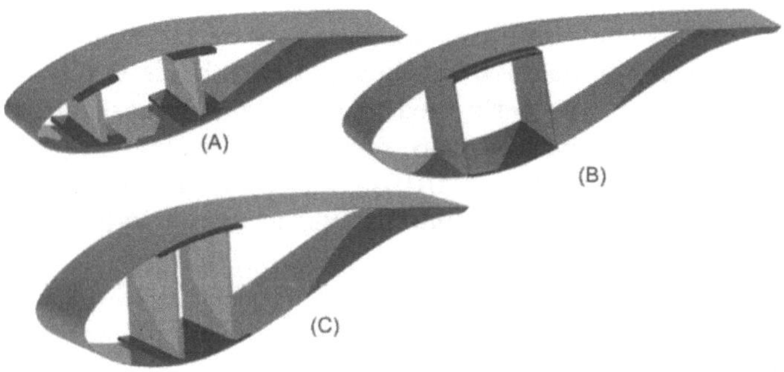

Figure 30. Cross section types. A: three cell with shear webs and spar caps; B: three cell box type, with single inter-web capping; C: three cell, with single capping extending fore and aft of the shear webs.

For conducting the transient simulation of all DLCs needed by the optimization procedures, it is necessary to automatically synthesize a pitch-torque controller capable of regulating each new instantiation of the machine across its entire operating envelope. This capability, which is crucial for the automation and robustness of the whole optimization procedure, is here implemented with the collective-pitch/torque linear quadratic regulator described in Bottasso et al. (2009).

4.3 Applications and Results

We first consider the optimization of a Class III A 2.8 MW machine, whose characteristics are given in Bottasso et al. (2010). The blade structural layout uses a box type three cell configuration, with a single cap confined within the two planar shear webs (see Figure 30(B)). The root region of the blade has caps that gradually extended to occupy a larger chord fraction, until they brace the full root circle. The thickness of the root section is set to 80 mm, to accommodate the connecting bolts. The structural optimization parameters $\boldsymbol{p}_s$ are defined as the common thickness of the two shear webs, the common thickness of the upper and lower caps, and the thickness of the external blade shell. The blade is made with two different materials, a unidirectional carbon fiber that is used for the caps, and a biaxial glass fiber for the external shell and the shear webs. Non-structural mass is accounted for with both surface-proportional and span-proportional quantities.

A family of three AEP-optimal blades was generated, each family member being indexed by a specific value of the taper measure $\tau = \int_0^1 c(\eta)\eta \, d\eta / S_b$, where $\eta \in [0, 1]$ is the non-dimensional span-wise coordinate, c the local blade chord and S_b the blade area. Next, for each blade family member an optimal weight sizing was performed. The principal characteristics of the computed blade family members are reported in Table 4.3 (further details are available in Bottasso et al. (2010)). The effect of the taper constraint on the extent of region II½ is apparent, and in turn it induces an effect on AEP.

Table 2. Principal characteristics of the blade family members for the 2.8 MW wind turbine.

τ	P_y^* [MWh]	C_P^{II}	λ^{II}	$V_{II^1/2}$ [m/sec]	V_r [m/sec]	w_b^* [Kg]
0.38	8.727e9	0.4910	10.05	7.4	12.9	17331
0.40	9.134e3	0.4911	9.01	8.4	12.0	11107
0.42	9.296e3	0.4882	8.15	9.2	11.4	10546
0.429[a]	9.312e3	0.4867	7.90	9.5	11.2	10261

[a] No constraint on τ.

Figure 31 shows the span-wise chord distribution of the blade family members; the taper constraint moves the blade area in-board, while clearly also affecting the solidity since the maximum chord constraint is active for all blade solutions.

It is interesting to observe that increased tapering comes with an increase in blade weight w_b^*. One might initially be tempted to think that the opposite should be true, since less tapered blades have more chord outboard along the span and hence should experience larger bending loads, and this in turn should induce a weight penalty. In reality, this blade is not sized by extreme loads, and in fact the stress and strain constraints (53e,53f) are mostly inactive at convergence. The blade sizing is driven in this case by the tip deflection and frequency placement constraints, that become harder to solve for more tapered blades, and this explains the lower weights of the less tapered solutions of Table 4.3.

Finally, the compromise between AEP and weight was analyzed by using stage two optimization, although in reality in this particular case the solu-

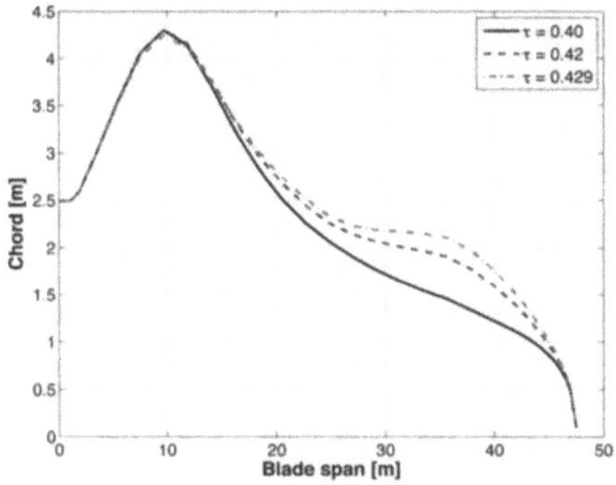

Figure 31. Chord shape of the blade family members for the 2.8 MW wind turbine.

tion is self evident, since increases in AEP are accompanied by decreases in weight (see Table 4.3). The behavior of the combined cost P_y^*/w_b^* vs. the taper parameter τ is given in Figure 32, at left. Clearly, the blade that achieves the best compromise is the one with a taper parameter $\tau = 0.429$, and it is therefore assumed as the final result of the present design process. Although the problem in this case does not present a maximum, the trend of Figure 32 (left) can be used for predicting possible further improvements in the design obtainable with reductions in tapering, i.e. for an increase in the τ parameter.

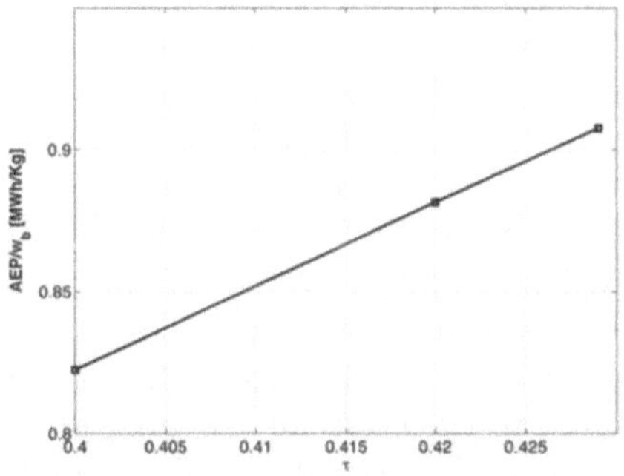

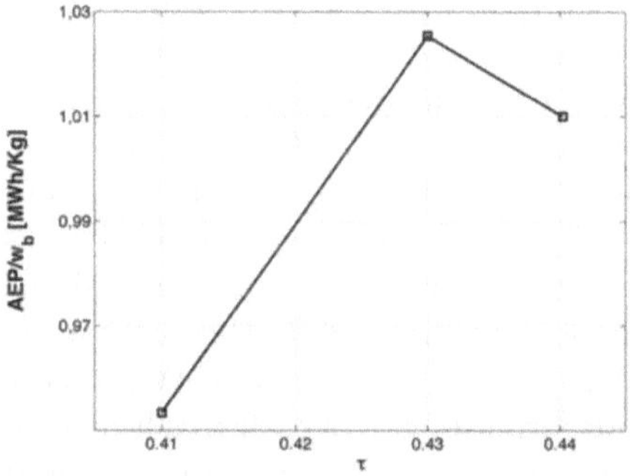

Figure 32. Combined cost AEP$/w_b$ vs. taper parameter τ, for the 2.8 MW (left) and the 3 MW (right) wind turbines.

To illustrate the complex and different effects that constraints can have on a design, we consider also the optimization of a 3 MW wind turbine, with a 53.2 m radius and a maximum chord of 3.9 m. The analysis was performed in a way very similar to the one of the previous example. In particular, the aerodynamic optimization was conducted for different values of the τ parameter, and for each of the members of the family of optimal aerodynamic designs a weight-optimal structural sizing was performed.

The behavior of the combined cost P_y^*/w_b^* vs. the taper parameter τ is given in Figure 32 at right. It appears that in this case the merit function has a maximum around $\tau = 0.43$. An analysis of the solution can help explain the different behavior than the one observed in the 2.8 MW case. In fact, the longer blade span and small maximum chord of the present design is penalized by excessive outboard chords, since this lowers the flap frequency and increases loads and in turn tip deflections, so that a compromise solution of intermediate taper emerges as the optimal one.

5 Wind Tunnel testing of Wind Turbines

Authored by A. Zasso and P. Schito

Since the first aerodynamical studies wind tunnel tests provided very helpful information to scientists and engineers. The advantage of wind tunnel testing is to allow measurement and visualization of the flow effects on the model in fairly well controlled and known conditions.

As far as concerns wind turbines, it is historically relevant to quote that the first quantitative design and optimization of a wind-mill has been attempted by the Danish scientist Poul La Cour in 1896, when he started to test small models of windmills in his own Wind Tunnel, probably the first such experiments in the world focusing on windmills. After a few weeks of experiments La Cour came to some general conclusions that are still accepted as far as concerns the overall architecture of a Wind Turbine.

Although being very interesting to note that the first quantitative approach to the wind turbine design relied strongly on Wind Tunnel scale model experiments, in the current development and optimization of modern Wind Turbines the role of Wind Tunnel experiments changed from the consideration of the overall machine essentially to the study of the wing profile performances. The design and prediction of the overall machine aerodynamic performances relies as a matter of fact almost entirely on numerical simulations.

Recently on the other hand a renewed attention has grown on possible valuable information coming from wind tunnel experiments, due to the complexity of the tasks involved evaluating the actual wind turbine behavior due to natural wind boundary layer, wind farms wakes effects, aeroelasticity issues.

For those reasons, the relevant role of the experimental approach is nowadays the allowance for validation of the numerical simulation techniques through direct comparison of the experimental measured data with the results predicted in the scale of the experiment by the same numerical simulation codes working full scale. of course the advantage relies on the controlled and known boundary condition of the experiment as well as experimentally known characteristics of the used wing profiles.

In the following paragraphs the fundamental concepts on which relies the wind tunnel experimental techniques are presented, with finally a focus on three test cases directly experienced at Politecnico di Milano Wind Tunnel.

A wind tunnel is generally composed of a inflow and outflow section (open circuit facility case), one or more fans to produce the flow and a test section. Of course there many important details in the design of the circuit as well as specific devices positioned along the flow (e.g. honeycomb / grids / vortex generators) allowing to realize the desired uniform / turbulent flow and to provide the most reliable results of the experiments. On the other hand for obvious reasons it is necessary to realize scale models of the true life Wind Turbines, with relevant problems concerning miniaturization , scaling effects, blockage effect and measuring issues.

5.1 Similitude Laws (and consequent limitations)

Wind tunnel testing needs the fulfilment of some basic similitude law. Not all laws must be observed in all wind tunnel tests, but it depends on the phenomenon under study. The definition of similitude is: *"Two phenomena are in similitude if, making adimensional each physical variable with respect to a suitable reference quantity, the mathematical relationships between them result identical"*. For example two phenomena are in dynamic similitude if all the dynamic quantities are in a constant relationship. Fluid dynamic similitude needs:

- **Geometrical Similitude** (geometrical scale)
- **Kinematic Similitude** (scale of velocities)
  - equal Reduced Velocities (*Strouhal Number*)
- **Dynamic Similitude** (scale of forces)
  - $\frac{Inertial forces}{Aerodynamic forces}$ (*Reynolds number*)
  - $\frac{Inertial forces}{Gravitational forces}$ (*Froude number*)

To formalize these similitudes it is necessary to define some scales:

$$LENGTHSCALE : \lambda_L = \frac{L_M}{L_R} \tag{55a}$$

$$VELOCITYSCALE : \lambda_V = \frac{V_M}{V_R} \tag{55b}$$

$$FREQUENCYSCALE : \lambda_f = \frac{\omega_M}{\omega_R} \tag{55c}$$

where the subscript M is referred to the model characteristics and the subscript R to the full scale object. In fluid dynamics some similitude numbers have been defined by Reynolds (Re), Froude (Fr) and Strouhal (St):

$$Re = \frac{inertialforces}{viscousforces} = \frac{\rho V L}{\mu} \tag{56a}$$

$$Fr = \frac{inertialforces}{gravitationalforces} = \frac{V^2}{Lg} \tag{56b}$$

$$St = \frac{aerodynamictime}{mechanicaltime} = f\frac{L}{V} = \frac{1}{V*} \tag{56c}$$

where ρ is the air density, V the wind speed, L the characteristic length, μ the dynamic viscosity, g the gravity acceleration and f the frequency. It can be seen that the Strouhal number is inversely proportional to the reduced velocity V^*. A very important similitude for wind tunnel tests is the geometric similitude. Another important similitude is the Reynolds similitude, but in cases where it is important to have a Reynolds and a Froude similitude a problem arises in the scaling procedure: the number of free parameters of the system is less than the non-dimensional numbers assumed as fundamental in the phenomenon. This does not allows to have a full similitude. Let's have an example: in a wind tunnel test it is possible to consider gravity g, viscosity μ and density ρ as fixed parameters so that it is shown in equations 57a and 57b that **a simultaneous Froude and Reynolds similitude is not allowed**.

$$Re_M = Re_R \Rightarrow (\frac{\rho V L}{\mu})_M = (\frac{\rho V L}{\mu})_R \Rightarrow \lambda_V = \frac{1}{\lambda_L} \tag{57a}$$

$$FR_M = Fr_R \Rightarrow (\frac{V^2}{Lg})_M = (\frac{V^2}{Lg})_R \Rightarrow \lambda_V = \sqrt{\lambda_L} \tag{57b}$$

Since the physics of a wind turbine requests a full fluid dynamic similitude, reliable wind tunnel tests should require:

- geometric similitude → geometric scale;
- kinematic similitude / wake similitude → equal Strouhal number or reduced velocity or tip speed ratio;
- dynamic similitude → equal Reynolds number.

The Reynolds similitude can be synthesized with a graphic representation as reported in figure 33. This relationship is then formalized in equation 58

Figure 33. Reynolds similitude sketch

$$Re_M = Re_R \Rightarrow (\frac{\rho VL}{\mu})_M = (\frac{\rho VL}{\mu})_R \Rightarrow \lambda_V = \frac{1}{\lambda_L} \quad (58)$$

For a HAWT it is possible to note in figure 34 and in equation 59 that Reynolds similitude permits to satisfy also the Strouhal similitude since the geometrical scale on R is balanced with the correct frequency scale on the rotor speed ω.

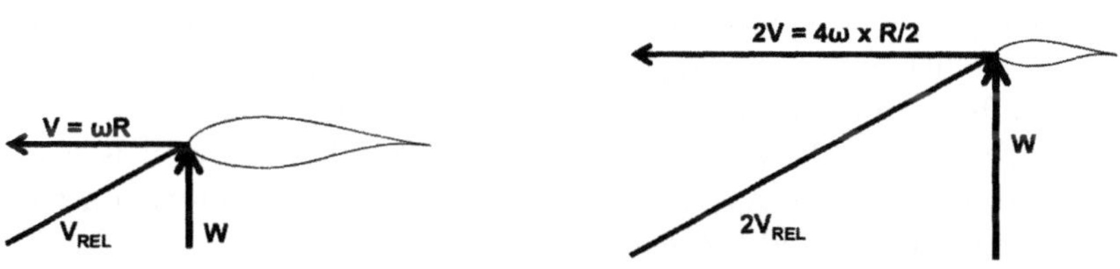

Figure 34. Kinematic similitude on a wind turbine profile airfoil

$$St_M = St_R \Rightarrow (f\frac{L}{V})_M = (f\frac{L}{V})_R \Rightarrow \lambda_f = \frac{1}{\lambda_L^2} \quad (59)$$

Full Scale	Wind Tunnel
Wind Speed $\approx$ 15 m/s Rotor Speed $\approx$ 40 rpm $Re \approx 10^6 - 10^7$	Wind Speed $\approx$ 750 m/s Rotor Speed $\approx$ 100.000 rpm $Re \approx 10^6 - 10^7$

Table 3. Examples of the assumptions that must be verified in case of full aerodynamic similitude

In this case the aerodynamic similitude is respected if the non-dimensional parameters are respected in full scale and in model scale. The biggest models that can be tested in wind tunnels have a geometrical scale $\lambda_L = 1/50$. Reynolds similitude needs a velocity scale $\lambda_V = 50$, while Strouhal similitude needs a rotational speed scale of $\lambda_f = 50^2$. Some problems arise in wind tunnel testing as can be seen in table 5.1.

The problem arises in technical difficulties in realizing a model that rotates so fast but specially in problems with the flow conditions: the very high Mach number leads to compressible flow with completely different aerodynamic behavior while the full scale wind turbine works at low Mach numbers and incompressible flow regime. It is clear that for this reason it is not possible to maintain the Reynolds similitude but another similitude must be chosen to define the velocity scale. For instance it can be based on the Froude number, using the same model scale of $\lambda_L = 1/50$, obtaining the velocity scale $\lambda_V = \sqrt{\lambda_L} \approx 1/7$ using equation 57b. In this case the Reynolds number between the full scale wind turbine ($Re_R \approx 10^6 - 10^7$) and the wind tunnel model ($Re_M \approx 10^4 - 10^5$) is completely different. This leads to a completely different behavior of the wing profiles in terms of flow, forces and efficiency as can be observed in figure 35 where it can be seen that lift and drag coefficients at the Re_R are far different from the Re_M ones: this causes a different performance in wind tunnel and full scale blades.

Using the similitude laws with a constant Froude number it is possible to derive the force scale λ_F, the mass scale λ_m, the acceleration scale λ_a, the damping scale λ_r and the stiffness scale λ_k as:

$$\left\{\begin{array}{l} \lambda_F = \frac{F_{aeroM}}{F_{aeroR}} = \frac{1/2\rho V_M^2 S_M C_X}{1/2\rho V_R^2 S_R C_X} = \lambda_V^2 \cdot \lambda_L^2 = \lambda_L^3 \\ \lambda_m = \frac{m_M}{m_R} = \frac{L_M^3 \rho_M}{L_R^3 \rho_R} = \lambda_L^3 \\ F_a = M \cdot a \Rightarrow \lambda_F = \lambda_m \lambda_a \Rightarrow \lambda_a = 1 \\ F_r = R \cdot V \Rightarrow \lambda_F = \lambda_r \lambda_V \Rightarrow \lambda_r = \lambda_L^5/2 \\ F_k = K \cdot L \Rightarrow \lambda_F = \lambda_k \lambda_L \Rightarrow \lambda_k = \lambda_L^2 \end{array}\right. \tag{60}$$

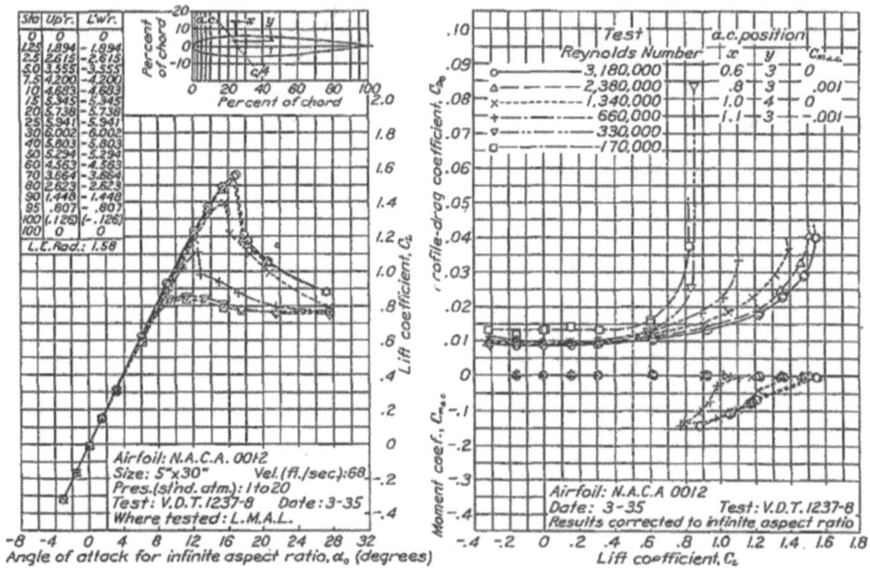

Figure 35. Lift and Drag curves for a NACA 0012 airfoil at different Reynolds numbers. Cfr. Jacobs and Sherman (1937)

At this point it is possible to make some considerations on the power production similitude following the Froude similitude (equation 57b) obtaining the power similitude λ_W reported in equation 61. It is possible to see that the scaled models give very low power.

$$\lambda_W = \lambda_V^3 \cdot \lambda_L^2 = \lambda_L^{7/2} \tag{61}$$

For instance the power generated by a $850kW$ full scale wind turbine should be, on a $1:50$ wind tunnel scaled model, something like $0.96W$, since the power scale is $\lambda_W = 1.14 \cdot 10^{-6}$. This leads to extremely low powers to be measured in the wind tunnel tests.

5.2 Natural Wind Boundary Layer Wind Tunnel simulation

Wind turbines wind tunnel testing, as already quoted at the beginning of the section, can be used for several different reasons: one can be the aerodynamic performance of the single blade or the entire rotor, but it is possible to study also the machine behavior in the natural atmospheric wind. The atmospheric wind does not presents uniform and smooth flow conditions, but has a non uniform flow in both space and time. The atmospheric wind presents different features depending on the terrain roughness (see figure 36), but also depending on the orography and conformation of the landscape.

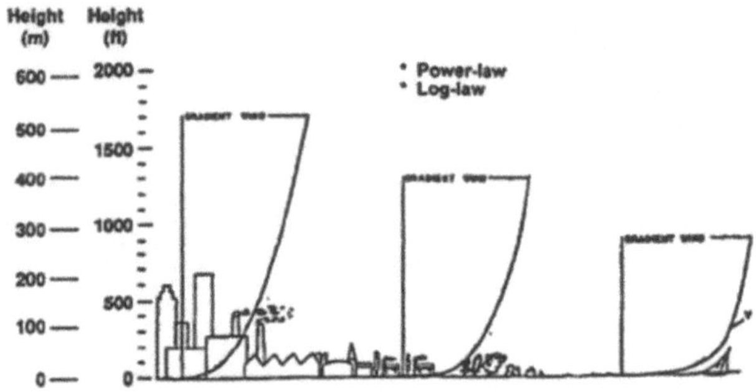

Figure 36. Different natural wind profiles over different terrains

Boundary layer wind tunnels have been built to reproduce with good accuracy the atmospheric wind. Usually these wind tunnels have a long test chamber to permit the development of the boundary layer. It is possible to reproduce the correct shape of the mean wind profile using passive generators of turbulence as, for instance, spires, bricks, floor roughness and other devices. These tools accurately combined allow to reproduce together with the mean wind profile also another important characteristic of the natural wind: the unsteadiness of the flow; this characteristic of the natural wind is usually presented in terms of turbulence intensity $I_{u,v,w}(z)$ and turbulence length scale $L_{u,v,w}(z)$ in the three space directions as a function of the height from the ground:

$$I_u(z) = \frac{\sigma_u(z)}{U(z)} \tag{62a}$$

$$L_u^x(z) = \int_0^\infty \rho_u(z, r_x) dr_x \tag{62b}$$

where $\sigma_u(z)$ is the standard deviation of the turbulent component of the wind, while $U(z)$ is the mean wind velocity vertical profile, ρ_u is the cross correlation function between the turbulence component in the u direction separated by a x-direction distance r. A simplified sketch to represent the natural turbulent wind can be seen in figure 37.

For structural and engineering purposes the characteristics of the natural wind boundary layer more detailed specifications can be found in the EURO

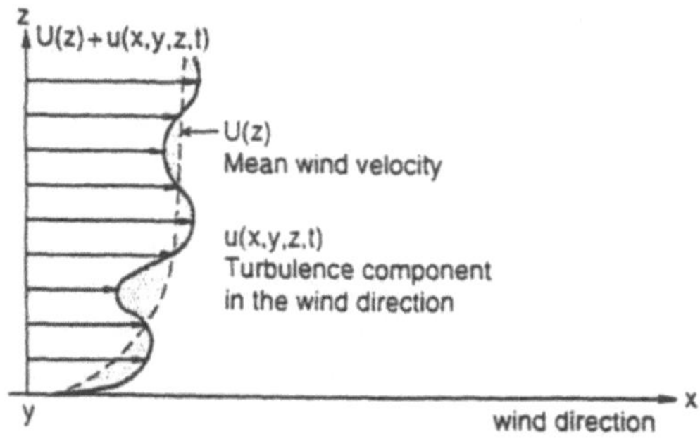

Figure 37. Definition of wind instantaneous wind profile for a turbulent flow condition

Code and in the ESDU Wind Engineering series. The effect of natural wind on structures can be seen in terms of actions and dynamic effects. Usually the actions on structures have a dependence in space and time, due to local environmental properties (i.e. wind intensity, turbulence, directionality), due to structure status (position and velocity) and to possible critical local effects. The dynamic effects depend on flow structure interaction involving aeroelastic effects and stability problems, as well as structural resonances in terms of fatigue effects and comfort. Boundary layer wind tunnel tests are used to perform many different studies: the pressure distribution can be measured on buildings to investigate the loads on the facades or entire zones of a city can be reproduced to study the flow, the heat and the pollution dispersion in urban areas, but also bridges and power lines can be tested as well as sailing boats, canopy roofs, all with or without the corresponding orography. As a conclusion it is possible to affirm that to grant a satisfying boundary layer wind tunnel tests it is necessary to ensure the desired flow quality in terms of mean wind and turbulence profile, the model dimension must be small compared to the test section in order to have negligible blockage effects. Depending on the test goals there must be a good accuracy in the model, not only from the geometrical point of view, but also in terms of velocities, frequencies, etc.

5.3 Test Case 1: Wind Tunnel test on 850kW Wind Turbine

This wind tunnel study has been performed to achieve some informations about the general performance of a set of three wind turbines. The analysis has been conducted on the flow condition and on the wake, with a particular interest on turbulence intensity, orography influence and wake interaction has been conducted. The measured quantity is the The test has been conducted in a 1:50 scaled model using the Froude similitude (equation 57b) for the velocity scale.

Setup and technical considerations A 1:50 scaled model of a wind turbine has been realized to perform the wind tunnel tests. The full scale machine is a $850kW$ horizontal axis wind turbine (see figure 38). The model reproduces the geometrical characteristics of the full scale turbine; the small space available on the nacelle has been used to install some devices in order to reproduce the correct blade pitch control, the rotor speed control and a torque measurement device.

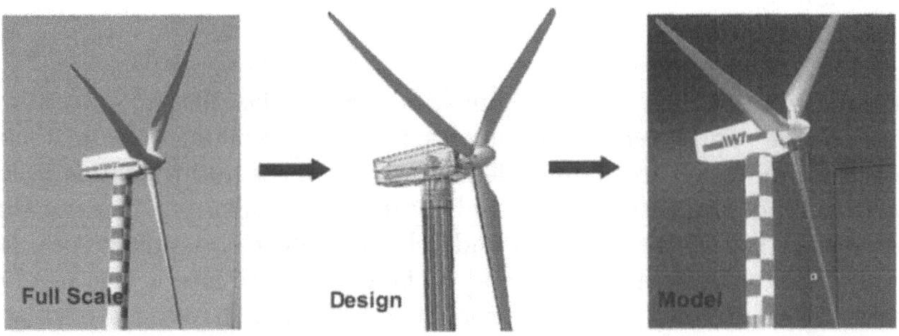

Figure 38. Full scale, CAD and wind tunnel model of the 850kW horizontal axis wind turbine

The pitch control is performed manually, realizing four different blade holders with which it is possible to give to the blades different pitch angles $(-1.1, 0.4, 7.5, 17.6$ degrees). The yaw angle of the wind turbine is controlled manually, placing the rotor supporting pylon on a rigid constraint, and aligning the rotor hub with the incident flow. The rotor speed control is given by a $20W$ brushless electric engine with gear, that realizes a maximum momentum ad the hub of $0.6Nm$; the rotational speed is measured through a 512 count per turn encoder; to reduce the power dispersion an elastic joint is used in order to reduce the mechanical friction. The electrical engine gives the possibility to read torque from the engine electrically, but this opportunity has not been used due to the low value of the generated power and the high mechanical friction. The aerodynamic torque on the nacelle has been measured by load cells on top of the tower. A good calibration of this device is necessary since the axial force has a great influences the momentum measurement.

Scaled model limits and applications As already introduced in section 5.1 and shown in figure 35 there are several critical scaling effects for wind tunnel testing of wind turbines: the most important is the mismatching Reynolds number between the full scale ($Re_R > 10^6$) and model scale turbine ($Re_M \approx 10^4 - 10^5$): the full scale Reynolds number states a fully turbulent flow around the blades, while Re_M has a laminar or transitional flow. The difference in lift and drag coefficients for a wind turbine blade profile at different Reynolds numbers is reported in figure 39.

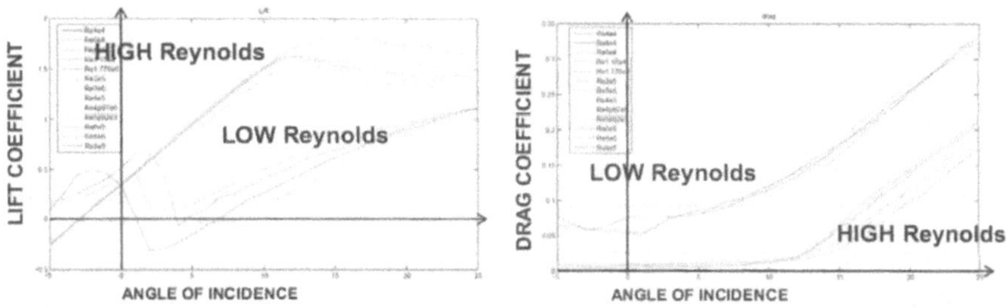

Figure 39. Lift and Drag coefficient for different Reynolds numbers

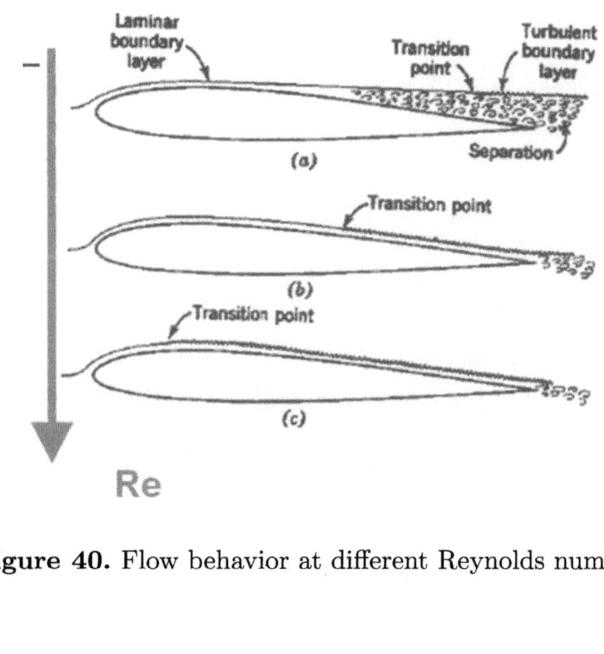

Figure 40. Flow behavior at different Reynolds numbers

The same conclusion can be reached considering the flow corresponding to the different Reynolds numbers as shown in figure 40: at lower Reynolds number (as happens for wind tunnel models) the flow around the airfoil is mainly laminar and separates at low angles of incidence, while for higher Reynolds numbers the transition region moves towards the stagnation point changing from laminar to turbulent flow, so detachment does not occurs and it is possible to obtain higher lift forces and lower drag as shown in figure 39, reaching better efficiency, which means higher driving forces and thus generating more power.

A solution that permits to force the flow transition is the introduction of trigger roughness called "*turbulators*"; despite its simplicity this device is not a low cost solution: it needs a long customized testing for every blade station profile in order to identify the correct height and shape for these turbulence generators.

Another approach that can be used to reproduce the correct aerodynamic forces is to neglect the geometrical blade similitude and choose an airfoil for the wind tunnel model that has the same scaled driving force of the full scale one at a different Reynolds number. This approach follows the similitude stated in the Buckingham Π Theorem. In this case the task is very difficult since it is quite impossible to find two equivalent wing profiles for Reynolds numbers with so high differences. It is possible to derive a new similitude regarding the power generation as expressed in equation 63. The power generation is a function of the geometry and of the velocity:

$$Power = f(geometry, velocity) = f(C_L, C_D, \alpha, chord, length, velocity) \tag{63}$$

Going back to equation 21a it is possible to see that the driving force is a function of the lift and drag forces: it is then possible to rewrite the equation for the power as:

$$Power = f(geometry, velocity) = f(C_T(\alpha), chord, length, velocity) \tag{64a}$$

$$\begin{aligned} Power &= f(geometry, velocity) \\ &= f(C_{TReal}(\alpha, Re_{Real}) \cdot g(chord, length, velocity) \end{aligned} \tag{64b}$$

It is possible to define the scaling in the driving coefficient C_{TModel} in the following equation:

$$C_{TModel}(\alpha, Re_{Model}) = \frac{C_{TReal}(\alpha, Re_{Real})}{n} \tag{65}$$

Wind Speed [m/s]	Pitch Angle	Rotation Speed ω [rpm]
8	-1.1	23.5
10.5	0.4	27
14	7.47	27
20	17.6	27

Table 4. Summary of the control parameters for the full scale parameters for some wind speeds

This similitude defines the power generation scaling as reported in equation 66; the only respected fluid dynamic similitude is the kinematic similitude, in terms of relative incidence angles ϕ and tip speed ratio λ.

$$Power = \frac{1}{n} C_{TReal}(\alpha, Re_{Real}) g(chord, length, velocity) \tag{66}$$

Scaled model performances The tested wind turbine is a variable speed controlled machine: the first step in the wind tunnel testing is the reproduction of the power generation performances. In figure 41 is reported the control law used at different wind speeds; the same data are summarized in table 5.3.

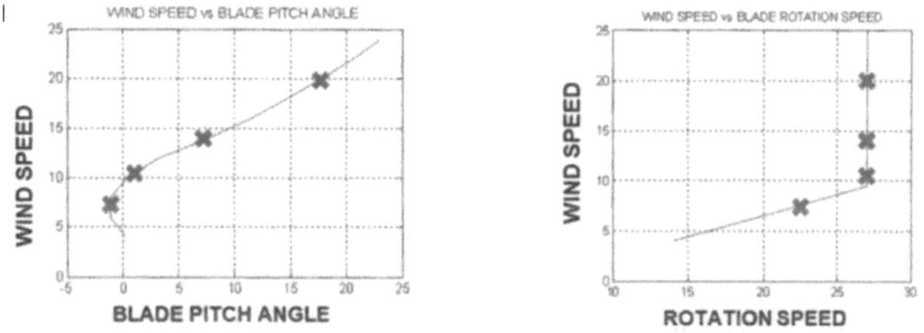

Figure 41. Control laws for blade pitch angle and rotation speed for the tested wind turbine

The power curve of the scaled model (figure 42) has been reproduced through the measurement of wind tunnel power curves at constant pitch. The data obtained have been interpolated and successively enveloped to get the correct power curve. This result has a correspondence with what reported in section 2.5 and marked out in figure 20.

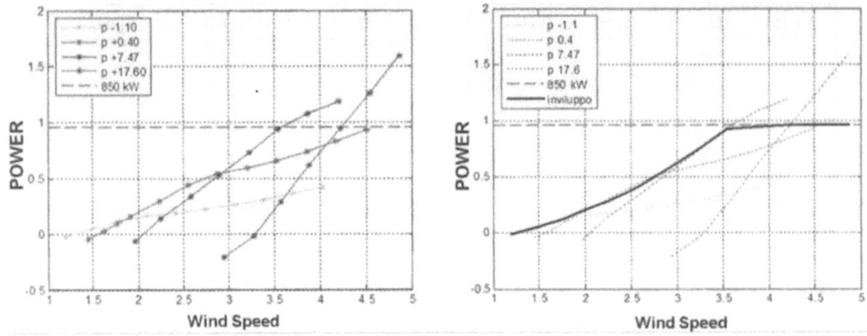

Figure 42. Measured power generated by the wind tunnel scaled model and envelop curve of the power generation of the wind turbine

In the following sections the influence of different parameters on this power curve will be presented.

Examples of boundary layer, orography and wake effects Wind tunnel tests have investigated different aspects of a wind turbine behavior, in particular:

- incoming wind profile effect;
- orography influence (see figure 43);
- wake interference between different wind turbines.

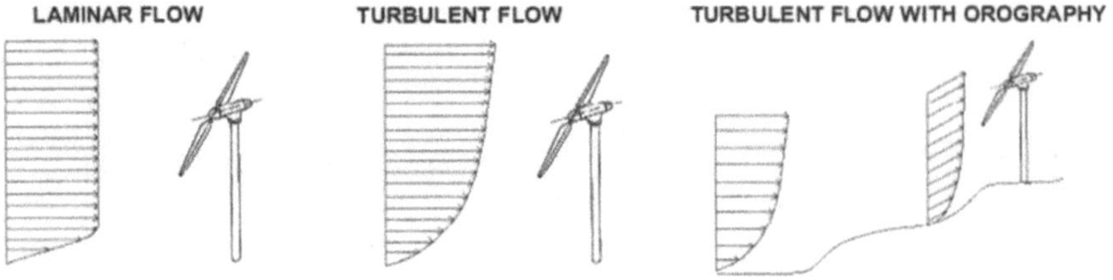

Figure 43. Sketch of the tested wind incoming conditions

In figure 44 is reported the incoming mean wind velocity profile and the turbulence intensity at the wind turbine location for the wind tunnel tests. The rotor position is highlighted inside the wind profile, and it is possible to compare the wind tunnel mean wind profile with full scale reference anemometers: mean wind vertical profile seems to fit very well the full scale data, while the turbulence intensity seems to be lightly underestimated in wind tunnel tests.

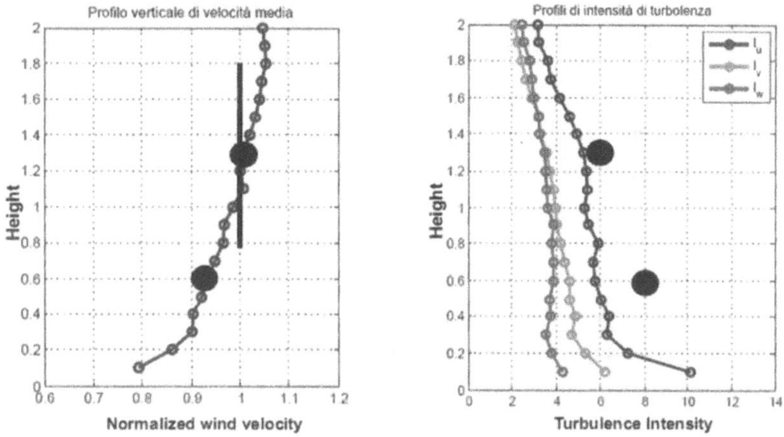

Figure 44. Vertical mean wind and turbulence intensity profiles. The vertical line on the left diagram represents the wind turbine rotor, the black dots two full scale reference anemometers

The measured power curve is reported in figure 45: the comparison between laminar and turbulent flow tests is done using the same wind speed at the rotor shaft height; the only difference between the laminar and the turbulent flow is the velocity distribution on the turbine disc. The wind turbine gives better performances in terms of power extraction when operating in turbulent wind condition: for this reason there is the need to correctly reproduce the atmospheric boundary layer in wind tunnel tests.

During the wind tunnel campaign another analysis has been performed to examine the influence of the orography of the wind turbine installation site on power generation. In this case the wind reference velocity has been measured far from the wind tunnel model in order to avoid unwanted speedup effects on the reference velocity. It is desirable that wind energy production is increased by the terrain structure: if speedup effects happen in the wind turbine location this may be a good place to install a machine. In this case three wind turbines have been placed on a hill in order to evaluate the interference of the landscape on power generation. The scenario can be seen in figure 46 and the result is presented only in terms of maximum power extraction for a constant air speed. In the graph it is possible to observe three different conditions of power production:

- for a wind incidence of around 150 $\approx$ 180 degrees there is a very low power production, especially for wind turbine 3: this can be ascribed

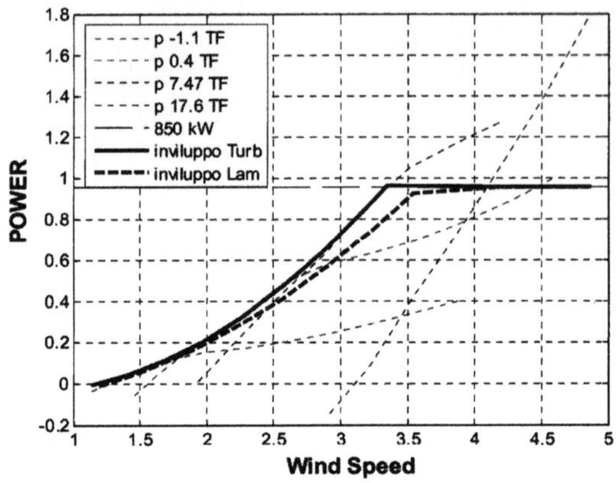

Figure 45. Envelop curve of the power generation of the wind turbine in turbulent flow. The dotted line is the power generation in laminar wind.

to the sheltering effect of the hill located south of the wind turbines;

- for a wind incidence of 60 and 240 degrees the power generation is very high, since the valley between the northern and southern hill causes a speedup in the air flow that permits to have an incoming wind with more energy;
- when the wind comes from a northern direction, the northern hill shelters again the wind turbines having a decrease in power generation.

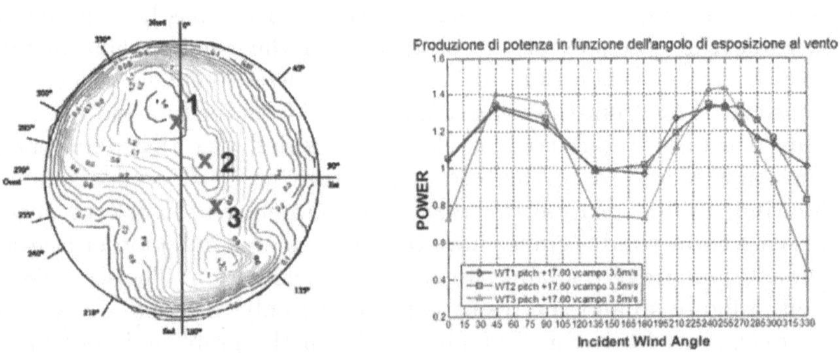

Figure 46. Orography with the detail of the three wind turbines location. Power curves as a function of the wind incidence angle for each wind turbine.

Examining these results it is possible to affirm that the most favorable winds for power generation in this site are eastern and western winds: a further meteorological analysis should be performed to see if the dominant winds blow from the directions that give higher energy production opportunities. The last important issue that has been studied in this campaign is a windfarm problem: the proximity between the wind turbines can cause some interference between the different machines; the test has investigated the wake interaction of three wind turbines aligned with the flow. As the simple Betz law states (see figure 1) the flow downwind a wind turbine is different than the free incoming air stream and, since a power extraction already occurred at the first turbine location, the available energy downstream is lower than the one disposable for the upwind turbine. The test has been conducted on three wind turbines, as can be seen in figure 47: it can be seen that the available power for the turbines in the wake decreases rapidly beyond the first one.

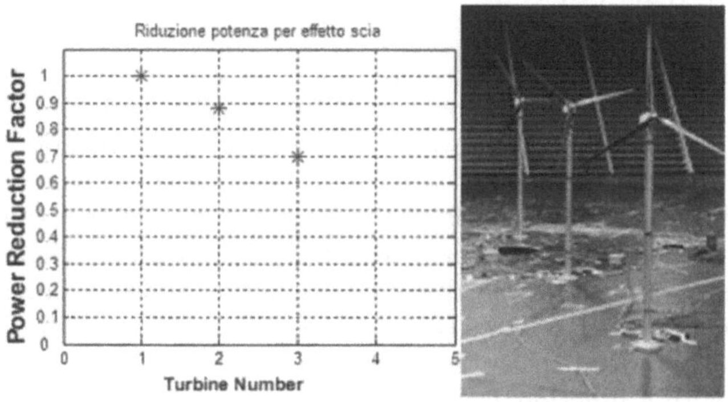

Figure 47. Available energy for three wind turbines aligned with the flow. The power of each machine has been divided with the power of the first one.

5.4 Test Case 2: Vertical Axis Wind Turbine

In this section will be presented the tests performed on a scaled model of a vertical axis wind turbine. The following discussion will assume steady state conditions when the inertial effects are negligible.

Experimental modeling considerations When designing an experimental campaign, it is important to define which is the target variable (the measured result of the test) and which are the design variables (the parameters that have an influence on the result). This process is especially

Figure 48. Wind tunnel model of the Vertical Axis Wind Turbine

important for scaled models because all the design variables must be correctly scaled in order to reproduce the physical phenomenon. In the case of a vertical axis wind turbine the target variable is the generated average power P while the design variables include the geometrical shape (airfoil), the geometrical dimensions (blade length L, chord c and radius R), the fluid characteristics (density ρ, the dynamic viscosity μ and the compressibility ϵ) and the cinematic parameters (fluid speed V and rotational speed $|Omega$). The relation between the target variable and the design parameter can be expressed with a function f:

$$P = f(airfoil, L, c, R, \rho, \mu, \epsilon, V, \Omega) \tag{67}$$

Considering the dimensional reference constituted by c, ρ, Ω that is commonly used in the aerodynamic field it is possible to normalize the function f to a dimensionless form. Applying this normalization and rearranging all the terms, a new dimensionless function Φ is obtained:

$$\frac{P}{\rho RLV^3} = \Phi(airfoil, \frac{L}{c}, \frac{C}{R}, \frac{\rho Vc}{\mu}, \frac{V}{\sqrt{\epsilon/\rho}, \frac{\Omega R}{V}}) \tag{68}$$

It is possible to write equation 68 in the final form using the typical aerodynamic parameters:

$$\eta = \Phi(airfoil, AR, c/R, Re, Ma, \lambda) \tag{69}$$

where η is the the turbine's efficiency, AR is the blade aspect ratio, c/R is the chord radius ratio, Re is the Reynolds number, Ma is the Mach number and λ the tip speed ratio. This dimensionless analysis confirms all the problems connected to the realization of a reliable analytical model of a vertical axis wind turbine as already mentioned in section 2.4. For these reasons an opportunity to design the experimental tests is given by the Riabucinski-Buckingham theorem, that states that two systems are similar if they produce the same values of the dimensionless design variables. For this reason it is necessary to assure that the results of the scale process are compatible with some experimental constraints. Usually, in the case of wind tunnel tests, the constraints that must be considered are the following:

- the dimensional scale factor is limited by the dimensions of the test chamber in order to avoid the interaction between the boundary layer of the walls and the model;
- the maximum fluid speed is limited to the capacities of the wind tunnel;
- the fluid must be air.

It is possible to apply the Riabucinski-Buckingham theorem to each dimensionless design variable of equation 69 hypothesizing a dimensional scale factor $\lambda_G = 1/10$ and neglecting the effect of the fluid compressibility (Mach number). In the following the variables related to the prototype and the model will be respectively indicated with the subscripts P and M. The parameters $airfoil$, AR, C/R are automatically assured with the geometrical scale factor and assuming the same shape. Requiring the similarity of the Reynolds number and the tip speed ratio, it is possible to obtain the velocity scale and the rotational speed scale:

$$Re_P = Re_M \Rightarrow V_M = \frac{\mu_M}{\mu_M} \frac{\rho_P}{\rho_M} \frac{C_P}{C_M} V_P = 10 \cdot V_P \tag{70a}$$

$$\lambda_P = \lambda_M \Rightarrow \Omega_M = \frac{V_M}{V_P} \frac{P_P}{R_M} \Omega_P = 100 \cdot \Omega_P \tag{70b}$$

In this case the similitude law requires a high fluid speed and an enormous rotational speed. The realization of a perfectly similar model can be therefore unfeasible for two reasons:

- the required fluid speed can exceed the maximum speed obtainable in the wind tunnel;
- the centrifugal loads connected to the high rotational speed of the model can break the structure.

These considerations lead to wind tunnel tests on vertical axis wind turbines that are commonly designed with distorted similarity between the prototype and the model.

Model Setup The experimental model reproduces the geometry of the prototype and is mainly constituted by a frame support and the rotor (figure 49).

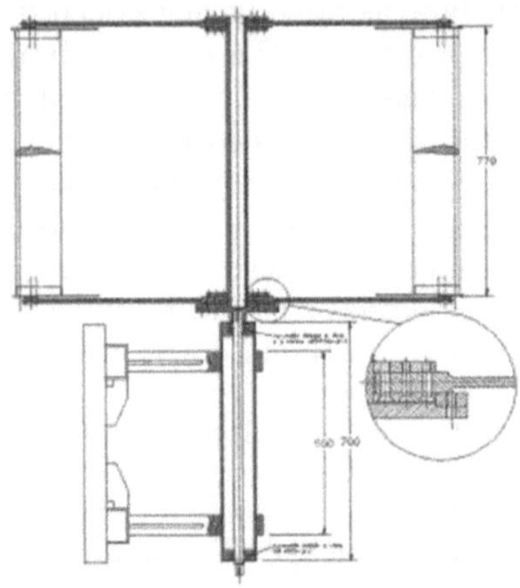

Figure 49. Section view of the model

The frame is composed by a steel structure and a stator that supports a vertical transmission shaft with two thrust bearing. The upper tip of the transmission shaft is connected to the rotor with a flange and the lower tip to a brushless electrical engine that is used to control the rotational speed. The rotor is constituted by a central axle with three blades, each supported by two radial arms. During the experimental campaign the attention has been focused on the following quantities of interest:

- the driving force of a single blade as a function of the azimuth angle;
- the average torque and power produced by the whole rotor.

The driving force is measured on board of the rotor and therefore, since a cable transmission is not possible, a wireless unit is used to send the measured data to an acquiring device. The driving force is obtained applying two strain gauge Wheatstone bridges to each arms in order to measure the bending moment in two different sections. Since the driving force acts like a shear load on each arms, it is calculated as the gradient of the bending moment. The position of the instrumented blade and the rotational speed are identified using a laser pointing to an eccentric cylinder inserted on the rotation axis of the turbine. The overall torque is determined with a torsiometer positioned between the transmission shaft and the electrical engine. Finally the power is not been directly measured but it is obtained as the product of the total torque times the rotational speed. In figure 48 is possible to see the model tested in the wind tunnel.

5.5 Test Case 3: Aeroelastic effects in the wake of a Horizontal Axis Wind Turbine

The last example of wind tunnel investigation on wind turbines is the interaction of the wind turbine with the surrounding landscape: this test case analyzes the interference of a horizontal axis wind turbine wake with a power line.

Model Setup The wind tunnel model reproduces a three blades $3MW$ wind turbine with a full scale diameter of $90m$. The tower has a height of $85m$ has a rotational speed of $8-20rpm$ full scale and the rotor has an inclination of 6 degrees from the vertical plane; the machine produces energy in the wind range of $4-25m/s$. The electrical power line has a length of $400m$ for a maximum camber of $15m$, a diameter of $40mm$ and a mass per unit length of $2.7kg/m$. The experimental tests will take into account the interference of the turbine with the cable placed at a distance of 1.5 or 2.5 rotor diameters. It is to note that the three cables of the power line are not standing at the same distance from the ground, but they have a medium height of $38.5-31-21.5m$ from the ground. The model is realized a 1:50 scale.

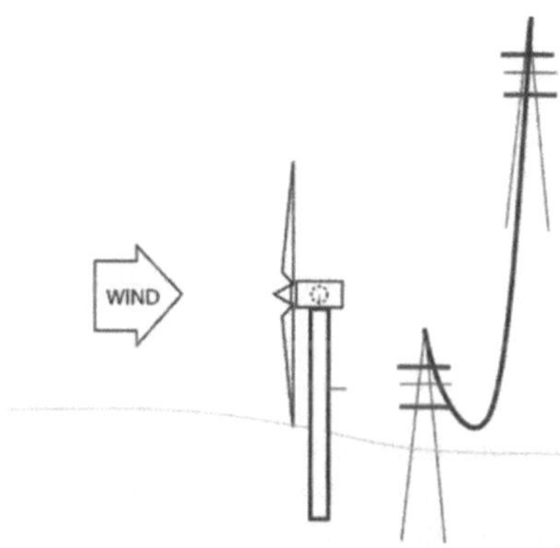

Figure 50. Sketch of the tested case: in figure are indicated the horizontal axis wind turbine and the electric power line downwind of the turbine.

The model setup, that can be seen in figure 50, has been built in order to reproduce the scaled wake, since the aim of the work is the interference of the wake with the objects placed downstream. The power line has been realized with an aeroelastic similitude, in order to reproduce the correct dynamic behavior, taking into account the correct scaled masses and frequencies. To ensure all these issues, several tests, measurements and visualizations have been performed downstream the model.

Results The first important issue to perform reliable results is the reproduction of the actual flow conditions around the models. Since in this test it is important to reproduce the correct wake of the wind turbine, it is important to correctly reproduce the atmospheric boundary layer in terms of mean velocity profile and turbulence intensity and length scale.

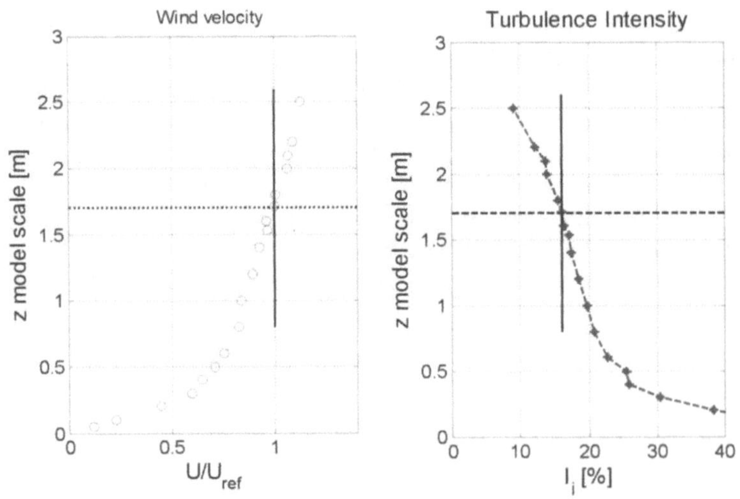

Figure 51. Mean velocity and turbulence intensity wind profiles measured in the wind tunnel test

A detailed analysis has been conducted measuring the wake shape at different distances downwind of the rotor. In figure 52 are reported the wake profiles for different distances from the rotor: it can be seen that the speed reduction near the hub is higher for small distances while the wake becomes wider with the increasing distance from the rotor; this result confirms Betz's theory assumptions as reported at the beginning of this contribution.

The tests have been conducted to analyze the interference of the wind turbine on the power line. In table 5.5 are reported the frequencies of the

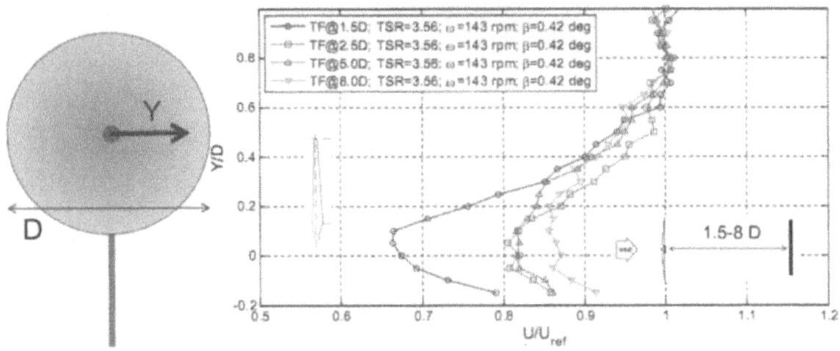

Figure 52. Mean velocity and turbulence intensity wind profiles measured in the wind tunnel test

Mode	Full Scale Frequency [Hz]	Model Scale Frequency [Hz]	Experimental Frequency [Hz]
1	0.14	0.98	0.96
2	0.28	1.98	2.12
3	0.42	2.96	2.89
4	0.56	3.95	4.19
5	0.71	5.02	5.01
6	0.84	5.93	6.19
7	0.99	7.00	7.11
8	1.13	7.99	8.40

Table 5. Full scale frequencies of the power line, scaled ideal frequencies and wind tunnel model frequencies

first 8 modes of the full scale, of the ideal and the realized model that regard the aeroelastic behavior of the power line: there is a quite good agreement between the realized model frequencies and the ideal model scaled frequencies. In this test the Froude Similitude (see equation 57b) has been maintained.

The measurement of the displacement of the power line has been performed using an optical system, composed by two infrared cameras and five reflective markers positioned on the line as indicated in figure 53. The principle for the estimate of the position of the markers is the triangulation: knowing the position of the two cameras it is possible to calculate the position of the markers. The image acquisition has been performed with a sampling frequency of $f_s = 18Hz$.

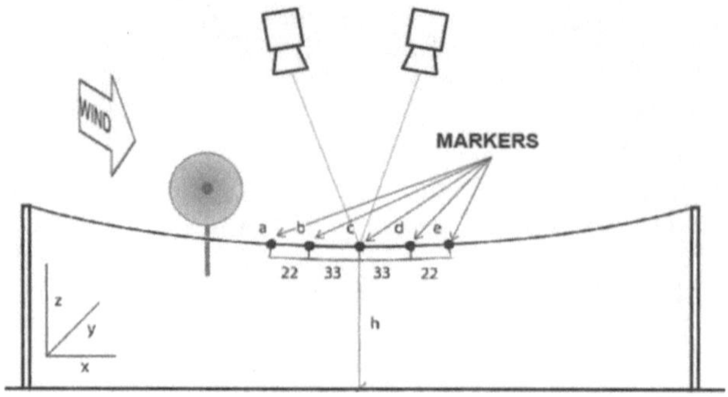

Figure 53. Power line data acquisition system: it is possible to see the position of the markers and the optical triangulation done by means of two infrared cameras

A first investigation has been performed on the central line ($h = 31m$) without and with the wind turbine and at different distances from the rotor to investigate the effect of the wind turbine on the power line. In figure 54 it is possible to see that the mean displacement of the markers on the power line is less in the presence of the wind turbine; another interesting result is that when there is a wind turbine upstream of the power line the displacement is higher when the machine is placed near the line. Considering the standard deviation the information provided by the tests is that that for bigger distances from the rotor the presence of the wind turbine does not affects very much the standard deviation in comparison to the configuration with the power line alone, while the proximity of the rotor affects very much the amplitude of the displacement of the line.

The study of the influence of the height of the power line has been conducted: in figure 55 it is possible to see the mean displacement and standard deviation of the markers for different heights of the line: lines at high quote present a lower displacement and also oscillations of lower amplitude.

The analysis of the dynamic performance of the power line has been carried out analyzing the power spectrum density (PSD) as can be seen in figure 56. The lower frequency contributions are ascribed to the transit of the blades, while the others are the natural frequencies of the line. It is interesting to note that lines placed at larger distance from the ground have the higher modes more excited than lines closer to the land.

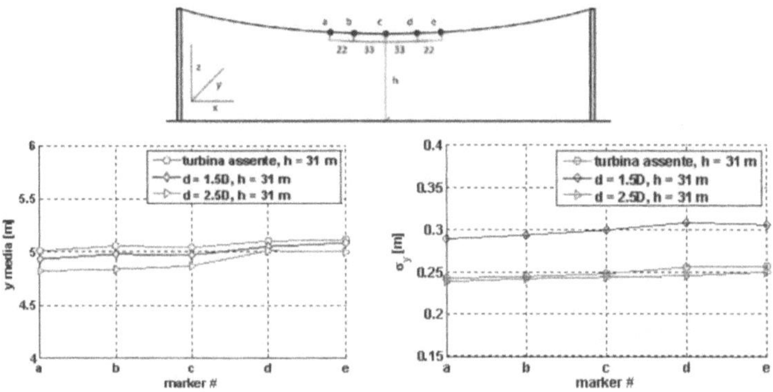

Figure 54. Displacement (left) and standard deviation of the displacement (right) of the power line in different conditions: with no turbine (red circle), 1.5 (black rhombus) and 2.5 (magenta triangle) rotor diameters in the wake of the wind turbine

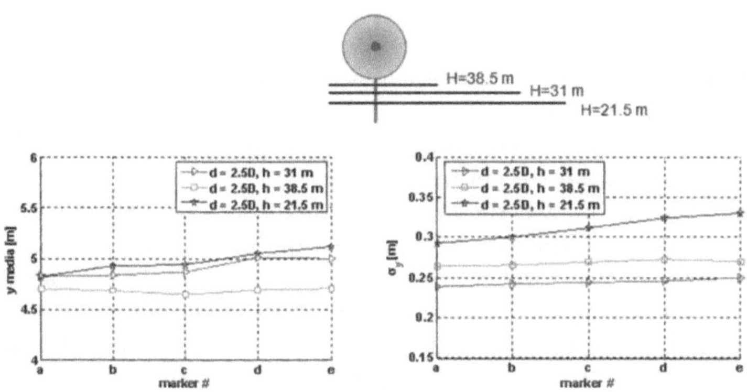

Figure 55. Displacement (left) and standard deviation of the displacement (right) of the power line at different height in the wake of the wind turbine.

6 Comments on cross validations of numerical modeling, wind tunnel and full scale testing

It is well known and accepted that in the nowadays practice of wind turbine modeling and design, the numerical simulation approach is by far the most reliable and diffused for predicting performances and wind loads of the machines. On the other hand it is a not obvious task to make available reli-

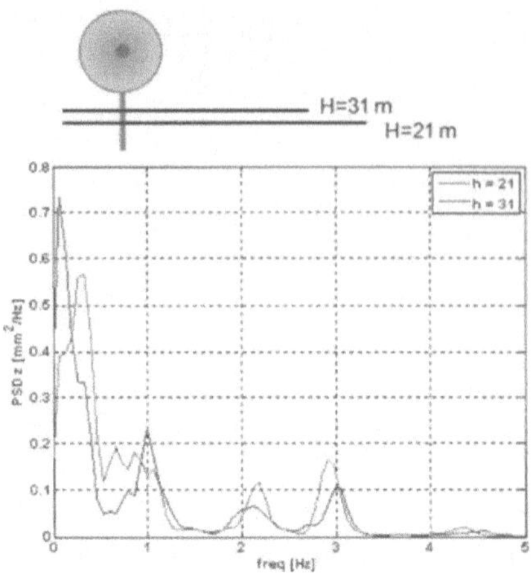

Figure 56. Power spectrum Density (PSD) of the vertical displacement of the power line at two different quotes: $21m$ (blue line) and $31m$ (red line)

able validation tools for the numerical approach, based on the experimental evidence. It is important to handle with care the comparative information that is possible to get from numerical simulation, experimental full scale and scaled model wind tunnel tests.

It is in fact well known the great complexity of a validation towards full scale data, due not to the lack of experimental information on performances and wind loads of real wind turbines, but due to the quite unreachable task of defining the boundary conditions and the space-time characteristics of the incoming turbulent wind that determined the experimentally measured quantities. On the other hand the characteristics of the wake generated by a full scale wind turbine are equally monitored with great difficulties due to obvious problems of dimensions and logistic difficulties of the measure, so that all the measured wake effects on downwind´structures or downwind wind turbines (i.e. wind farms) cannot be validated with a known description of the inflow conditions.

On the other hand the great advantage of an experimental wind tunnel approach on scaled models is obviously the allowance for well controlled laboratory boundary conditions as well as measurable smooth-turbulent flow time-space distribution. Similarly measures of wind loads, steady and unsteady mechanical effects on the turbines due to the wind interaction are clearly measured with extremely less demanding efforts in the wind tunnel case compared to large field environment.

Moreover it is clear that the direct control permitted by the wind tunnel experiment on the flow characteristic, allows to make whatever test in extreme wind conditions as well as sensitivity tests on the effect of different flow/structure arrangement parameters, without having to rely on unpredictable field experiment meteorological conditions.

As an example wind tunnel test concerning parametrical studies on the incoming flow turbulence as well as on wake effects on a wind farm due to the upwind turbines are easily arranged with full control of the flow parameters at a relatively low cost.

On the other hand it is true that the wind turbine scaled model cannot match exactly the full scale aerodynamic characteristic of the corresponding machines, due to unresolvable Reynolds number problems, in particular as has been described in the previously wind tunnel quoted experiences, the efficiency of the blade airfoil on the scaled model will always be lower than the very high efficiency airfoil characteristics of the full scale wind turbine. As a consequence the wind tunnel experiment will always suffer for a lack in the extracted energy from the incoming flow compared to the field prototype.

Nevertheless being all the actual aerodynamic and mechanical characteristics of the scaled model known and allowed to be modeled in the numerical simulation codes, the wind tunnel experiment can find a precious role as a fully suitable validation tool of the numerical simulation procedures. The proposal could be in fact to use the wind tunnel experiment for a validation "model - towards model" (numerical towards wind tunnel experimental). The numerical results can be compared with the experimental ones, under the condition of sharing among the wind tunnel experiment and the numerical simulation procedures, the same aerodynamic/mechanical/geometrical wind turbine characteristics, although different from a real full scale case, of course under the same incoming flow boundary condition. In this case it is possible to say that the wind tunnel tests can be used to validate the numerical models of the scaled models, in order to obtain a reliable and validated numerical simulation instrument to be used with confidence for the prediction of the full scale wind turbines performances and mutual effects.

The numerical code, validated through the wind tunnel scale model experimentation even in scaled extreme wind / response conditions, not realistic to be experienced in full scale in a reasonable time, could be finally used with more confidence for predicting the full scale wind turbine performances and wind loads on its structure.

Bibliography

Anonymous. Wind turbines — Part 1: design requirements; Part 2: design requirements for small wind turbines; Part 11: acoustic noise measurement techniques. International Standard IEC 61400, 2005–2006.

Anonymous. `Adams`, MSC.Software Corporation, 2 MacArthur Place, Santa Ana, CA 92707, USA www.mscsoftware.com.

Anonymous. `ECN BOT`, ECN Wind Energy, P.O. Box 1, 1755 ZG Petten, The Netherlands, epos.ecn.nl.

Anonymous. `HAWC2`, Risø National Laboratory for Sustainable Energy, Technical University of Denmark, Frederiksborgvej 399, P.O. Box 49, DK-4000 Roskilde, Denmark, www.risoe.dtu.dk.

Anonymous. `RotorOpt` perfects rotor design. *LM Glasfiber News Letter*, 5, 2007.

Anonymous. `Mecano`, SAMTECH, Liege Science Park, Rue des Chasseurs-Ardennais, 8, B-4031 Liége (Angleur), Belgium, www.samcef.com.

Anonymous. `Simpack`, SIMPACK AG, Friedrichshafener Strasse 1, 82205 Gilching, Germany, www.simpack.com .

O.A. Bauchau, C.L. Bottasso, and Y.G. Nikishkov. Modeling rotorcraft dynamics with finite element multibody procedures. *Mathematics and Computer Modeling* 33:1113–1137, 2001.

O.A. Bauchau, C.L. Bottasso, and L. Trainelli. Robust integration schemes for flexible multibody systems. *Computer Methods in Applied Mechanics and Engineering*, 192:395–420, 2003.

O.A. Bauchau, A. Laulusa. Review of contemporary approaches for constraint enforcement in multibody systems. *Journal of Computational and Nonlinear Dynamics*, 3:011005, 2008.

O.A. Bauchau, A. Epple, and C.L. Bottasso. Scaling of constraints and augmented lagrangian formulations in multibody dynamics simulations. *ASME Journal of Computational and Nonlinear Dynamics*, 4:021007, 2009.

O.A. Bauchau, and J. Rodriguez. Formulation of modal based elements in nonlinear, flexible multibody dynamics. *Journal of Multiscale Computational Engineering*, 1:161–180, 2003.

V. Bertogalli, S. Bittanti, and M. Lovera. Simulation and identification of helicopter rotor dynamics using a general-purpose multibody code. *Journal of the Franklin Institute*, 336:783–797, 1999.

P. Betsch, and S. Leyendecker. The discrete null space method for the energy consistent integration of constrained mechanical systems. Part II: multibody dynamics. *International Journal for Numerical Methods in Engineering*, 67:499–552, 2006.

S. Bittanti, and P. Colaneri. Invariant representations of discrete-time periodic systems. *Automatica*, 36:1777–1793, 2000.

M. Borri, L. Trainelli, and C.L. Bottasso. On representations and parameterizations of motion. *Multibody Systems Dynamics*, 4:129–193, 2000.

M. Borri, L. Trainelli, and A. Croce. The embedded projection method: a general index reduction procedure for constrained system dynamics. *Computer Methods in Applied Mechanics and Engineering*, 195:6974–6992, 2006.

E.A. Bossanyi. `GH Bladed` theory manual. *Garrad Hassan and Partners Limited*, Document No. 282/BR/009, Bristol, UK, 2008.

C.L. Bottasso, O.A. Bauchau, and A. Cardona. Time-step-size-independent conditioning and sensitivity to perturbations in the numerical solution of index three differential algebraic equations. *SIAM Journal on Scientific Computing*, 29:397–414, 2007.

C.L. Bottasso, F. Campagnolo, and A. Croce. Computational procedures for the multi-disciplinary constrained optimization of wind turbines. *Scientific Report DIA-SR 10-02*, Dipartimento di Ingegneria Aerospaziale, Politecnico di Milano, January 2010.

C.L. Bottasso, A. Croce. Advanced control laws for variable-speed wind turbines and supporting enabling technologies. *Scientific Report DIA-SR 09-01*, Dipartimento di Ingegneria Aerospaziale, Politecnico di Milano, January 2009.

C.L. Bottasso, A. Croce, C.E.D. Riboldi, and Y. Nam. Power curve tracking in the presence of a tip speed constraint. *Renewable Energy*, under review, 2009. Also: *Scientific Report DIA-SR 09-04*, Dipartimento di Ingegneria Aerospaziale, Politecnico di Milano, March 2009.

C.L. Bottasso, D. Dopico, L. Trainelli. On the optimal scaling of index three DAEs in multibody dynamics. *Multibody Systems Dynamics*, 19:3–20, 2008.

O. Brüls, P. Duysinx, and J.C.Golinval. The global modal parameterization for non-linear model-order reduction in flexible multibody dynamics. *International Journal for Numerical Methods in Engineering*, 69:948–977, 2007.

T. Burten, D. Sharpe, N. Jenkins, and E. Bossanyi. *Wind Energy Handbook.* John Wiley & Sons Ltd, West Sussex, England, 2001.

A. Cardona. *An Integrated Approach to Mechanism Analysis.* PhD thesis, Université de Liège, Belgium, 1996.

N.P. Duineveld. `FOCUS5`: an integrated wind turbine design tool. In *Proceedings of the 2008 Wind Turbine Blade Workshop*, Sandia National Laboratories, Albuquerque, NM, USA, 12–14 May, 2008.

J. Fehr, and P. Eberhard. Error-controlled model reduction in flexible multibody dynamics. *Journal of Computational and Nonlinear Dynamics*, 5:031005, 2010.

P. Fuglsang, and H.A. Madsen. Optimization method for wind turbine rotors. *Journal of Wind Engineering and Industrial Aerodynamics*, 80:191–206, 1999.

L. Fuglsang. Integrated design of turbine rotors. In *Proceedings of the European Wind Energy Conference & Exhibition EWEC 2008*, Brussels, Belgium, 31 March – 3 April, 2008.

C. Gear, B. Leimkuhler, and G. Gupta. Automatic integration of Euler-Lagrange equations with constraints. *Journal of Computational and Applied Mathematics*, 12–13:77–90, 1985.

M. Gèradin, and A. Cardona. *Flexible Multibody Dynamics: a Finite Element Approach.* John Wiley & Sons Ltd, West Sussex, England, 2001.

V. Giavotto, M. Borri, P. Mantegazza, and G. Ghiringhelli. Anisotropic beam theory and applications. *Computers & Structures*, 16:403–413, 1983.

E. Hairer, and G. Wanner. *Solving Ordinary Differential Equations II: Stiff and Differential Algebraic Problems* (2nd edn). Springer-Verlag, 1996.

M.O.L. Hansen. *Aerodynamics of Wind Turbines* (2nd edn). Earthscan, London, UK, and Sterling, VA, USA, 2008.

J.M. Jonkman. NREL structural and aeroelastic codes. In *Proceedings of the 2008 Wind Turbine Blade Workshop*, Sandia National Laboratories, Albuquerque, NM, USA, 12–14 May, 2008.

J.M. Jonkman, and M.L. Buhl Jr. FAST User's Guide. NREL Techical Report, NREL/EL-500-38230, Golden, CO, USA, August, 2005.

J.M. Jonkman, and M.L. Buhl Jr. Development and verification of a fully coupled simulator for offshore wind turbines. In *Proceedings of 45th AIAA Aerospace Sciences Meeting and Exhibit, Wind Energy Symposium*, Reno, Nevada January 8–11, 2007.

M. Jureczko, M. Pawlak, and A. Mezyk. Optimization of wind turbine blades. *Journal of Material Processing Technology*, 167:463–471, 2005.

D. Laird. NuMAD: blade structural analysis. In *Proceedings of the 2008 Wind Turbine Blade Workshop*, Sandia National Laboratories, Albuquerque, NM, USA, 12–14 May, 2008.

A. Laulusa, and O.A. Bauchau. Review of classical approaches for constraint enforcement in multibody systems. *Journal of Computational and Nonlinear Dynamics* 3:011004, 2008.

K. Lee, W. Joo, K. Kim, D. Lee, K. Lee, and J. Park. Numerical optimization using improvement of the design space feasibility for Korean offshore horizontal axis wind turbine blade. In *Proceedings of the European Wind Energy Conference & Exhibition EWEC 2007*, Milan, Italy, 7–10 May, 2007.

M. Lehner, and P. Eberhard. A two-step approach for model reduction in flexible multibody dynamics. *Multibody Systems Dynamics*, 17:157–176, 2007.

K.Y. Maalawi, and M.A. Badr. A practical approach for selecting optimum wind rotors. *Renewable Energy*, 28:803–822, 2003.

J.F. Manwell, J.G. McGowan, and A.L. Rogers. *Wind Energy Explained — Theory, Design and Application.* John Wiley & Sons Ltd, West Sussex, England, 2002.

J. Méndez, and D. Greiner. Wind blade chord and twist angle optimization using genetic algorithms. In *Proceedings of the Fifth International Conference on Engineering Computational Technology*, Las Palmas de Gran Canaria, Spain, 12–15 September, 2006.

N. Orlandea, M. Chace, and D. Calahan. A sparsity oriented approach to the dynamic analysis and design of mechanical systems. Part I and II. *ASME Journal of Engineering for Industry*, 99:773–784, 1977.

S. Øye. FLEX 4 simulation of wind turbine dynamics. In *Proceedings of the International Energy Agency, Annex XI, 28th Meeting of Experts*, Lyngby, Denmark, 11–12 April 1996.

P. Passon, M. Kühn, S. Butterfield, J. Jonkman, T. Camp, and T.J. Larsen. OC3-benchmark exercise of aero-elastic offshore wind turbine codes. In *Proceedings of the Science of Making Torque from Wind, Journal of Physics: Conference Series* 75:012071, 2007.

D.A. Peters, and C.J. He. Finite state induced flow models — Part II: three-dimensional rotor disk. *Journal of Aircraft*, 32:323–333, 1995.

L. Petzold, and P. Lötstedt. Numerical solution of nonlinear differential equations with algebraic constraints II: practical implications. *SIAM Journal on Scientific and Statistical Computing*, 7:721–733, 1986.

S.R.J. Powles, the effects of tower shadow on the dynamics of a horizontal-axis wind turbine. *Wind Engineering*, 7:26–42, 1983.

A.A. Shabana. *Dynamics of Multibody Systems* (2nd edn). Cambridge University Press, 1998.

K.A. Stol, and G.S. Bir. User's guide for SymDyn, Version 1.2. NREL Techincal Report, NREL/EL-500-33845, Golden, CO, USA, November 2003.

W. Xudong, W.Z. Shen, W.J. Zhu, J.N. Sørensen, and C. Jin. Blade optimization for wind turbines. In *Proceedings of the European Wind Energy Conference & Exhibition EWEC 2009*, Marseille, France, 16–19 March, 2009.

V. Vaughn. *Wind Energy — Renewable Energy and the Environment.* CRC Press, 2009.

M.O.L. Hansen. *Aerodynamics of Wind Turbines* (2nd edn). Earthscan, London 2009.

E.N. Jacobs, A. Sherman. *Airfoil section characteristics as affected by variations of the Reynolds number* NACA Report n-586, Langley Research Center, Hampton VA 1937.

Support Structures of Wind Energy Converters

Peter Schaumann[1], Cord Böker[2],
Anne Bechtel[1] and Stephan Lochte-Holtgreven[1]

[1] Institute for Steel Construction, Leibniz University Hannover, Hannover, Germany
[2] REpower Systems AG, Hamburg, Germany

1. Introduction

The wind energy market is one of the most promising markets of renewable energies. Besides biomass, photovoltaic, geothermal, and ocean energy especially the offshore wind energy will deliver the biggest part in renewable electricity. Regarding National and European demands for 2030, 25% of the required electricity will result from renewables. The biggest player will be the wind energy. To reach this aim a significant installation of offshore and onshore wind energy turbines is necessary. Figure 1 shows the estimated annual installation and cumulated capacity of onshore and offshore wind energy in Europe.

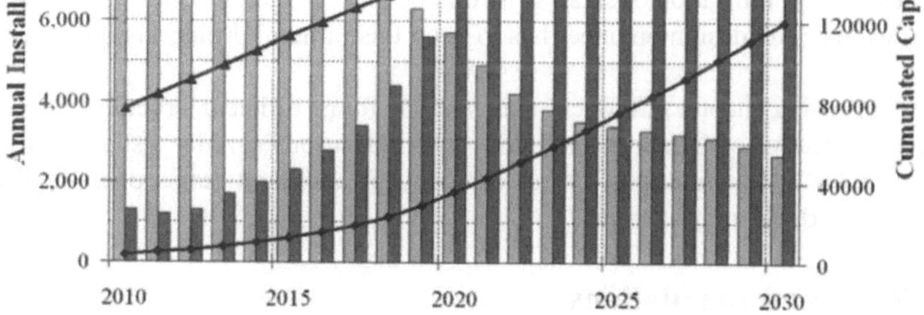

Figure 1. Annual and cumulated capacity of onshore and offshore wind energy (© European Wind Energy Association)

It can be seen that in 2020 annual installation of offshore wind energy will exceed the part of onshore wind. In 2030 nearly 300 Gigawatt (GW) of electrical power will be produced by wind energy. Compared to 2010 this means a three-fold rise of installed capacity. Especially for offshore wind energy the demands lead to an annual installation of approximately 7 to 9 GW. The large number of offshore wind turbines cannot be located near shore. Finally the installation areas and estimated water depths will be significantly different to today's offshore wind parks. To minimize maintenance costs reliable turbines with nearly fatigue resistant support structures have to be installed. Additionally to offshore wind parks onshore location will become rare. Turbine heights and rotor diameters will turn into be higher and larger to increase the amount of wind power. For these locations reliable and economic support structures have to be built.
This chapter introduces the different types of support structures for onshore and offshore wind energy converters. Furthermore basic design criteria like material, corrosion protection or connections regarding different limit design states are presented. Latest developments in offshore support structures are shown finally.

2. Structural Design of Wind Energy Converters

The reliability of the structural design of support structures for Wind Turbines is essential to guarantee the functionality and serviceability during the service-life. Additionally to the calculation of action and action effects other mechanical engineering and electrical engineering specific requirements have to be considered in the structural design of Wind Turbines.

Regarding possible series production techniques of support structures for Wind Turbines, an optimised design leads to an increase of the economic efficiency. Already little optimisation of small structural details can increase the total efficiency significantly. The design engineer has to keep this in mind during the design process.

The following chapter introduces main aspects for an efficient structural design of support structures of Offshore and Onshore Wind Turbines. After a short introduction of rules and regulations, the safety concept and limit states of valid Wind Turbine standards are presented.

2.1 Rules and Regulations

Several design codes for Wind Turbine structures including support structures and foundations have been implemented in the last decades. Most of the international rules for Onshore Wind Turbines originate from codes for towers, mast and chimneys. The codes for support structures for Offshore Wind Turbines have their origins in the standards for structures of oil and gas platforms.

The international standards are modified and completed by national annexes. Most of the international standards for the turbine design orient on the recommendation of the International Electrotechnical Commission (IEC). Table 1 exemplarily shows parts of the IEC 61400 for Wind Turbines.

In addition to the IEC-rules, national codes for the structural design and certification of support structures of Wind Turbines have been implemented. For instance, in Germany the code of the DIBt[1] has to be considered. This code transfers the international recommendations of the IEC to national aspects. Additionally, the recommendations of the Germanischer Lloyd[2] can be taken into account.

Table 1. IEC-Guidelines for Wind Turbine design

Number	Content
IEC 61400-1	Wind Turbine Safety and design
IEC 61400-2	Small Wind Turbine Safety
IEC 61400-3	Design requirements for offshore wind turbines
IEC 61400-11	Noise measurements
IEC 61400-12	Power performance
IEC 61400-13	Mechanical Load Measurements
IEC 61400-21	Power Quality
IEC 61400-22	Wind Turbine Certification
IEC 61400-23	Blade Structural Testing

For Offshore Wind Turbine Structures national codes for the structural design define which design rules and codes have to be used. The German design standard of the Federal Maritime and Hydrographic Agency (BSH) allows the use of different international technical codes of practice for the design of Offshore Wind Turbine structures in German territorial water[3]. Load assumptions and design regulations for steel support structures of Offshore Wind Turbines can be derived by the following standards:

- DIN 1055-4: 2005-03
- DIN 18800-1 to 4
- EN 1993-1
- IEC 61400-3

[1] DIBt (2004)
[2] GL-Onshore Guideline (2003)
[3] BSH (2007), p. 15

- API RP 2A-LRFD
- GL Guideline for the Certification of Offshore Wind Turbines
- DNV-OS-J101

The regulation of the BSH as mentioned above is only valid for Germany. For other European countries national regulations differ.

Generally the structural design of steel tower sections is based on mechanical principles, so that the country-specific regulations only influence the load and safety levels.

2.2 Safety Concept and Safety Factors

The design of Wind Turbine structures is based on the verification of the structural safety of all structural components of a wind energy converter. For the design the stress levels of every member have to be calculated taking several load cases into account. The comparison of resulting stresses to permissible material strengths displays the main task of the design. For the design process loads and material strengths are taken into account on a design level. Within this, an appropriate safety level shall be assured. The design loads and material strengths are based on the so-called 'probabilistic design concept'. This design concept includes partial safety factors for loads and material strengths in different limit states. Partial safety factors cause increased characteristic load levels. Based on statistical approaches, load factors consider how often a load level will occur during a design life of a Wind Turbine structure. Within the safety factors an exceedance of the real load shall be avoided. The design load can be calculated to:

$$F_d = \gamma_F \cdot F_k \tag{1}$$

with the design load value F_d, the partial safety factor for the load γ_F and the characteristic load value F_k. If there is a linear relationship between the loads which acts on the structure, all single design loads can be combined to one load effect S_d:

$$S_d = \sum_{i=1}^{n} \gamma_{F,i} \cdot F_{k,i} \tag{2}$$

While the loads are increased by the partial safety factors, characteristic material strengths are reduced by partial material safety factors. The material safety factors reduce the nominal resistances of the materials. Hence, material uncertainties due to local cracks or installation errors are regarded.

Design resistance values are derived by division of characteristic strength values through the appropriate safety factor of the considered material:

$$R_d = R_k / \gamma_M \tag{3}$$

where R_d is the design resistance, R_k is the characteristic resistance, and γ_M is the partial safety factor for the material.

Finally the design value of the load effect has to be less or equal than the resistance value (see eq. (4)).

$$S_d \leq R_d \tag{4}$$

Figure 2 shows distribution functions for loads and material strengths. The structural design is carried out for the overlapped region of the two bell-shaped curves. For high load effects and small material resistances a structural failure might occur. The larger the overlapping area is, the earlier the failure occurs.

The crossover of effect and resistance is regulated in the design codes. Some standards define a target safety[4], which considers a nominal annual probability of failure.

In special cases the safety level and survival probability can be modified in accordance with the certification institution and owner of the Wind Turbine. Due to economic reasons an increase of the safety level is not recommendable.

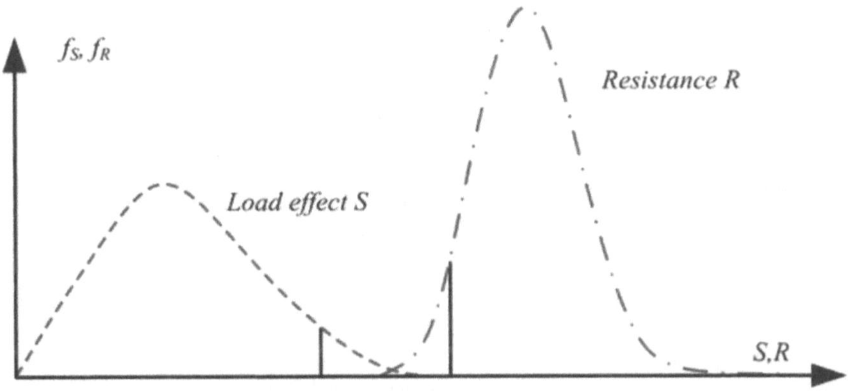

Figure 2. Safety concept for Wind Turbine structures

[4] cf. DNV-OS-J101 (2007), C200, Sec.2 – Page 18

Onshore Foundations. For Onshore Wind Turbine structures the partial safety factors can be taken from the GL-Guideline for the Certification of Wind Turbines[5]. This guideline states the partial safety factors for loads and materials for three limit states. For varied types of loadings and design situations the partial safety factors have to be considered. Table 2 presents partial safety factors for loads in the Ultimate Limit State (ULS).

For the Serviceability Limit State (SLS) a partial safety factor of $\gamma_{F,SLS} = 1.0$ for loads has to be considered. A value of $\gamma_{F,FLS} = 1.0$ has to be assumed for the Fatigue Limit State (FLS) for normal and abnormal design situations.

The partial material safety factor for structural steel components is of about $\gamma_{M,ULS} = 1.1$ in the ULS. The prove of serviceability can be achieved by a partial material safety factor of $\gamma_{M,SLS} = 1.0$.

Table 2. Partial safety factors for loads γ_F[6]

Source of loading	Unfavourable loads				Favourable loads
	Type of design situation				All design situations
	N Normal	E Extreme	A Abnormal	T Transport and erection	
Environmental	1.2	1.35	1.1	1.5	0.9
Operational	1.2	1.35	1.1	1.5	0.9
Gravity	1.1/1.35*	1.1/1.35*	1.1	1.25	0.9
Other inertia forces	1.2	1.25	1.1	1.3	0.9
Heat influence	-	1.35	-	-	0.9
* in the event of the masses not being determined by weighing					

Compared to the partial material safety factors in the ULS and SLS the choice of the partial material safety factors in the FLS has to include possible consequences of failure of structural members. Besides a total breakdown of the component, it has to be considered, whether the operation of the Wind Turbine is interrupted or not. Table 3 shows different partial material safety factors in the FLS for structural steel components of Onshore Wind Turbines.

[5] GL-Onshore Guideline (2003)

[6] GL-Onshore Guideline (2003), p. 4-20

Table 3. Partial material safety factor γ_M for fatigue design[7] (steel, onshore)

Inspection and accessibility	Component failure results in destruction of wind turbine or endangers people	Component failure results in wind turbine failure or consequential damage	Component failure results in interruption of operation
Periodic monitoring and maintenance, good accessibility	1.15	1.0	0.9
Periodic monitoring and maintenance, poor accessibility	1.25	1.15	1.0

Offshore Foundations. For different limit states the partial safety factors have to be considered in the calculation in the same manner as described for onshore towers. Partial safety factors for loads of Offshore Wind Turbines are equal to the safety factors for Onshore Structures (cf. Table 2). For the Serviceability and Fatigue Limit State partial safety factors for loads can be taken as mentioned above.

Partial material safety factors in the FLS have to be chosen considering inspection and accessibility of the structural member. Table 4 shows partial material safety factors γ_M for the FLS of steel members of offshore structures.

Structural members as tower of Offshore Wind Turbines are always members which are non 'fail-safe' structural members. Bolted connections and butt welds in the tower are easy to access so that they can be considered as 'fail-safe'. Grouted connections are comparably bad accessible. A periodic monitoring of them can be excluded. They can be considered as non 'fail-safe'.

In contrast to Table 3 and Table 4 in EN 1993-1-9 (2005) a partial material safety factor[8] of $\gamma_{FLS, EC3} = 1.35$ for structural members with high consequences of failure is recommended. This factor is approximately 10% higher than the required partial material safety factors of the Wind Turbine codes.

[7] GL-Onshore Guideline (2003), Table 5.3.1

[8] EN 1993-1-9 (2005), Table 3.1

Table 4. Partial material safety factors γ_M for fatigue design[9] (steel, offshore)

Inspection and accessibility	part of a non 'fail-safe' structure	part of a 'fail-safe' structure
Periodic monitoring and maintenance, good accessibility manufacturing and installation surveillance	1.15	1.0
No periodic monitoring and maintenance possible or poor accessibility	1.25	1.15

2.3 Ultimate Limit State (ULS)

The ultimate limit state (ULS) corresponds to maximum load levels from wind and wave actions. This limit state includes[10] the rupture of critical parts of the structure including components, cross-sections, and connections by:

- Fracture and exceeding of ultimate strength
- Stability failure
- Fatigue

The strength analysis can be carried out acc. to GL-Onshore Guideline (2003) for Onshore Wind Turbines structure or GL-Offshore Guideline (2005) and DNV-OS-J101 (2007) for Offshore Wind Turbine structures, respectively. National regulations have to be taken into account.

2.4 Serviceability Limit State (SLS)

The serviceability limit state (SLS) depends on the functionality of the structure. Within the SLS no loss of stability or breakdown of the structure will occur, but the serviceability of the structure can be reduced significantly. Within the SLS the following topics are checked and limited[11]:

- Deformations
- Vibration amplitudes
- Crack widths

[9] GL-Offshore Guideline (2005), p. 6-36

[10] GL-Offshore Guideline (2005), p. 1-13

[11] GL-Offshore Guideline (2005), p. 1-13

- Stresses and strains
- Water tightness

2.5 Fatigue Limit State (FLS)

Temporal fluctuating and frequently repeating loadings lead to material damage due to material fatigue phenomena. Preferable on defects, notches and cross-sectional variations incipient cracks arise due to fluctuating loads. According to an increasing number of stress cycles, the cracks dilate and result in a residual fracture. The damage-causing load respectively has a smaller value as the static strength requires. Consequently, the verification of fatigue strength requires an important design part especially for high dynamic loaded structures and structural components. This procedure evaluates the endurance of the construction. In order to define the fatigue strength, different approaches can be used. In general, the fatigue design procedure is based on the global concept depending on nominal stress S-N curves, which are established on published test results. Limitation is given to this global concept due to the failure criterion of total damage.

An initial crack is the trigger which starts the crack propagation in a cross section under amplitude loadings. The damage process starts with the crack initiation phase followed by the crack propagation process. Figure 3 shows different stages of fatigue. The crack propagation starts with an initial crack on the tension side.

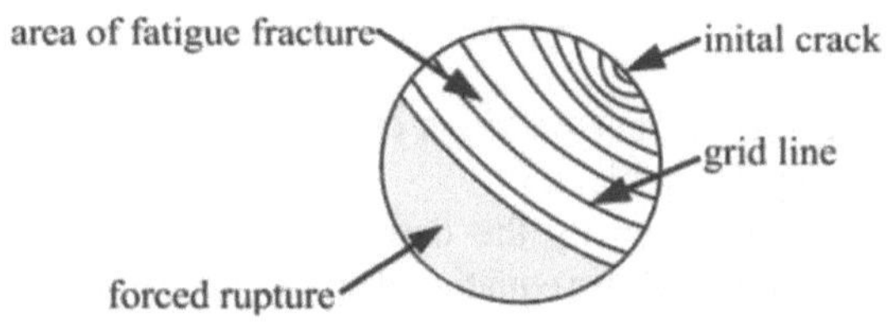

Figure 3. Fracture areas

The stage of 'initial crack' is followed by the 'crack propagation'. Under cyclic loads the stress concentrations at the crack tip leads to a local exceedence of yield strength, so that the crack tip opening displacement (CTOD) is increased. Every single stress cycle exceeding a particular stress level leads to crack propagation. The number of load cycles causing crack propagation can be seen in the number of grid lines. The more stress cycles loading the component and the more the crack is opened, the faster is the crack propagation.

The forced rupture occurs when the remaining area is smaller than the area which is necessary for the stress transfer.

Nominal stress approach. The fatigue strength evaluation of structural steel components is normally performed by the nominal stress approach if a nominal cross section is defined and nominal stress S-N curves are available for the particular construction detail. First, occurring fatigue loadings have to be described by load functions. Statistical evaluation of the loading function results in the stress spectrum. Generally, the nominal stress approach proceeds from these nominal stress amplitudes compared with the S-N curve of the permissible nominal stress amplitudes. The nominal stress S-N curve for welded constructions comprises the influence of material, notch class and weld quality.

The service life results from the nominal stress S-N curve and the nominal stress spectrum according to damage accumulation hypotheses. Several hypotheses exist whereas the Miner`s rule is used most frequently.

The evaluation of the durability generally results from the linear damage accumulation hypothesis. The literature recommends different hypotheses based on the nominal stress and in accordance to the basic Miner`s rule. The linear damage accumulation encompasses the damage ratio proceeding from the number of load cycles at one level compared to the number of cycles to failure on the corresponding level.

$$D = \sum_{i=1}^{I} \frac{n_i}{N_i} \leq 1 \tag{5}$$

with the damage rate D, the number of stress cycles n_i in block i and N_i as the number of endured stress cycles in block i.

Proof to the service life is given by the damage ratio not exceeding a value of 1.0. The introduced Palmgren/Miner equation displays the original one published 1945 by Miner. Several researches established further damage accumulation rules as the elementary and the modified Palmgren/Miner rule. The elementary Palmgren/Miner rule developed by Cortan and Dolan additionally regards the damage beneath the endurance strength. Refinement of this method combines the modified Palmgren/Miner rule by Haibach.

The number of load cycles can be derived by counting the load cycles between the cusp points of the load time function. Possible counting methods are the rainflow cycle counting or the reservoir counting method (see Figure 4).

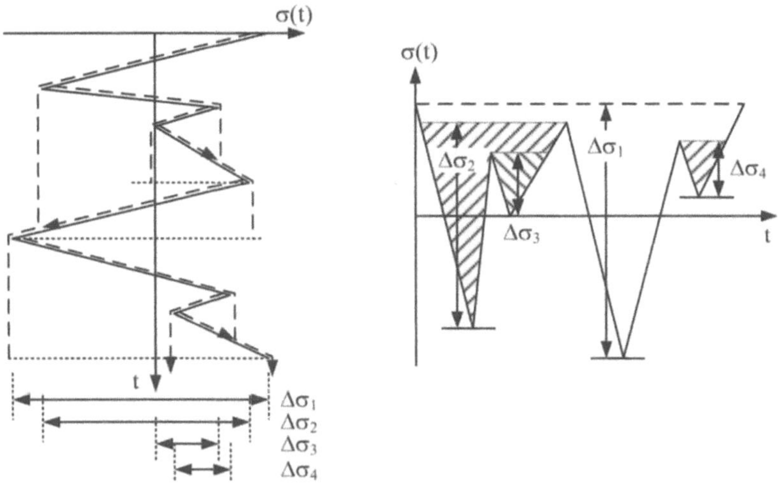

Figure 4. Counting methods (left: rainflow, right: reservoir)

For multi-axially stressed steel components, the Offshore-Guidelines[12] recommend considering the complex stress conditions in a realistic manner. Therefore the time series of the fatigue load spectra have to be taken into account. It is also possible to transform the multi-axial stress state to a mono-axial stress condition. This can be done by the application of the equivalent stress hypothesis. In some cases this approach may lead to non-conservative design due to the dependency only on the mean stress.

Definitions. For different load levels in the design life of the turbine the stress for the fatigue design in the nominal stress approach can be calculated according to the 'von-Mises'-stresses as shown in eq. (18) to (21).

The fatigue of steel components under cyclic loadings is significantly related to the mean stress level and the stress amplitude per load cycle. Figure 5 shows the basic definitions for alternating stress levels under cyclic loadings.

[12] GL-Offshore Guideline (2005), p. 5-10

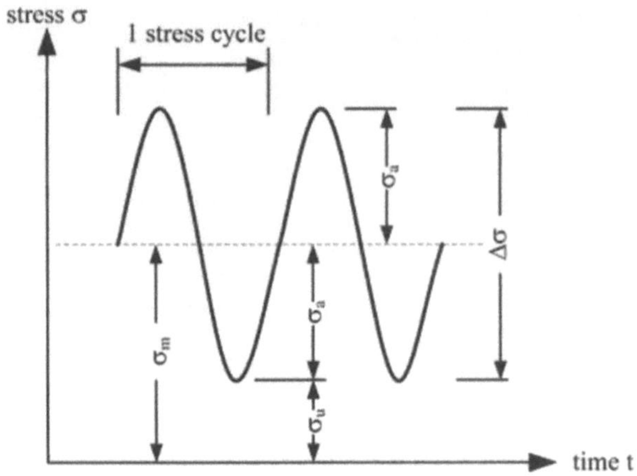

Figure 5. Definitions

The stress amplitude σ_a can be determined to:

$$\sigma_a = \frac{1}{2}(\sigma_o - \sigma_u) = \frac{1}{2}\sigma_o(1-R) = \sigma_m \frac{1-R}{1+R} \qquad (6)$$

with:

σ_o	maximum upper stress of stress cycle		
σ_u	minimum lower stress of stress cycle		
σ_m	mean stress:	σ_m	$= (\sigma_o + \sigma_u)/2$
σ_a	stress amplitude:	σ_a	$= (\sigma_o - \sigma_u)/2$
$\Delta\sigma$	applied stress range:	$\Delta\sigma$	$= 2 \cdot \sigma_a$
R	stress ratio:	R	$= \sigma_u/\sigma_o$

A stress ratio R = -1 describes a stress state where the mean stress level is equal to zero. Stress ratios of $0 \leq R \leq 1$ describe a stress state and predominant tensile stress and pure tension, respectively. If the stress ratio R exceeds a value of ±1 compression stress are prevailing.

S-N curves and detail categories. The design S-N-curves for the calculation of the cumulative damage ratio acc. to (5) can be calculated for different stress ranges. Figure 6 depicts the S-N curves for different detail categories for welded joints. S-N curves can be calculated according to valid offshore guidelines[13].

[13] GL-Offshore Guideline (2005) or DNV-OS-J101 (2007)

According to GL-Offshore Guideline (2005)[14] the S-N curves for welded joints can be calculated by the following equation taking the detail category into account:

$$\log N = 6.69897 + m \cdot \left(\log\left(\Delta\sigma_R / \Delta\sigma \right) - 0.39794/m_0 \right) \tag{7}$$

where m is the inverse slope of the S-N Curve and m_0 is the inverse slope of the S-N curve in the range $N \leq 5 \cdot 10^6$ (see Figure 6)
The S-N-curves of the GL-Offshore Guideline (2005) represent the lower limit of the scatter band of 95% of all test results available. In comparison to the GL-guideline the S-N curves in the DNV-OS-J101 (2007)[15] correspond to the 2.3% quantile of N. A comparison of S-N curves from different standards is quite difficult because in addition to loads the partial safety factors have to be compared carefully.

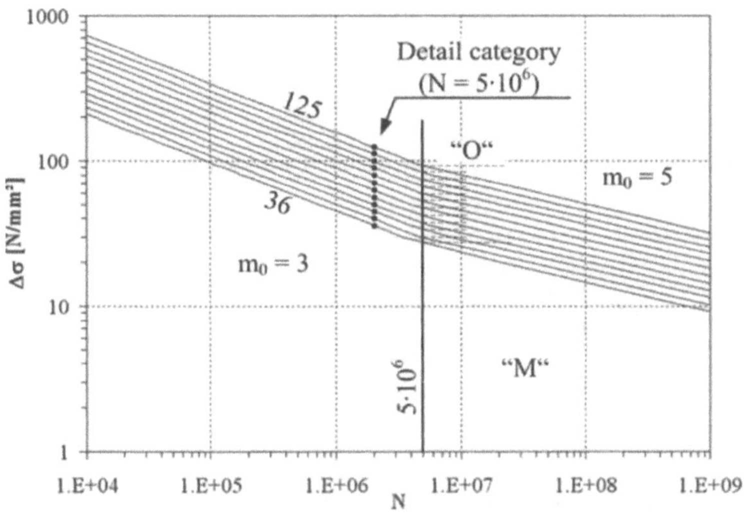

Figure 6. S-N curves for welded joints[16]

For base materials and welded connections the plate thickness has to be taken into account. Acc. to GL-Offshore Guideline[17] displayed S-N curves are applicable for base materials with thicknesses equal or smaller than 25 mm. For base materials with

[14] GL-Offshore Guideline (2005), p. 6-40

[15] DNV-OS-J101 (2007), Sec. 7 – page 78

[16] GL-Offshore Guideline (2005), p. 6-42

[17] GL-Offshore Guideline (2005), p. 6-42

thicknesses higher than 25 mm the permissible stresses e.g. the permissible load cycles have to be reduced. Therefore the following approach can be used:

$$f_t = \left(\frac{t}{25}\right)^n \tag{8}$$

with the thickness effect f_t, the plate thickness t in [mm] and the exponent n. For cruciform joints, transverse T-joints, plates with transverse attachments, as welded, n can be considered to 0.3. At the toe ground the exponent has to be reduced to 0.2 for theses joints. For butt welds the thickness effect is about 0.1.

Steel substructures for Offshore Wind Energy Converters are complex structures. Exact detail category classification of structural components might be difficult. As a simplification different Offshore-Guidelines[18] have implemented a catalogue where examples for several detail categories are shown.

Type No.	Structural detail	Description, Requirements and Remarks (St. = steel, Al. = Aluminium)	Detail category	
			Steel	Al
216	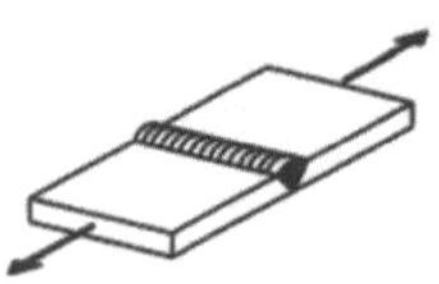	Transverse butt welds welded from one side without backing bar, full penetration root controlled by NDT no NDT Misalignment < 10%	71 28	28 12

Figure 7. Exemplary detail category for transverse butt welds[19]

Local approaches and fracture mechanics. The prediction of fatigue life can be done by using nominal stress approaches as shown above. In some cases the nominal approach underestimates the real stress distributions in the component due to discontinuities in the structure or residual stresses at welded connections. Detail categories sometimes do not reflect the stress distributions satisfyingly. For high stressed connections or constructional details it is recommended to use local ap-

[18] For example GL-Offshore Guideline (2005),

[19] GL-Offshore Guideline (2005), p. 6-62

proaches for the fatigue design of dynamically loaded support structures of Wind Turbines.

The following topics reflect some local approaches for the fatigue design. Local concepts should be used when the component cannot be described by the detail categories in a satisfying way.

Structural stress approach (Hot-Spot). The structural stress approach is based on stress concentration resulting from local structural discontinuities. Maximum stresses due to welded connections are taken into account by suitable S-N curves. The structural stress approach describes the macrostructural behaviour without consideration of local notch effects[20].

Using linear or quadratic extrapolation methods for the calculation of stress distributions at discontinuities the structural stress can be derived at the so-called 'hot spot'. The ratio between stress at the 'hot spot' and the nominal stress results the stress concentration factor (SCF):

$$SCF = \frac{\sigma_{structural}}{\sigma_{nominal}} \tag{9}$$

Figure 8, top shows the location of the structural stress at the 'hot spot'. It can be seen that the notch stress at the 'hot spot' is slightly higher than the interpolated structural stress.

The strains at the extrapolation spots A and B can be derived by measurements. Therefore strain gages are applied at the component in specific distances from the weld toe (see Figure 8, bottom). In most cases the strains can be evaluated by numerical investigations. For the derivation of local stresses or strains by measurements it is important to evaluate the influence of 'local' notch stresses and 'global' structural stresses.

[20] cf. Radaj and Sonsino (2006), pp. 33 ff.

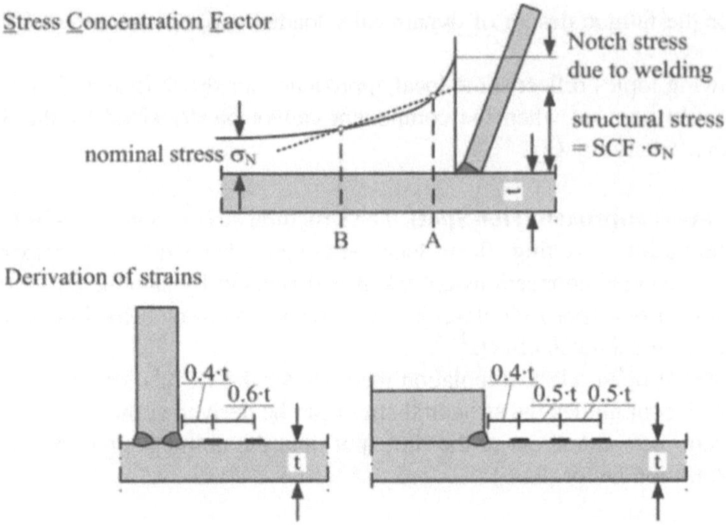

Figure 8. Hot-spot stress[21] (top), derivation of strains by measurements (bottom)

For the calculation of the 'hot spot' stresses different extrapolation methods can be used. Table 5 shows the location of the extrapolation points for linear extrapolation by two grid points. The given distances are related to the weld toe. Alternatively three extrapolation points can be used (see Figure 8).

Table 5. Recommended location of extrapolation points

Guideline	Distance from the weld toe (extrapolation points)	
	A	B
DNV[22]	0.5 · t	1.5 · t
IIW[23]	0.4 · t	1.0 · t

The structural approach can also be used for weld seams of tubular joints. Under consideration of the location of the investigated 'hot-spot' (chord or brace), it is recommended to evaluate the notch stress with numerical methods and to calculate

[21] taken from Schaumann et al. (2010)
[22] cf. DNV-OS-J101 (2007), App. D - p. 119
[23] cf. IIW (2007), p. 29

the 'hot-spot' stress subsequently. The methods can be taken from DNV-OS-J101 (2007) or IIW (2007). For an accurate stress evaluation it may be useful to consider the weld geometry in the numerical model (see Figure 9, left).

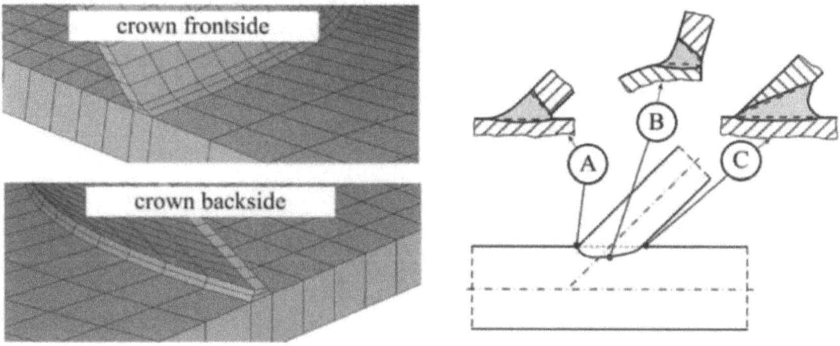

Figure 9. Numerical model with different weld geometries

If the weld seam is modelled in the numerical simulation, the location dependant weld geometry must be considered (see Figure 9, right). This leads to an immense modelling effort but also to most realistic results.

For the fatigue design with structural stresses, the structural stress S-N curve is equal to the nominal stress S-N curve for same unnotched materials[24]. Notches lead to material and stress discontinuities in the structural component - the so-called 'notch' effect which describes stress concentration and reduced permissible strengths caused by notches. Due to the fact that the fatigue strength of a structural steel member depends on its notch effect, it is recommendable to use local approaches taking the notch effect into account.

3. Steel Support Structures for Onshore Wind Energy Converters

Onshore Wind Energy Converters with steel support structures are mainly build as steel towers or lattice structures. The load-bearing behaviour of tubular towers and lattice structures is different. This section describes the load carrying capacity of these two support structure variants. Certain structural members as ring flange or bolted connections are described in more detail. The stress and fatigue design of the

[24] cf. Radaj and Sonsino (2006), pp. 33. ff

different steel sections or structural members can be carried out as mentioned above.

3.1 Tubular Towers

The most popular support structure for Onshore Wind Energy Converter is the tubular tower. Besides the simple structural system of cantilever beam, the cross-sections and their section properties can be calculated very easy.
Figure 10 shows a wind energy converter with tubular steel tower.

Figure 10. Tubular steel tower (©REpower) and cross section

Geometrical parameters. For the design of the tubular members the structural properties for tubular sections have to be determined. The cross section of a tubular tower section can be calculated to:

$$A = 2\pi \cdot r \cdot t \tag{10}$$

with A as the cross section, r as the mean radius of the cylinder and t as the wall thickness (see Figure 10, right). The secondary moment of inertia I and the moment of resistance W results to:

$$I = \pi \cdot r^3 \cdot t \tag{11}$$

$$W = \pi \cdot r^2 \cdot t \tag{12}$$

In case of torsion the torsional moment of inertia can be evaluated to:

$$I_T = 2\pi \cdot r^3 \cdot t \tag{13}$$

Eigenfrequency of tubular sections. The calculation of the eigenfrequencies of Wind Energy Converters is crutial for the design. If numerical methods are used the eigenfrequencies can be determined easily. In some cases it is recommendable to calculate the fundamental frequency with empirical calculation formulae. One simple approach is the approach based on Morleigh[25]:

$$\omega^2 = g \cdot \frac{\sum Q_i \cdot \eta_i}{\sum Q_i \cdot \eta_i^2} \tag{14}$$

with the eigenfrequency ω in [1/Hz], the gravity force g, the own weight Q_i at the centre of gravity of section i, and ordinate of the deflection curve η_i of section i. Figure 11 schemes a discretization of a tubular tower in several sections.

In a first step the self weight of every single section is calculated and afterwards applied as vertical loads in the centre of gravity of every section. Afterwards the deflection curve is calculated taking the dead loads into account. Finally the eigenfrequency can be determined acc. to eq. (14).

Additionally, the influence of the foundation stiffness has to be included. Therefore the approach acc. to Dunkerley can be used. First the eigenfrequency of the totally constrained cantilever beam is calculated (eq. (14)). Then the boundary condition for bending is modified from total constraint to flexible. The mass m of the tower is concentrated in the centre of gravity (see Figure 12).

The eigenfrequency of simplified rigid bodies can be calculated acc. to Schaumann et al. (2007-01) to:

$$f_{K_\phi} = \frac{1}{2\pi} \sqrt{\frac{K_\phi}{m \cdot l_S^2}} \tag{15}$$

[25] see Petersen (2001)

The combination of the eigenfrequency $f_{K\phi}$ of the rigid body and acc. to Morleigh leads to the determination of the eigenfrequency of the flexible tower:

$$f = \sqrt{\left(\frac{1}{f_{Morleigh}^2} + \frac{1}{f_{K_\phi}^2}\right)^{-1}} \tag{16}$$

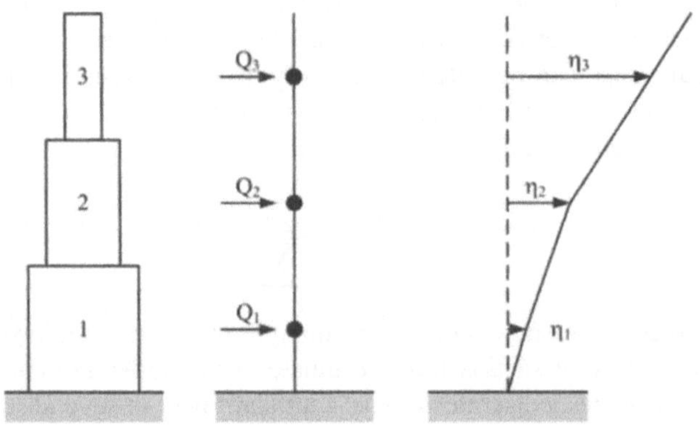

Figure 11. Discretization of sections of a tubular tower (cf. Schaumann et al. (2007-01))

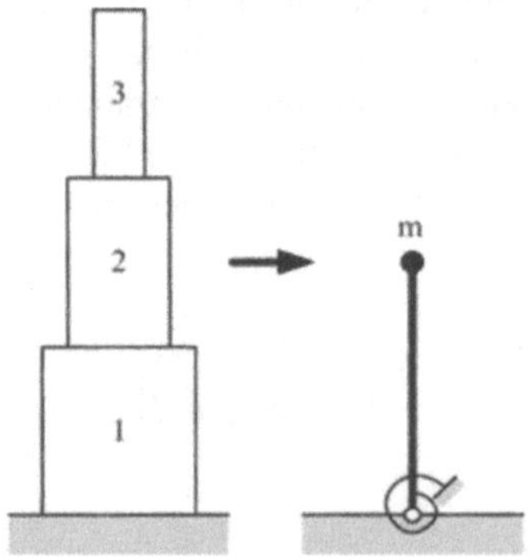

Figure 12. Tower as rigid body with concentrated mass m

Materials and coatings. Design standards for Onshore Wind Turbines specify valid steel materials and give recommendations regarding possible coatings of the weathered steel sections.

For tubular towers of Onshore Wind Turbines the used steel materials have to fulfil requirements regarding strength, toughness, cold deformability, and suitability for welding[26]. Steel material according to EN 10025 and fine grained steel acc. to EN 10113 can be used for Onshore Wind Turbine structures. Additionally, cast steel and stainless steel can be applied. Normally steel grades S 235 and S 355 are used.

Table 6 shows appropriate steel grades for Wind Energy Converters with yield and ultimate tensile strength.

Coatings for tubular tower sections have to be chosen in accordance to the ISO 12944-5. Coatings of weathered steel sections must be monitored at regular intervals.

Table 6. Nominal values of yield strength f_y and ultimate tensile strength f_u for hot rolled structural steel[27]

Standard and steel grade	Nominal thickness of the element t [mm]			
	$t \leq 40$ mm		$40 \text{ mm} < t \leq 80$ mm	
	f_y [MPa]	f_u [MPa]	f_y [MPa]	f_u [MPa]
EN 10025-2				
S 235	235	360	215	360
S 275	275	430	255	410
S 355	355	510	335	470
S 460	460	550	410	550
EN 10025-3				
S 275 N/NL	275	390	255	370
S 355 N/NL	355	490	335	470
S 420 N/NL	420	520	390	520
S 460 N/NL	460	540	430	540

26 cf. GL-Onshore Guideline (2003), p. 3-6 f.

27 Table taken from EN 1993-1-1 (2005), Table 3.1, first part

EN 10025-4 S 275 M/ML S 355 M/ML S 420 M/ML S 460 M/ML	275 355 420 460	370 470 520 540	255 335 390 430	360 450 500 530
EN 10025-5 S 235 W S 355 W	235 355	360 510	215 335	340 490
EN 10025-6 S 460 Q/QL	460	570	440	550
N: normalised / M: thermomechanically / Q: quenched and tempered				

Stress design for plain steel sections. The design of plain steel sections are based on the comparison of occurring stresses to permissible strengths. For plain steel sections stresses can be evaluated according to EN 1993-1-1 (2005) or DIN 18800-1 (2008). The occurring axial stresses can be calculated to:

$$\sigma_d = \left|\frac{N_d}{A}\right| + \left|\frac{M_d}{W}\right| \tag{17}$$

with: σ_d design value of the axial stress
N_d design value of the axial force
A cross sectional area
M_d design values of the bending moments
W moments of resistance of the y-axis

The shear stresses due to shear forces result to:

$$\tau_d = \frac{2 \cdot V_d}{A} \tag{18}$$

where V_d is the design value of the shear force. Additionally shear forces due to torsion can be calculated by the 1st Bredt formula to:

$$\tau_d = \frac{M_{z,d}}{2\pi \cdot r^2 \cdot t} \tag{19}$$

where $M_{z,d}$ is the design value of the torsional moment, r is the mean radius of the cylinder section, and t is the wall thickness of the cylinder.

If both ratios between stress and strength exceed a value of 0.5 the equivalent stress has to be calculated. Therefore the "von-Mises"-stress distribution can be considered:

$$\sigma_v = \sqrt{\sigma_x^2 + \sigma_y^2 + \sigma_z^2 + \sigma_x \cdot \sigma_y + \sigma_x \cdot \sigma_z + \sigma_y \cdot \sigma_z + 3\tau_{xy}^2 + 3\tau_{xz}^2 + 3\tau_{yz}^2} \tag{20}$$

The calculation formula for equivalent stress σ_v can be simplified to:

$$\sigma_v = \sqrt{\sigma^2 + 3\tau^2} \tag{21}$$

The resistance stresses can be determined acc. to eq. (22) and (23)

$$\sigma_{R,d} = \frac{f_y}{\gamma_M} \tag{22}$$

$$\tau_{R,d} = \frac{f_y}{\sqrt{3} \cdot \gamma_M} \tag{23}$$

with $\sigma_{R,d}$ as the normal stress resistance, $\tau_{R,d}$ as the shear resistance, f_y as the yield strength of the used steel material, and γ_M as the partial material safety factor. Table 6 shows nominal values for the yield and tensile strength of different steel grades.

Ring flange connections. The connection of tower sections in tubular towers is realised by ring flange connections with highly preloaded bolts (HV) with diameters between M30 and M48. Every ring flange connection consists of 70 to 120 bolts. For modern wind turbine structures also HV-Sets of M64 with a nominal weight per bolt of 20 kg are used. For the design of highly preloaded bolts only the maximum loaded bolt is considered (see Figure 13).

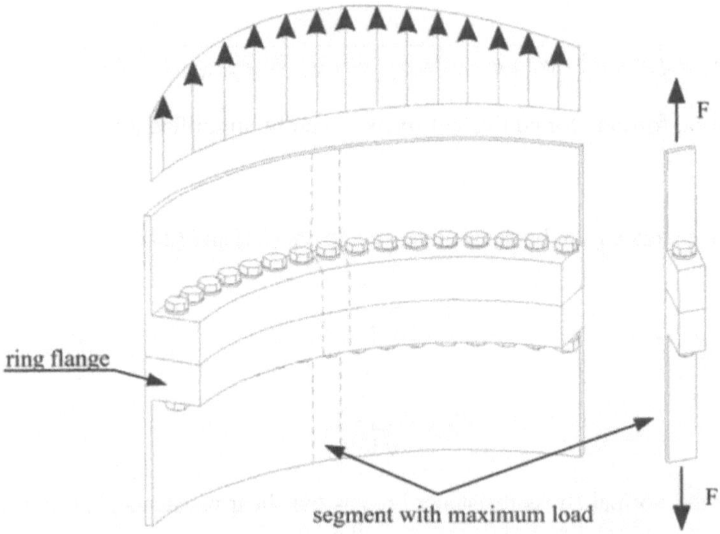

Figure 13. Common design approach for ring flange connections[28]

The reduction of the whole connection to the maximum loaded segment leads to a simple numerical model and a small calculation effort. On the other side the supporting effect of the tubular structure is neglected and cannot be reproduced. Additionally, shell imperfections of the ring flange are not taken into account which influences the load-carrying behaviour of the connection significantly.

According to DIN 18800-7 (2008) ring flange connections of wind turbines have to be preloaded. In the ultimate limit state the preload of the bolt has to be neglected, that means that the design process of the bolt can be carried out equally to normal bolted connections. Local plasticification of structural steel members of wall or flange can be taken into account.

In the fatigue limit state preloads of the bolted connections have to be taken into account under consideration of the pressures in the flanges. The bolt forces have to

[28] cf. Seidel (2001), p. 7

be calculated considering a non-linear bolt-force relation. Due to the eccentricity between bolt axis and the axis of the steel shell thickness, the bolt force is significantly influenced by the opening of the ring flange gap under increasing loads. Figure 14 shows the bolt force relation of a preloaded bolt in a ring flange connection of a tubular tower.

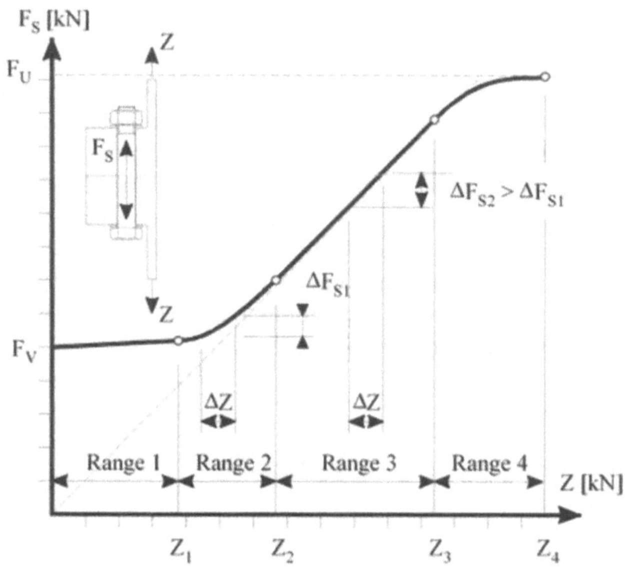

Figure 14. Bolt force relation for HV-sets in ring flange connections[29]

The considerable preload level of the bolt depends on the service intervals of the structure. If the bolts of the ring flange connections in a wind turbine structure are installed and maintained minimum two years later, the preload of the bolt can be determined to:

$$F_v = 0.7 \cdot A_s \cdot f_{y,b,k} \tag{24}$$

where A_s is the tensile stress area and $f_{y,b,k}$ is the yield strength of the bolt material. If the bolt force F_v is controlled during the first half-year after erection, but not immediately following the commissioning, the factor of 0.7 can be increased to 0.9.

The bolt force relation of preloaded bolted connection consists of four significant stress ranges. In the beginning (range 1, left) the bolt is loaded by the preload F_V.

[29] taken from Seidel (2001)

The bolt force increases linearly with tensile forces Z. Due to occurring gaps on the outside of the ring flange the bolt force relation becomes non-linear (range 2). An increasing gaping joint leads to a large increase of the bolt force until the ring flange is nearly totally opened. Range 3 displays a total gaping joint. Only the inner edge of the ring flange and the bolt can transfer the tensile forces Z. If the tensile stresses in the bolt exceed the value of the tensile strength, the connection will plasticize until failure (range 4).

Due to the highly non-linear bolt force relation it is recommended to use numerical models for the design of preloaded bolts in ring flange connections taking gaps and imperfections of the ring flange into account. Otherwise non-linear bolt force relations acc. to Seidel (2001) or Schmidt/Neuper (1997) can be used. The letter one is generally used in practice.

Due to non-linear bolt force relations and spatial bearing behaviours numerical models are often used for the design of the ring flange connections. Nevertheless, easy pre-design values for geometric dimensions are helpful. Suggested relationships between wall and flange thickness are given in ENV 1993-3-2 (1997). Notations can be seen in Figure 15. According to Seidel (2001) the following relationships can be considered for a simple pre-design:

$$b + s/2 \leq 2d \tag{25}$$

$$a \geq 1.45 \cdot (b + s/2) \tag{26}$$

$$c/d \leq 2 \cdot d/s \leq 10 \tag{27}$$

$$\begin{aligned} & t \geq 1.5 \cdot d \\ & with: t \geq 4 \cdot s \;\; for \;\; r/s \leq 50 \\ & \qquad t \geq 3 \cdot s \;\; for \;\; r/s = 100 \\ & \qquad t \geq 2 \cdot s \;\; for \;\; r/s \geq 200 \end{aligned} \tag{28}$$

with the flange thickness t, the shell thickness s, the diameter of the bolt d, the distance b between inner surface of the shell and the axis of the bolt, and the distance a between axis of the bolt and edge of the ring flange.

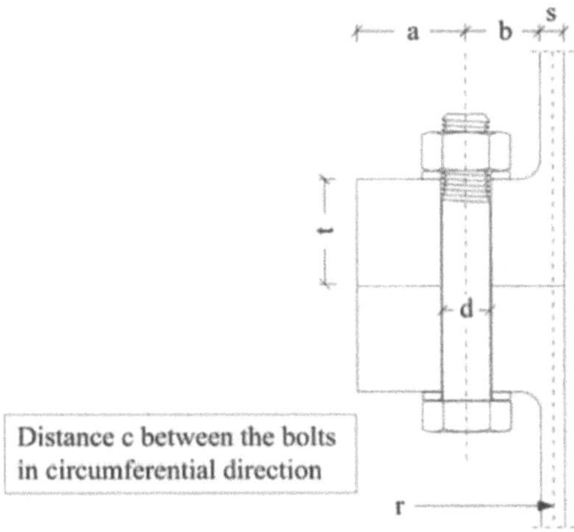

Figure 15. Notations for recommended flange dimensions[30]

All notations have to be considered in [mm]. Minimum values for dimensions of ring flanges can be taken from Seidel (2001) (see Table 7).

Table 7. Minimum values for dimensions of ring flanges in [mm]

	M16	M20	M22	M24	M27	M30	M36	M39	M42	M48
min (b-s/2)	30	35	40	40	45	45	50	55	55	65
min c	45	55	60	65	70	75	90	95	100	110

Welded connections. The design of welded connections in Wind Energy Converters is based on the comparison of occurring weld stress with permissible strength of the welded connection. In relation to the used steel material and weld geometry the permissible strength can be calculated according to EN 1993-1-8 (2005).

For the design of welded connections in the ULS the following equation has to be fulfilled:

$$F_{w,Ed} \leq F_{w,Rd} \tag{29}$$

where $F_{w,ED}$ is the design value of the weld force per unit length and $F_{w,Rd}$ is the design weld resistance per unit length. The resistance for fillet welds can be deter-

[30] taken from Seidel (2001), p. 2

mined to:

$$F_{w,Rd} = a \cdot f_{VW,d} = a \cdot \frac{f_u / \sqrt{3}}{\beta_w \cdot \gamma_{M2}} \tag{30}$$

with the design shear strength $f_{VW,d}$, the effective throat thickness a, nominal ultimate tensile strength f_U of the weaker part joined, the appropriate correlation factor β_W, and the partial material safety factor γ_{M2}. For butt welds the weld geometry has to be considered in the calculation of the throat area.

The partial material safety factor γ_{M2} can be assessed to 1.25.[31]. In accordance with the used steel material the correlation factor β_W is defined in Table 8 (only steel grades for wind turbine structures are shown. Further steel grades can be taken from EN 1993-1-8 (2005).).

Table 8. Correlation factor β_W for fillet welds[32]

standard and steel grade **EN 10025**	**correlation factor β_W**
S 235, S 235 W	0.8
S 275, S 275 N/NL	0.85
S 355, S 355 N/NL S 355 M/ML, S 355 W	0.9
S 420 N/NL, S 420 M/ML	1.0
S 460 N/NL, S 460 M/ML, S 460 Q/QL /QL1	1.0

3.2 Lattice Towers

Lattice towers for Onshore Wind Turbines (see Figure 16) are comparably lightweight support structures. The load bearing behaviour is dominated by the method of attack with several drifts.

The onsite installation of the single structural components of lattice towers realise larger turbine heights by comparable low weights. The highest wind turbine with lattice support structure is located in Laasow, Brandenburg, Germany with a nacelle height of 160 m (see Figure 16, middle and left).

[31] see EN 1993-1-1 (2005), Sec. 6.1, p.45
[32] taken from EN 1993-1-8 (2005), Table 4.1

Figure 16. Lattice towers (Osterkappeln ©Schaumann / Laasow ©Wind Kraft Journal)

The dimensions of the lattice tower result from the loadings on the structure. The lattice tower in Laasow has got a tower width of 2.9 m at the tower top and of 29 m at the bottom.

Variants and structural components. Lattice towers for wind turbines can be built with different numbers of struts. Figure 17 shows different variants of lattice towers of Onshore Wind Turbines.

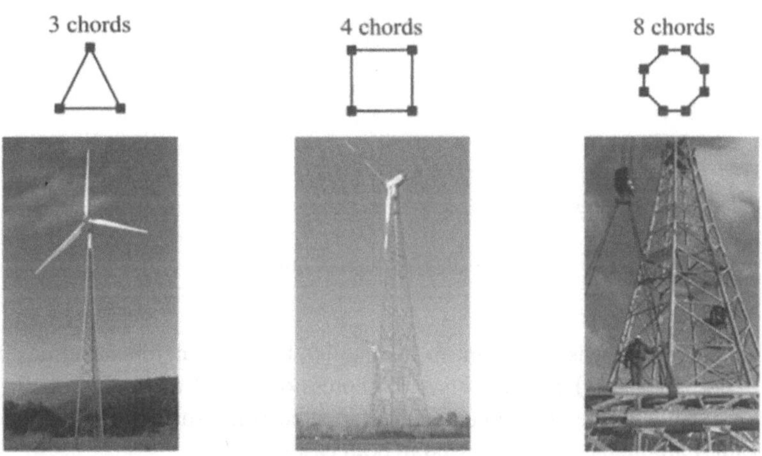

Figure 17. Lattice tower variants

Lattice towers with four chords are commonly used. Towers with three or eight struts have been realised, too. Theses support structures consist of different structural components. The chords transfer the loads from top to the bottom. The framework is realised by diagonal and horizontal bracings. Alternatively K-bracings can be installed. The load bearing behaviour of the lattice tower is dominated by the chosen type of bracing (see Figure 18).

Load bearing behaviour. Besides K-braced lattice towers, also towers with parallel chords can be designed (see Figure 19). The bearing capacity of K-braced lattice towers is mainly influenced by the geometric conditions.

The load bearing behaviour of the trestle is dominated by the inclination of the chords. Diagonal members are not necessary. The internal forces of the steel members are in balance. Values for compression and tension forces are equal. The vertical loads lead to an overturning at the foundation.

Parallel chord lattice towers transfer the horizontal life loads, especially wind loads, by couples of forces as compression or tension forces from tower top to tower foundation.

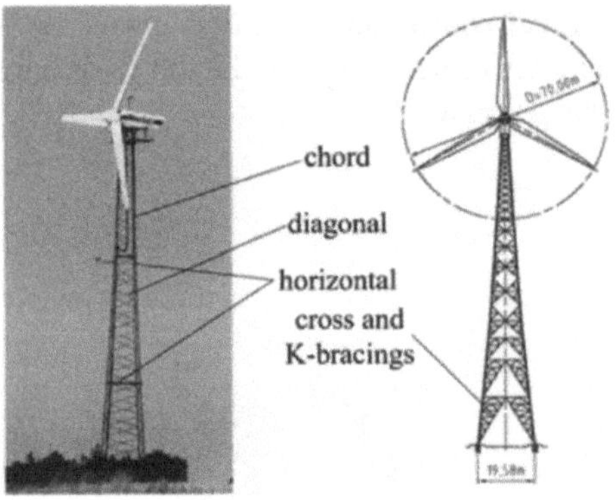

Figure 18. Structural components

For both systems it is important to distinguish between members in tension and those in compression. For members in compression lateral torsional buckling may become decisive. In members in tension the load bearing capacity of the bolted joint is relevant. Fatigue has to be assessed in both cases.

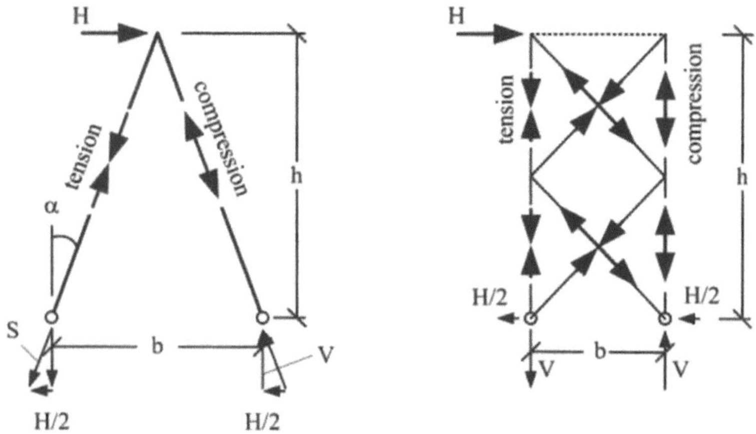

Figure 19. Load bearing behaviour: trestle and parallel chord lattice tower

Generally the overall structural system adopts the cross sections acc. to the increased bending moments from tower top to foundation. Due to necessary clearance of the tower the width of the tower sections near the rotor blades is limited. Inclined flow and gust on the rotor blades lead to torsional moments along the tower height. Lattice towers have smaller torsional stiffness and load bearing capacity than tubular towers.

Usually angle sections with equal legs or unequal legs acc. to EN 10056-1 (1998) are used. Figure 20 shows different types of unequal or equal, single our double angel sections. For lattice towers the following assessments have to be carried out:

- Stress design
- Bar buckling
- Design of bolted connections
- Fatigue

The stress design for angle sections can be carried out as for tubular sections.

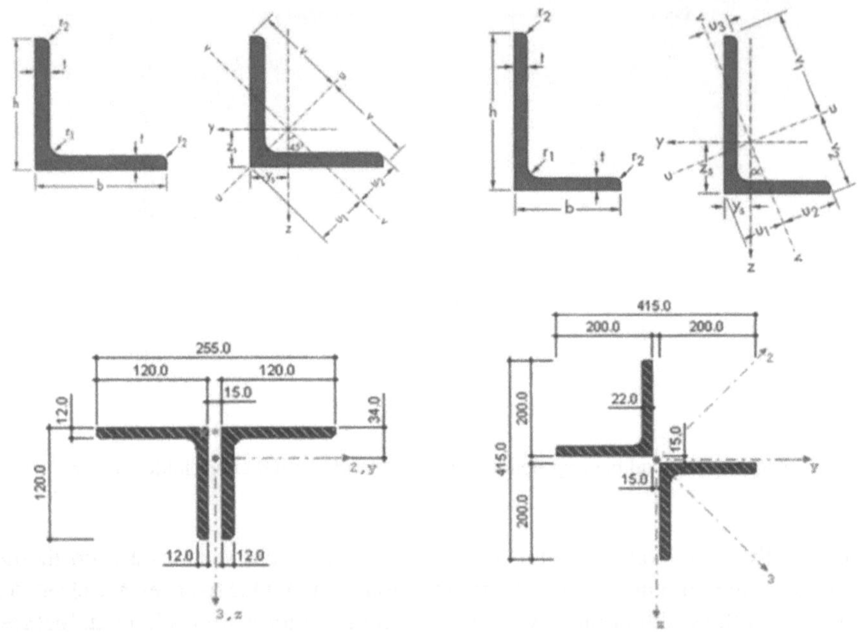

Figure 20. Equal and unequal angle sections, double angle sections

Stability. In addition to the stress design the stability of single or double angle sections have to be checked. Buckling analyses for bars can be carried out acc. to. DIN 18800-2 (2008).

Under pure axial pressure the buckling failure can be excluded if the following equation is fulfilled:

$$\frac{N_d}{\kappa \cdot N_{pl,d}} \leq 1 \tag{31}$$

where N_d is the design member force, $N_{pl,d}$ is the plastic axial force, and κ is the reduction factor acc. to DIN 18800-2 (2008), Sec. 3.2. The reduction factor has to be chosen in accordance with the existing cross sections (see Figure 21).

5	U-, L-, T- und Vollquerschnitte 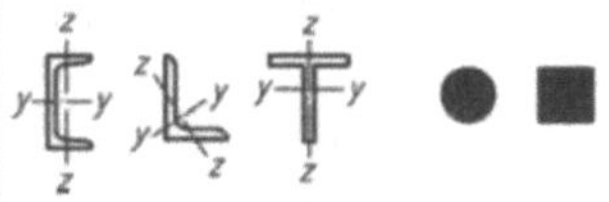 und mehrteilige Stäbe nach 4,4	y – y z – z	c

Figure 21. Buckling curve classification[33]

The slenderness ratio can be calculated as follows:

$$\bar{\lambda} = \frac{\lambda}{\lambda_a} \tag{32}$$

$$\lambda = \frac{s_K}{i} \ and \ \lambda_a = \pi \cdot \sqrt{\frac{E}{f_y}} \tag{33}$$

where i is the radius of inertia and s_k is the buckling length. The buckling length and within this buckling resistance of axial compressed beams is influenced by the support conditions at the ends. It has to be distinguished between rigid and hinged connections. For full constraint beam elements the slenderness ratios can be modified.

4	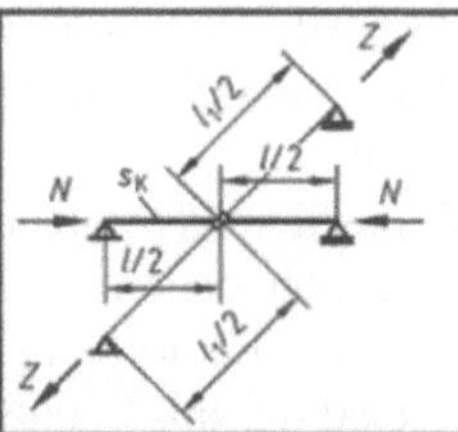	$s_K = l\sqrt{1 - 0{,}75\frac{Z \cdot l}{N \cdot l_1}}$ but $s_K \geq 0{,}5\ l$
5	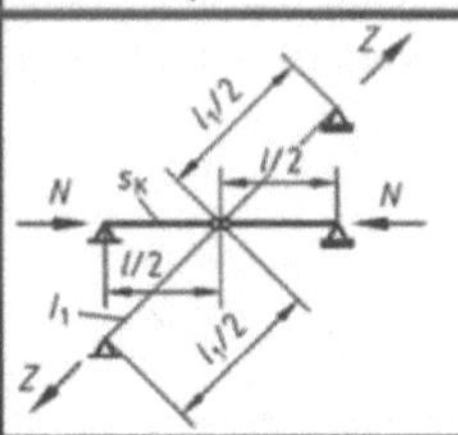	$s_K = 0{,}5\ l$ if $\frac{N \cdot l_1}{Z \cdot l} \leq 1$, or if valid $(E \cdot I_1)_d \geq \frac{3Z \cdot l_1^2}{4\pi^2}\left(\frac{N \cdot l_1}{Z \cdot l} - 1\right)$

Figure 22. Buckling length of lattice bars[34]

[33] cf. DIN 18800-2 (2008), Table 5

Besides the boundary conditions the eccentricities of the compact elements are taken into account. The calculation of the slenderness ratio can be taken from DIN 18800-2 (2008).

For crossed diagonal members the buckling length can be reduced. Therefore the stress state in the supporting member has to be taken into account (see Figure 22).

Connections. Single or double angle sections in lattice towers are connected by bolted connections. In contrast to tubular towers bolted connections in lattice structures are installed in transverse direction of the load, so that the bolts are predominantly loaded by shear forces. Figure 23 shows a bolted connection of two single sections with inner and outer steel plates.

Figure 23. Bolted connection in a chord of a lattice structure

The dash-dot line marks the centroidal axis of the steel sections. The outer and inner steel plates lead to an equidistant load transmission in every bolt. In cases of single section connected by only an inner or an outer steel plate the connection has to be considered as unsupported. The one sided load inducement leads to small eccentricities and additional bending moments in the bolts. The interaction of shear and bending moment has to be regarded in the design of the bolt.

In Figure 24 three types of unsupported bolted connections loaded by shear forces are depicted. Type a) represents the simplest type of shear connection. The bolt is not preloaded. The load transmission is realised by the bolt. Due to bad shearing transferability properties of shear forces under dynamic loading, this connection is not suitable for dynamic loaded connections in lattice towers of wind turbines.

[34] taken from DIN 18800-2 (2008), Table 15

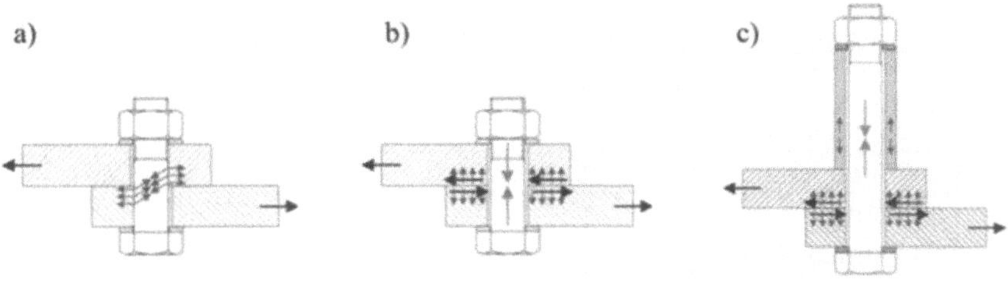

Figure 24. Types of bolted connections: a) shear b) slip-resistant c) expansion sleeve

Type b) displays a slip-resistant preloaded bolted connection which is preferably used in dynamic loaded lattice towers. The load transmission between the steel sections is realized by friction and contact pressure between the preloaded steel members. Under perfect conditions the bolted connection is only loaded by preloads. For this type of connection normally no fatigue analysis is necessary.

In case of dynamic loadings small movements of the transverse loaded bolts can occur. Movements lead to local settlements due to abrasion of the corrosion protection between bolt washer and steel section. The settlements lead to a loss of preload in the bolt which can reduce the design life of the connection significantly[35]. The shorter the clamping length of the connection the higher is the influence of loss of preload. The installation of the expansion sleeve (Figure 24; Type c)) increases the clamping length of the bolted connection and decreases the influence of local settlements simultaneously. Alternatively spring washers can be installed which also increase the clamping length and ensure the preload even if small settlements occur.

The shear load can be determined acc. to DIBt (2004)[36] to:

$$F_{s,Rd} = 0.9 \cdot \frac{\mu}{\gamma_{M,3}} \cdot F_V \qquad (34)$$

where $F_{s,Rd}$ is the design slip resistance, μ is the friction coefficient ($\mu \leq 0.5$), $\gamma_{M,3}$ as partial safety factor for loadings, and F_V as the nominal preload of the bolt. The partial safety factor $\gamma_{M,3}$ has to be determined in accordance to the combinations of actions. Table 9 shows the values for the partial safety factor for different operating conditions.

[35] see Schaumann et al. (2007-02)

[36] cf. DIBt (2004), Sec. 12.3

Table 9. Partial safety factor $\gamma_{M,3}$

Operating conditions	$\gamma_{M,3}$
1) Full operation	
2) Full operation and failure occurrence	1.25
3) Start	
4) Normal stop	
5) Emergency stop	
6) Standstill or idling	1.1
7) Standstill after failure occurrence	
8) Transport, installation, maintenance or repair works	

Transition piece. The connection between the lattice tower and the nacelle of the wind turbine is realised by a massive transition piece. Figure 25 shows the transition piece of the lattice tower of the wind turbine located in Laasow, Brandenburg, Germany.

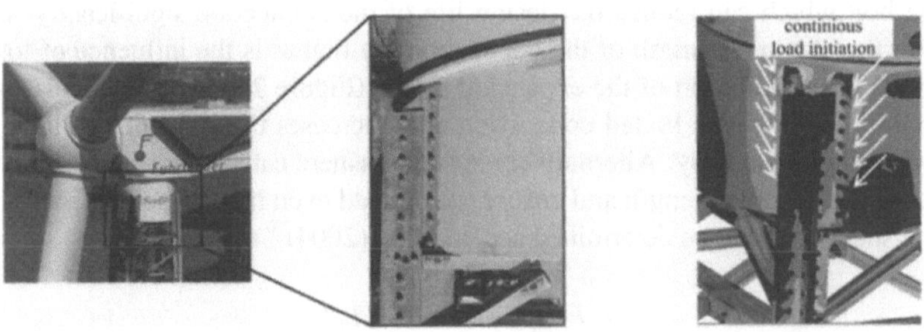

Figure 25. Transition piece for lattice towers: wind turbine Laasow (©SeeBa Technik)

The transition piece between the nacelle and the lattice tower always represents a point of structural and geometrical discontinuity which leads to local stress concentrations at single points. The plot of the numerical model of the transition piece (see Figure 25, right) shows the equivalent stresses. The bright areas reflect high stresses.

In addition to local stresses the number of bolts is limited which results from the maximum stresses of the exterior bolts. Figure 26 shows the bolt forces related to their location in the connection. If more then eight bolts in a row are installed, the bolt forces of the outer bolt connections may exceed the allowed bolt strength.

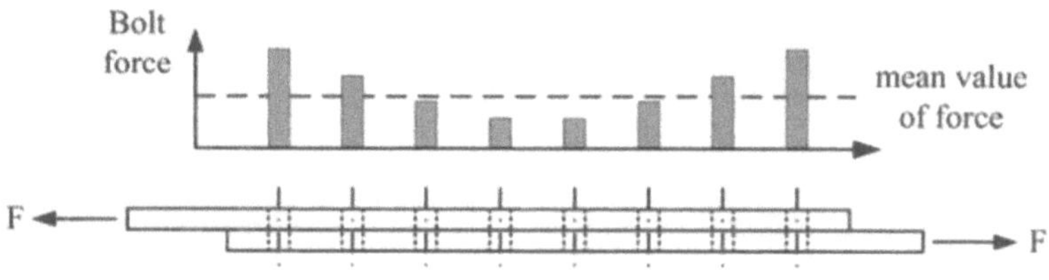

Figure 26. Maximum number of bolts in a row

If more than eight bolts are necessary a continuous load transfer has to be guaranteed. In this case it is recommended to calculate the bolt forces by using numerical models (see Figure 25, right).

3.3 Production and Erection

The prefabrication level of tubular towers for wind turbine structures is high. Most of the works can be carried out in the workshop. In a first step steel plates are rolled in a steel mill. After the transportation to the tower manufacturer the steel plates are cut to size according to the necessary geometric dimensions. The edges of the steel plates are prepared at the same time. For V- or K-welds the edges will be chamfered. Normally a 3-point-bending machine shapes the steel plates to steel cylinders with the required diameters (see Figure 27, left). The longitudinal gaps are welded so that finally several single tubular sections can be assembled to a tower section. The ends of the tower sections will be equipped with ring flanges. Every single tower section is finally coated by a corrosion protection system according to the on-site requirements (see Figure 27, right).

The production of lattice towers for wind turbine structures is shaped by comparably low prefabrication level. Most of the structural components can be prefabricated by steel producers. The cost intensive installation and assembly of the single components has to be done on-site.

The single or double sections are rolled at the steel mill and the bolt holes will be drilled. The structural members will be protected against corrosion by hot-dip galvanization. The friction surfaces are additionally coated by an ASI-friction coating (ASI = alcali - zinc-dust - silicate -coating). Single parts of the braces and chords are pre-assembled. The length of the pre-assembled parts depends on the transportation length.

Afterwards all members including transition piece are transported to the construction site. The members will be assembled on the ground and then lifted in the final position (see Figure 28).

Figure 27. Tubular tower: 3-point bending machine, coating (©SIAG)

Figure 28. Erection of a lattice tower

4. Steel Support Structures for Offshore Wind Energy Converters

Compared to the structural design of onshore structures the design of steel support structures of offshore wind turbines is more complex due to the offshore conditions.

Besides the additional turbulent wave loadings the structure is exposed to high corrosive environmental conditions. Additionally, the turbines are located far away from shore so that the effort for maintenance is very high. Offshore structures have to be more reliable than onshore.

This chapter presents the additional load case "wave loadings" in more detail. Different offshore steel materials and corrosion protection systems are introduced. Finally, different types of support structures with constructional details are shown.

4.1 Wave Loadings

Offshore Wind Energy Converters (OWECs) are exposed to wind, operational and particularly wave loads. Wave loads result from the water flow past and around the support structure of the OWEC. The real turbulent sea states can be described by a superposition of many different single sinusoidal waves or waves systems. The systems take different wave heights, periods, lengths, directions, and shapes into account. The waves interfere and superimpose each other.

Wave loads of slender structures can be described by the Morison's equation. Slender structures are structure with a small ratio between diameter and wave length. For non-slender structures the diffraction theory has to be used for the calculation of wave loads. Support structures of OWECs especially for lattice support structures like Jackets the Morison equation can be used. The empirical calculation formula takes structural information and the kinematics of the water particles into account. The particle kinematics can be derived from wave theories. Besides the regular wave theories, the kinematics can be described by sea state models.

Regular sea states. Wave theories describe the kinematics of regular, monochromatic, two-dimensional waves mathematically. In cases where the wave height is small compared to the water depth, the linear Airy wave theory may be used. This assumes a sinusoidal shape of the surface elevation and movement of the water particles along circular paths, varying over depth. With the circular frequency of the wave ω, and the wave height H, the surface elevation η, can be expressed as follows:

$$\eta = \frac{H}{2}\cos(\omega \cdot t) \qquad (35)$$

with: $\omega = \dfrac{2\pi}{T}$

In deep water, the particle kinematics result to:

$$v_h = \frac{H}{2} \cdot \omega \cdot e^{k \cdot z} \cdot \cos(\omega \cdot t) \tag{36}$$

$$v_v = \frac{H}{2} \cdot \omega \cdot e^{k \cdot z} \cdot \sin(\omega \cdot t) \tag{37}$$

$$a_h = \frac{H}{2} \cdot \omega^2 \cdot e^{k \cdot z} \cdot \sin(\omega \cdot t) \tag{38}$$

$$a_v = \frac{H}{2} \cdot \omega^2 \cdot e^{k \cdot z} \cdot \cos(\omega \cdot t) \tag{39}$$

with the indices ‘h’ for the horizontal and ‘v’ the vertical velocity v and acceleration a respectively. The wave number k can be determined to:

$$k = \frac{2\pi}{L} \tag{40}$$

The particle kinematics becomes non-linear with increasing wave steepness. In that case different nonlinear wave theories can be used. Figure 29 shows appropriate wave theories related to water depth and wave height.

The different theories all provide approximate solutions on the same differential equations with appropriate boundary conditions.

All wave theories compute a waveform that is symmetric about the crest and propagates without changing shape. The theories differ in their functional formulation and in the degree to which they satisfy the non-linear kinematics and dynamic boundary conditions at the wave surface.

Waves of small height in deep water are approximately linear in nature. Regular waves in this region are sinusoidal in shape and may be modelled using linear Airy wave theory or a low order stream function solution.

As the wave height is increased or the water depth reduced, wave profiles become steeper sided and the height of the wave crest above the still water level becomes greater than the depth of the trough below the same datum. The wave profile and water particle kinematics can no longer accurately be described using linear wave theory. Stream function theory can be suitably applied over a wide range of depths if the correct choice of solution order is made. Stokes 5th theory may be used to model steep waves in deep water.

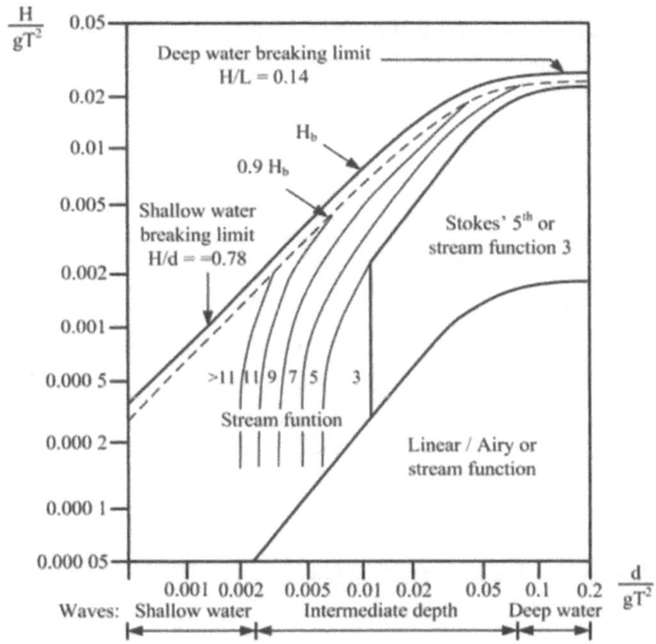

Figure 29. Regular wave theory selection diagram (IEC 61400–3 (2009))

As wave height is further increased or the water depth further reduced, the horizontal velocity of water particles in the wave crest will at some point exceed the wave celerity and the structure of the wave will break down. Water particles are ejected forward from the crest and the wave is said to break.

Wave theories for regular waves are valid only up to the still water level. In order to calculate the kinematics for parts of the structure above the still water level the wave kinematics as obtained from one of the wave theories described above needs to be stretched.

A common approach is the method suggested by Wheeler[37], the so-called Wheeler Stretching:

$$z = \frac{z_S - \eta}{1 + \frac{\eta}{d}} \tag{41}$$

[37] Cf. DNV-RP-C205 (2007), p. 28 f.

with η as the surface elevation, d as the water depth, and z_S as the z-coordinate of the point that the wave kinematics shall be stretched. Figure 30 shows a schematic of this stretching method.

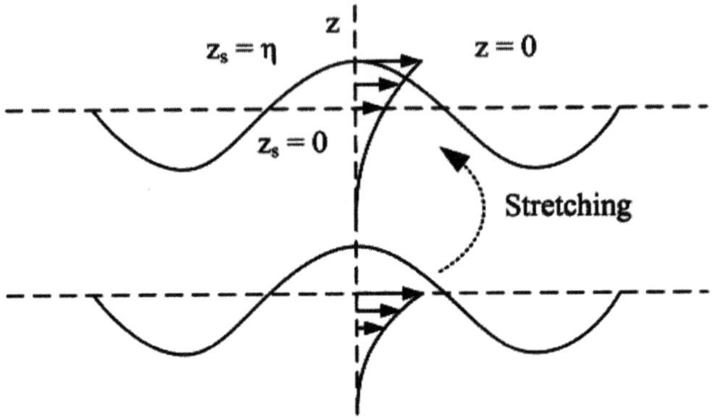

Figure 30. Wheeler stretching (taken from the DNV-RP-C205)

Irregular sea states. Real sea states can be described as stochastic processes that are stationary over a certain period of time. Usually, the assumption is made that over a time span of three hours a sea state is stationary. The characteristics of a stationary sea state can be modelled with wave energy spectra, i.e. the spectral density function of the surface elevation in that sea state. If no measured wave spectra are available for a given site, models of wave spectra formulations can be used that depend on characterizing parameters of a sea state, the significant wave height and the zero-crossing period. The explanations in this section base on the summary given in DNV-RP-C205[38].

The mean value of the wave height of the 1/3 biggest wave heights observed in a time series of recorded wave heights is called the significant wave height and is usually denominated H_S. Another definition of H_S can be derived from the spectral moment m_0:

$$H_S = 4\sqrt{m_0} \tag{42}$$

[38] Cf. DNV-RP-C205 (2007), pp. 31 ff.

with the spectral moments m_0 defined as:

$$m_0 = \int_0^\infty f^n \cdot S(n)\,df \tag{43}$$

where f is the wave frequency and n = 0, 1, 2, ...

The zero-crossing period T_Z is the mean period of successive up-crossings of the zero-water level of the water surface and can be described in a spectral form:

$$T_Z = \sqrt{\frac{m_0}{m_2}} \tag{44}$$

The wave energy spectra usually depend on the peak period T_P, i.e. the peak period of the spectral density function. The relation of T_Z and T_P depends on the shape of the spectrum.

The Pierson-Moskowitz (PM-) spectrum[39] has been implemented for fully developed sea states with unlimited fetch and unlimited duration of wind exposure. It is defined as given in (eq. (45)), using a formulation in angular frequency:

$$S_{PM}(\omega) = \frac{5}{16} \cdot H_S^2 \cdot \omega_P^4 \cdot \omega^{-5} \exp\left(-\frac{5}{4}\left(\frac{\omega}{\omega_P}\right)^{-4} \right) \tag{45}$$

with ω_P as the angular spectral peak frequency. The JONSWAP spectrum takes developing sea states with limited fetch and duration into account. It is based on the PM-spectrum, extended by the shape parameter γ:

$$S_{JS}(\omega) = nf \cdot S_{PM}(\omega) \cdot \gamma^{\exp\left(-\frac{1}{2}\left(\frac{\omega-\omega_P}{\sigma\cdot\omega_P}\right)^2\right)} \tag{46}$$

with: nf normalizing function $nf = 1 - 0.287\ln(\gamma)$

γ shape parameter

σ bandwidth parameter: $\sigma = \sigma_a$ for $\omega \le \omega_P$

$\sigma = \sigma_b$ for $\omega > \omega_P$

The normalizing function nf ensures that the same significant parameters will be reproduced as in the PM-spectrum. For $\gamma = 1$, the JONSWAP spectrum is equal to the PM-spectrum.

The DNV-RP-C205 (2007) exemplarily gives data resulting from experiments for the calculation of the JONSWAP - spectrum. Average values are $\gamma = 3.3$, $\sigma_a = 0.07$, and $\sigma_b = 0.09$.

[39] Cf. Pierson W J and Moskowitz L (1964)

The relationship between zero crossing period and peak period can only be established by approximate means. IEC 61400-3 (2005) recommends the following correlation:

$$T_p = T_z \cdot \sqrt{\frac{11+\gamma}{5+\gamma}} \tag{47}$$

The equation yields the following values for the spectra considering the average values mentioned above:

- $T_p = 1.41\ T_z$ for the Pierson-Moskowitz spectrum
- $T_p = 1.31\ T_z$ for the JONSWAP spectrum

Figure 31 shows a comparison of the Pierson-Moskowitz and the JONSWAP - spectrum for $H_s = 4$ m and $T_z = 8$ s.

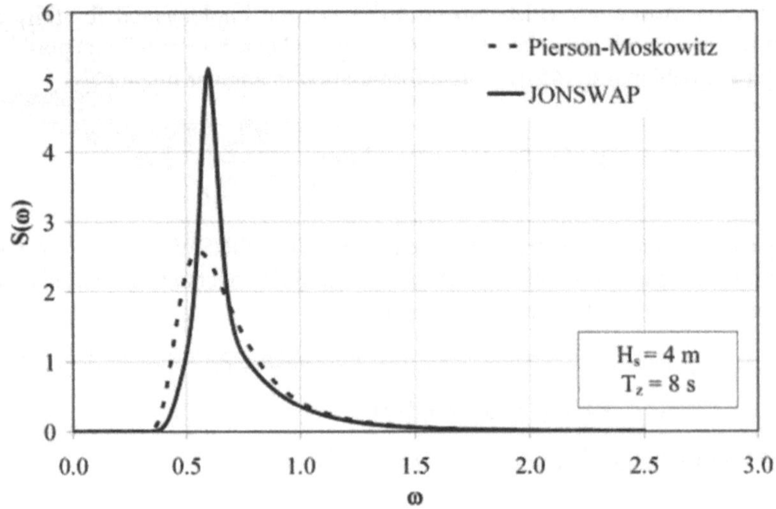

Figure 31. PM and JONSWAP spectrum for $H_s = 4$m and $T_z = 8$s

Real waves vary in directions even within short term sea states, too. Therefore, a wave spreading function D can be applied to take this directionality into account.

$$S(\omega,\vartheta) = S(\omega)\cdot D(\vartheta) \tag{48}$$

with ϑ as the angle which is relative to the main direction of the unidirectional sea state, as it is defined by the wave energy spectrum S(ω). The spreading function

D(ϑ) has to fulfil the following condition:

$$\int_{\vartheta} D(\vartheta)\,d\vartheta = 1 \tag{49}$$

Different approaches for the spreading function can be found in the literature, e.g. refer to Mittendorf (2006)[40] and DNV-RP-C205 (2007). Figure 32 illustrates the difference in the appearance of the sea surface for a linear wave (left) and a sea state with a spreading function applied (right).

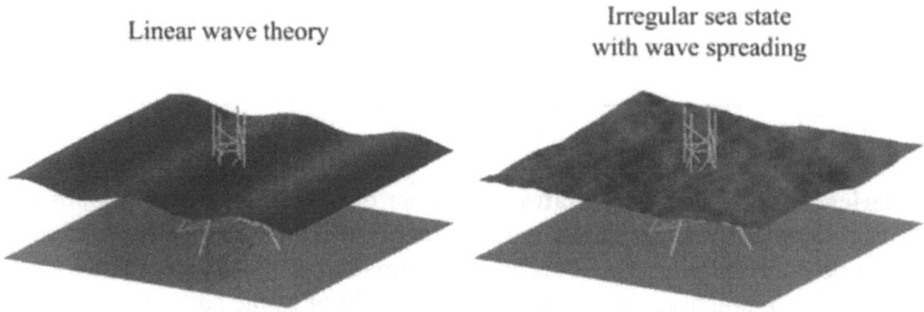

Figure 32. Illustration of the difference in the appearance of the sea surface[41]

Long term distribution of short term sea states. Long term statistics of sea states can be found in so called scatter diagrams that provide information of the relative frequency of occurrence of a sea state of a given class. Sea state classes are usually characterized by zero-upcrossing period T_z and significant wave height H_S. Additionally, often the directional dependency is considered in scatter diagrams, too. Table 10 gives an example of a non-directional scatter diagram for the North Sea.

Data for scatter diagrams can be generated by simulation, measurements, or from literature (see Kleineidam P (2005)).

[40] Mittendorf K (2006), pp.35 ff
[41] taken from Böker (2009), p. 13

Table 10. Scatter diagram for the North Sea[42]

$H_{s,i}$ [m]	$T_{0,i}$ [s] < 4	4 .. 5	5 .. 6	6 .. 7	7 .. 8	8 .. 9	9 .. 10	10..11	Total
0..1	19	86	94	41	10	2			252
1..2	3	49	121	99	40	10	2		324
2..3	1	17	63	73	40	13	3	1	211
3..4		6	27	39	26	10	3	1	112
4..5		2	11	19	14	6	2	1	55
5..6		1	4	9	7	4	1		26
6..7			2	4	4	2	1		13
7..8			1	2	2	1	1		7
8..9				1	1	1			3
9..10				1	1				2
total	23	161	323	288	145	49	13	3	1005

Numerical simulation of sea states. A number of models have been proposed to simulate irregular sea states based on wave energy spectra numerically. In principle this can be done by an inverse Fourier transformation of the spectrum back to the time domain. In practice, a sea state is often thought of as the sum of many partial waves with different amplitudes, periods, and phase angles. This model is the so called superposition model.

In this superposition model of sea states, the surface elevation becomes:

$$\eta(t) = \sum a_i \cdot \cos(\omega_i \cdot t + \varphi_i) \tag{50}$$

with the amplitude of the partial waves a_i, the angular frequency of the partial waves ω_i, and randomly distributed phase angles φ_i that are uniformly distributed in the interval $(0, 2\pi)$.

The same superposition method is valid for the superposition of the velocities and accelerations of the linear partial waves.[43]

In order to obtain the angular frequencies ω_i of the partial waves, the wave energy spectrum needs to be discretized into slices of $\Delta\omega_i$. Several methods exist carrying out this discretization.

The simplest approach is to divide the spectrum in to slices of constant thickness $\Delta\omega$. This results in time series that are periodic with $T = 2\pi/\Delta\omega$. This is undesirable due to the necessity to increase the number of partial waves in order to obtain longer random time series. This leads to unacceptable numerical costs.

[42] cf. Hapel K-H (1990), p. 214

[43] Mittendorf K (2006), p. 126

Another approach is to define the frequency intervals such that they become irrationally distributed[44]:

$$\Delta\omega_i = \sqrt{\frac{i}{i+1}}\frac{\omega_{\max}}{\sum\sqrt{i/\sqrt{i+1}}} \tag{51}$$

Time series generated by this approach are no longer periodic. Alternatively, the intervals can be chosen such that all partial waves have the same amplitude a_i:

$$a_i = \sqrt{\frac{2\cdot m_0}{N}} \tag{52}$$

with N being the total number of partial waves used to simulate the sea state. Descriptively, this means that all intervals cover the same area under the wave energy spectrum. Therefore, this method leads to a rather coarse resolution of the spectrum at the tail of higher frequencies, which can yield inaccurate results in dynamic time domain simulations if structural parts have resonance frequencies in that range. Kleineidam P (2005) therefore suggests using a combination of eq. (51) and (52) for different regions of the wave energy spectrum[45].

If the angular frequencies of the partial waves are known, the amplitudes can be calculated as follows:

$$a_i = \sqrt{2\cdot S(\omega)\cdot\Delta\omega_i} \tag{53}$$

Morison's equation. With the particle kinematics it is possible to calculate wave loads on hydrodynamically transparent structures using Morison's equation, which is an empirical formula widely used in the offshore industry.

The requirement for the applicability of Morison's equation is that the structures are hydrodynamically transparent, i.e. the structure itself influences the water flow only locally rather than obstructing the free flow on a global scale. It is assumed that this is true if the structure's diameter D does not exceed one fifth of the wave length λ:

$$\frac{D}{\lambda} \leq 0.2 \tag{54}$$

For structures that cannot be defined as slender in the meaning of eq. (54), other methods of wave load calculation have to be taken into account to consider the

[44] Mittendorf K (2006), p. 127

[45] Kleineidam P (2005), p. 122

diffraction of the flow. For more detailed information on diffraction theory refer to IEC 61400–3 (2009).

For static structures, i.e. in cases where the structural movements can be neglected, Morison's equation is:

$$f = f_d + f_m = \frac{1}{2} C_d \cdot \rho \cdot D \cdot v \cdot |v| + C_m \cdot \rho \frac{\pi D^2}{4} a \tag{55}$$

where: f force per unit length of the member
- f_d drag term of the wave force
- f_m inertia term of the wave force
- C_d hydrodynamic drag coefficient
- $C_m = (1+C_a)$ hydrodynamic inertia coefficient
- C_a added mass coefficient
- ρ water density
- D diameter of the member in the respective section
- v velocity of the flow normal to the member force
- a acceleration of the flow normal to the member force

It can be seen that the forces acting on members in the water consist of a drag term f_d and an inertia term f_m, both of which depend on the hydrodynamic coefficients C_d and C_m.

The drag term includes viscous force and friction effects and depends nonlinearly on the particle velocity.

The inertia term includes the so called Froude-Krylov force, which results from the integration of the pressure field of the undisturbed wave over the surface of the member, and the hydrodynamic mass force. For cylindrical bodies, the hydrodynamic mass is equal to the displaced water mass.

The Froude-Krylov force is independent of the structural answer. However, the hydrodynamic mass portion of the inertia term in eq. (55) is not. Considering the relative velocities and accelerations between structural movement and particle kinematics yields a modified formulation of Morison's equation, called the relative velocity formulation[46]:

$$f = \frac{1}{2} C_d \cdot \rho \cdot D \cdot v_r \cdot |v_r| + C_m \cdot \rho \frac{\pi D^2}{4} \cdot a_w - C_a \cdot \rho \cdot \frac{\pi D^2}{4} \cdot a_s \tag{56}$$

where: v_r relative velocity of the flow normal to the member surface
- a_w water particle acceleration normal to the member
- a_s structural acceleration normal to the member

[46] DNV-RP-C205 (2007), p. 53

The consideration of relative kinematics effectively takes hydrodynamic damping into account implicitly.

The hydrodynamic coefficients C_d and C_m can only be determined by experiments. They depend on the flow conditions around the structure, the structural shape, and the roughness of the structure. The particular conditions for a given site may be characterized by the Reynolds number Re, the relative roughness, and the Keulegan-Carpenter number KC. However, significant variations of coefficients that have been measured experimentally for seemingly identical conditions prove that there is still a considerable portion of uncertainty in the determination of hydrodynamic coefficients. The abovementioned standards as well as many other sources give guidance on the selection of C_d and C_m. As standard values for the coefficient, they can be taken to $C_d = 0.7$ and $C_m = 2.0$. These are average values proposed by several authors.

Joint distribution of wind and waves. Since Offshore Wind Turbines are loaded by wind and wave loads it is necessary to find suitable means of combining both. This can be done by applying the joint probability function for the simultaneous occurrence of a certain wind speed and sea state. If no measured or hindcast data on combined wind and wave conditions for a site is available, e.g. if the Offshore Wind Turbine is to be designed according to one of the wind turbines classes, the GL-Offshore Guideline (2005) gives guidance on choosing H_s and T_p based on the hourly mean wind speed at 10 m above the sea surface[47]:

$$T_p = \frac{u_{10}}{g} \cdot \frac{1}{0.16} \approx 0.6371 \cdot g \qquad (57)$$

$$H_s = 0.0094 \cdot 0.16^{-5/3} \cdot \frac{u_{10}^2}{g} \approx 0.02032 \cdot u_{10}^2 \qquad (58)$$

4.2 Material and Components

Offshore wind turbine support structures are exposed to extreme climate and loading conditions. Materials in OWECs have to fulfil requirements regarding the structural safety and reliability. Beside extreme loadings the materials have to be suitable regarding their mechanical properties. Therefore offshore standards[48] define requirements for metallic materials of OWEC for example strength, toughness also for

[47] GL-Offshore Guideline (2005), p. 4-73

[48] GL-Offshore Guideline (2005), page 3-6

low temperatures, cold deformability, and suitability for welding. The material has to be chosen considering the importance of the structural member, the load and stress level, and other specific material and structural properties.

Valid standards distinguish between component categories in OWEC support structures. The materials have to be chosen for the appropriate component category.

Component categories. In relation to the importance of the structural member and on the type of load and stress level different component categories exist. According to GL-Offshore Guideline (2005) a structure can be subdivided in three different component categories.

The first component category is for special structural members which are essential to the overall integrity of the structure. They are exposed to particularly arduous conditions like stress concentration or multi-axial stresses. The second category is valid for primary structural members, which are members participating in the overall integrity of the structure or which are important for operational safety and exposed to calculated load stresses comparable to the special structural members. The third category considers secondary structural members which are of minor significance and are exposed to minor stresses.

Steel strength classes. Material strengths of metallic materials are subdivided in three different strength classes. Steel materials with yield strengths of 285 MPa can be assigned to "normal strength". If materials are used with yield strengths between 285 and 380 MPa, these materials have to be associated to "higher strength" class. For yield strengths above 380 MPa the "high strength" class has to be evaluated. For special and primary steel structures, the GL-Offshore Guideline (2005)[49] recommends using fine grained structural steels suitable for welding with nominal yield strengths not exceeding 355 MPa due to reasons of resistance to fatigue.

Table 11 summarises appropriate steel grades for plates and sections. For tubular sections different steel grades have to be considered. Detailed information for tubular members can be taken from the standards.

[49] GL-Offshore Guideline (2005), page 3-7

Table 11. Appropriate steel for plates and sections[50]

Structural member category	Steel strength class	Standard and/or rules	Designation of material
Special	High	EN 10225 EN 10025-3/-4	S 460 + S420 of group 2+3 S 460 NL, S 460 ML S 420 NL, S 420 ML
	Higher	EN 10225 EN 10025-3/-4	S 355 of group 2 + 3 S 355 NL, S 355 ML
	Normal	EN 10025-3/ -4	S 275 NL, S 275 ML
Primary	As for category “special” and additionally:		
	High	EN 10025-3/ -4	S 460 N + NL, S 460 M + ML S 420 N + NL, S 420 M + ML
	Higher	EN 10225 EN 10025-3/-4 EN 10025-2	S 355 group 1 S 355 N + NL, S 355 M+ ML S 355 J2 + K2
	Normal	EN 10025-3/-4 EN 10025-2	S 275 N + NL, S 275 M + ML S 275 J2 + K2 S 235 J2 + K2
Secondary	As for the category “special” and “primary” and additionally:		
	Higher	EN 10025-2	S 355 J0
	Normal	EN 10025-2	S 275 JR, S275 J0 S 235 JR, S 235 J0

Additional requirements. Valid offshore codes have additional requirements regarding the material thickness, the carbon equivalent used, the yield strength ratio, and the impact energy. Further information can be taken from GL-Offshore Guideline (2005).

Corrosion protection and coatings. Due to the marine climate corrosion protection is necessary for support structures of Offshore Wind Turbines.

Different corrosion protection systems are possible. Within this subchapter only a

[50] GL-Offshore Guideline (2005), page 3-73

short overview of possible corrosion protection systems is given. Further details have to be taken from the guidelines[51].

Offshore wind turbine components are exposed to aggressive environmental conditions and not easily accessible. Because of the operational conditions, in many cases repeated protective coating is not possible. Special importance therefore attaches to the design, choice of material and corrosion protection measures. The corrosion protection can be chosen according to the following codes.

For Coatings:

- EN ISO 12944: "Corrosion protection of steel structures by protective paint systems"
- NORSOK Standard M-501: "Surface preparation and protective coating"

For Cathodic Protection Systems:

- EN 12945: "Cathodic protection for fixed steel offshore structures"
- NORSOK Standard M-503: "Cathodic Protection"
- NACE Standard RP 0176: "Control of Corrosion on Steel Fixed Offshore Platforms"

Table 12 summarises three possible corrosion protection systems and reflects the possibility of corrosion allowance. In most cases support structures of Offshore Wind Turbines are protected by a combination of different corrosion protection systems.

For structural members of Offshore Wind Turbines the corrosion protection is significant. Corrosion initiates small cracks in the material which reduce the fatigue resistance of the structure significantly[52].

Typical coating systems for the passive corrosion protection can be taken from EN ISO 12944. Offshore steel support structures are classified in the corrosion class C5-M acc. to EN ISO 12944-2. The appropriate coating system can be chosen according to EN ISO 12944-5 subsequently. For an evaluated design life (class H = high), systems with different sequence coatings are possible (see Table 12).

The active corrosion protection by sacrificial anodes can be chosen acc. to EN 12495. A design for the galvanic system has to be carried out.

[51] GL-Offshore Guideline (2005) or DNV-OS-J101 (2007)

[52] Bignonnet, A (1987)

Table 12. Corrosion protection[53]

Corrosion Protection	Description
Coatings	Coating is a collective term for one or several coherent layers on a base material, which are made from non preformed materials and the binding agents of which mostly are of an organic nature. Coatings can be selected acc. to EN ISO 12944-5.
Sacrificial Anode	An anode is an electrode in a galvanic cell or the part within a corrosion cell which emits a direct current in the form of positively charged ions, mostly with anode substance dissipation. Anode materials are alloys on aluminium, zinc, and magnesium basis.
Cathodic protection	The cathodic protection is a protective method by which the material to be protected is made a cathode, by sacrificial anodes or impressed current systems. This method is intended to prevent corrosion in the submerged zone electrochemically.
Corrosion allowance	In addition to constructional corrosion protection systems, corrosion allowance can be considered. The average value of allowed corrosion per year can be estimated to 0.3 mm/a. The value has to be evaluated for every single wind park location.

Table 13. Coating systems for offshore steel structures acc. to EN ISO 12944-5

low alloyed steel, surface preparation SA 2 ½						
protection system	**base coating**			**Top Coating**		
	bonding agent	**layer**	**per layer µm**	**bonding agent**	**layer**	**per layer µm**
A5M.02	EP, PUR	1	80	EP, PUR	3-4	320
A5M.07	EP, PUR, ESI	1	60	EPC	3-4	400
EP: epoxy resin; PUR: polyurethane; ESI: ethyl silicate						

4.3 Types of Offshore Support Structures

Within the last decades several types of support structures have been investigated. The support structure of an Offshore Wind Turbine consists of the tower and the

[53] GL-Offshore Guideline (2005), pp. 3-65 ff

substructures. The substructure includes all structural components below the tower including the foundation (see Figure 33).

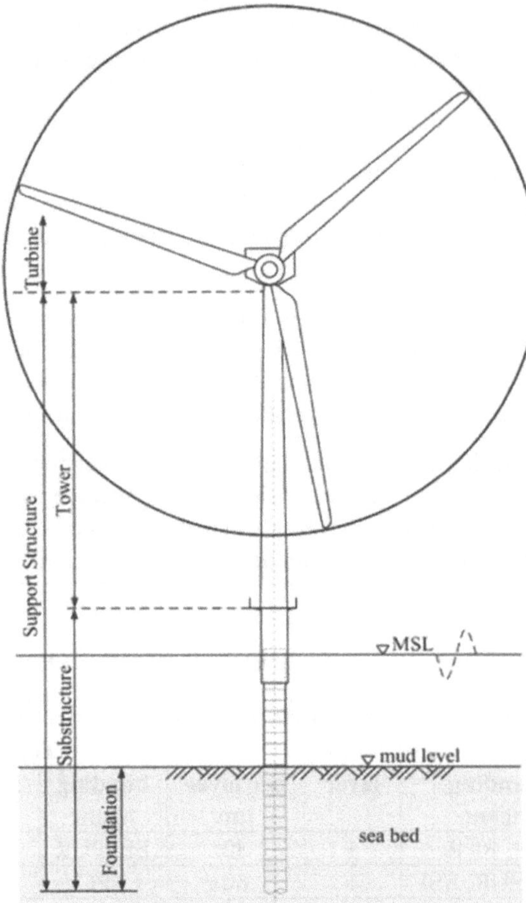

Figure 33. Components of an Offshore Wind Turbine with Monopile foundation

Depending on the water depth, the size of the turbine, and the local conditions different types of substructures have been developed. This chapter describes several types of support structures. Most offshore wind parks in Europe are supported by Monopile foundations.

Monopile. This type of foundation is a simple construction and extends effectively the turbine tower under water and into the seabed (see Figure 33). The Monopile consists of a steel pile with a diameter between 3.5 and 5.5 meters driven or drilled 10 to 40 meters into the seabed. This cylindrical tube is connected to the steel tower by a so-called transition piece. This transition piece unifies the tower pile with the foundation pile. Within this unit, the foundation tube overlaps the tower tube and the annulus between the tubes is filled with high performance grout. This type of connection is also called grouted joint. Monopiles are used extensively in the near- and offshore environment up to water depths of 25 m. For the installation of Monopiles only small preparations of the seabed are necessary. Heavy piling equipment drills the foundation into the seabed. Monopiles are not suitable for locations with many large boulders.

Jacket. The Jacket resembles with a lattice tower (see Figure 34, left). As a result of the spatial carcass the steel amount and therefore the material costs experience a reduction. Compared to Monopile structures a Jacket substructure consists of 40 % to 50 % less steel. To the top of the Jacket construction, the tower including the turbine is mounted. The anchorage is formed by piles at all four structure feet according to the Tripod construction. Conveniently, the specific details are of comparatively small dimensions affording a production standardized largely. All mentioned facts lead to marginally increasing project costs with significantly increasing water depth. According to the suitability to deep water, this structure might become favorable for wind energy farms in the outer sea. Jackets can be used in water depths up to 50 m.

Tripod. Tripod foundations are multiple-pile constructions because of their three-legged foundation (see Figure 34, middle). This structure consists of a central steel cylinder connected with a steel lattice to three foundation piles symmetrically arranged around the central pile. The three legs transfer the forces from the tower into the foundation piles. The three piles are driven 10 to 20 meters into the seabed depending on soil conditions. The installation needs marginal seabed prearrangement what leads to an expeditious assembling. Though, the seabed needs to be free of boulders. Compared to the dimensions of the Monopile, the pile diameter of the three legs is significantly reduced. Tripods can be used for water depths up to 50 m.

Tripile. The Tripile foundation is a development by BARD Engineering (see Figure 34, right). The Tripile is suitable for water depths of 25 to around 40 meters and is more compact, lighter and cheaper than other offshore support structures. The supporting crosspiece and struts are welded from flat steel elements. On the other hand the torsional stiffness is smaller. This influence has to be considered thoroughly.

The joints between the supporting crosspiece and the three foundation piles are glued with special cement. All the assembly work can be done with a heavy-duty crane on a construction vessel.

The structural members of the Tripile are produced in series. For different soil conditions the lengths of the three piles can be varied. The design process of the crosspiece considers different boundary conditions so that the structural design is valid for several boundary conditions.

The three piles are driven into the seabed. The necessary length of the piles depends on the soil conditions. The driving of the three piles is done by using a template. The crosspiece is put on top of the three piles. The gaps between piles and crosspiece are grouted with a high performance mortar.

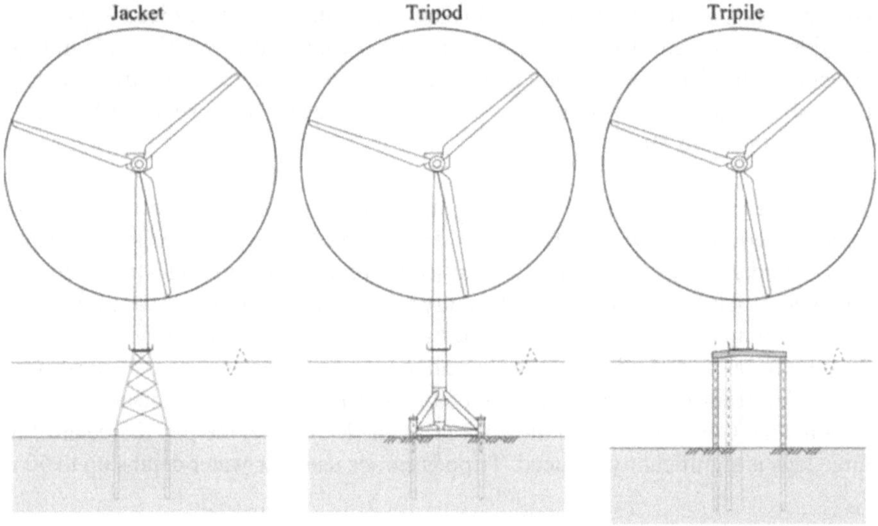

Figure 34. Support structures of Offshore Wind Energy Converters (Jacket, Tripod, Tripile)

Suction bucket. Suction buckets are tubular steel foundations that consist of an upside down cylinder (see Figure 35, left). They are installed by sealing the top and applying suction inside the bucket. The hydrostatic pressure difference of the outer and the inner side and the deadweight cause the bucket to penetrate the soil.

This benign installation procedure allows the bucket to be connected to the rest of the structure before installation, enabling a reduction in steps of the installation procedure. The bucket technology has been widely used in the construction of offshore oil and gas platforms. Concerning Wind Energy Converters there is little knowledge by now. Compared to other support structures advantage is given by the installation without drilling procedures and the small amount of steel.

Floating structures. Floating structures are one of the latest developments for Offshore Wind Turbine support structures (see Figure 35, middle). Suitable for extreme large water depths between 150 and 700 metres, these types of substructures are kept in position by mooring chains and anchors. Additionally, the chains have the advantage that they contribute to dampen the motions of the floater[54]. At the bottom of the hull of the floater, a stabiliser is placed to further reduce roll. The installation is simple since the structure can be towed to the site and then be connected by the chains to the anchors. The anchors can be fluke anchors, drag-in plate anchors and other plate anchors, suction anchors or pile anchors, depending on the actual seabed conditions. When the anchors have been installed, the chains can be attached and tightened and hook-up cables can be assembled.

The HYWIND project in Norway is the first Offshore Wind Turbine which is supported by a floating substructure.

Tension leg platform. The tension leg concept is a foundation concept which is well known from the oil and gas industries. This type of foundation is suitable for water depths of 300 to 1500 metres. The tension leg support platform is a floater submerged by means of tensioned vertical anchor legs[55] (see Figure 35, right). The base structure helps dampen the motions of the structural system.

The installation is simple since the structure can be towed to the site and then be connected to the anchors. When anchors such as anchor piles have been assembled and steel legs have been put in place, the hook-up cable can be installed. The platform is subsequently lowered by use of ballast tanks and/or tension systems. The entire structure can be disconnected from the tension legs and floated to shore in case of major maintenance or repair of the wind turbine.

Table 14 shows the water depths for different types of support structures.

[54] DNV-OS-J101 (2007), Sec.1 – p. 17

[55] DNV-OS-J101 (2007), Sec. 1 – p.17

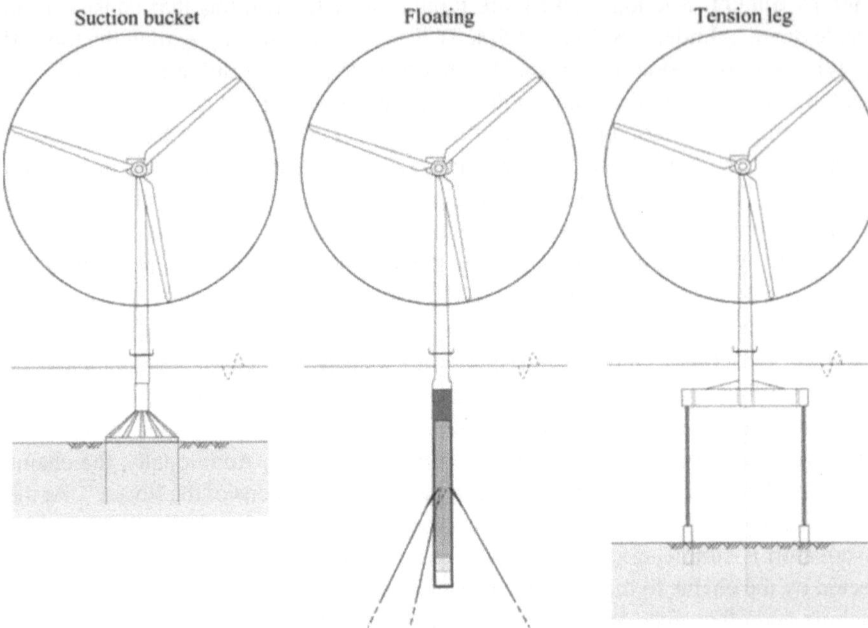

Figure 35. Support structures of Offshore Wind Energy Converters (Suction bucket, Floating, Tension leg)

Table 14. Appropriate water depth for different types of support structures

Concept	water depth small 0 to 20 metres		medium 20 to 50 metres		large > 50 m	
Monopile	suitable	suitable	suitable			
Jacket		moderate	suitable	suitable		
Tripod		moderate	suitable	suitable		
Tripile	suitable	suitable	suitable	suitable		
Suction bucket	suitable	suitable	suitable			
Floating						suitable
Tension Leg			moderate	suitable	suitable	suitable

Legend: dark shading = suitable; light shading = moderate

4.4 Construction Details

Compared to Onshore Wind Turbines support structures of Offshore Wind Turbines use additional types of joints. This chapter gives a short overview of constructional details.

Grouted joints. The grouted joint is used for the connection between substructure and tower of the Offshore Wind Turbines. It consists of the top of the foundation pile and a transition piece, which is slid over the pile. The annulus between pile and transition piece is filled with a high performance grout.
Figure 36 schemes the components of a grouted joint.

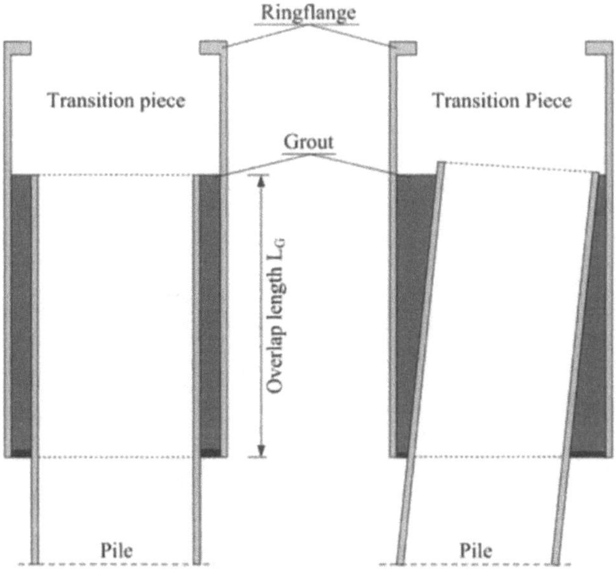

Figure 36. Components of a grouted joint

Within the grouted joint it is possible to overcome pile eccentricities and inclinations of the driving process. The load transmission is realised by contact and friction between the steel and the grout surfaces. Latest investigations show that the application of shear keys increase the load capacity and stiffness of the connection significantly. Shear keys are weld beads or steel bars in circumferential direction. They are adjusted on the opposite surfaces of the steel members. The authors strongly recommend the use of mechanical measures for shear in order to improve the fatigue resistance of this kind of joint.

Valid standards recommend an overlap length L_G of 1.5-times the diameter of the pile. The used grout material have compressive strengths of 100 MPa and higher. Grouted connections have to be designed within a numerical simulation under consideration of non-linear contact and non-linear brittle material for the grout. Besides the ultimate limit state the fatigue limit state has to be investigated. Therefore concrete codes for high performance grout materials have to be considered.

Boat landing, work platform and J-tubes. Any Offshore Wind Turbine has to provide good accessibility for maintenance and service. Mostly this is achieved by boat landing facilities. The boat landing consists of a ladder with fenders and a shock cell to absorb ship impact (see Figure 37).

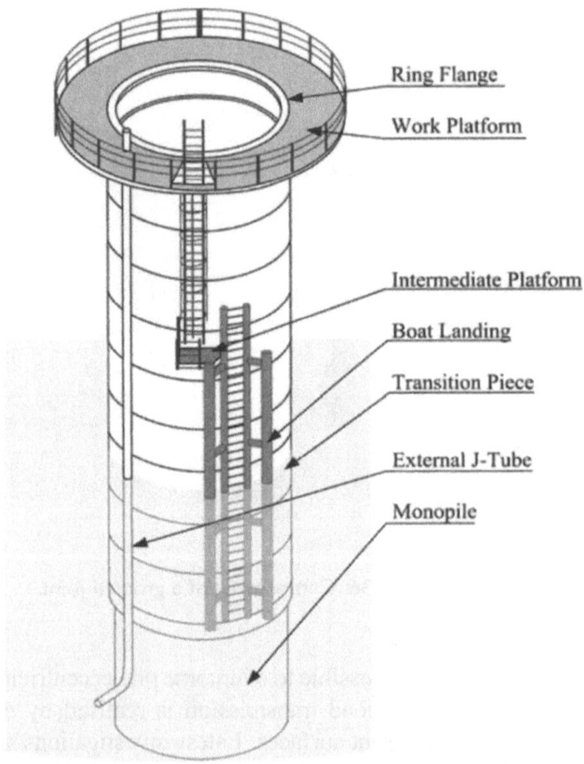

Figure 37. Transition piece of a Monopile with secondary items

For instance, the boat landing facility is installed at the transition piece of a Monopile foundation. The ladder leads to a work platform at the tower base. The service platform is normally above the splash zone. In some cases an intermediate platform has to be installed.

Modern Offshore Wind Turbine support structures are assembled with a helicopter deck. This is useful, when the turbine structure is very high.

Besides the work platform and boat landing J-Tubes are preinstalled at the transition piece. The metallic tubes are used for the connection of the single wind turbine to the grid. Cables for the power supply are routed from the seabed up through the J-Tubes. J-Tubes can be built internal or external of the transition piece. Platforms, landing facilities and J-Tubes belong to secondary structural members acc. to GL-Offshore Guideline (2005)[56].

4.5 Erection of Offshore Wind Turbine Structures

The erection of OWEC support structures including substructure and tower is a big challenge. In comparison to the installation of wind turbines onshore, offshore constructions have to consider weather conditions in more detail. For the North Sea approximately 100 to 120 days a year[57] are suitable for the installation of an OWEC support structure.

The installation of foundations and substructures of Offshore Wind Turbines are mainly influenced by the wave action. Today's offshore installation techniques allow maximum significant wave heights of 2.6 m and water depths up to 45 m.
For the installation of OWECs mobile platforms are used. These so called jack-up rigs stand still on the sea floor resting number of supporting legs. Jack-up rigs are sometimes used for transportation of foundation piles, tower tubes, or nacelles with blades.

For the installation of a wind turbine structure with Monopile foundation the following steps are carried out:

- Drilling or driving of the foundation piles
- Installation of the transition piece with grouting
- Installation of the tower sections
- Installation of the nacelle and rotor blades
- Grid connection
- Scour protection

Mostly the preparation of the seabed is not necessary.

[56] GL-Offshore Guideline (2005)p. 3-7
[57] Greenpeace (2000), Sec. 3.5.1

5. References

Almar-Naes (1985): *Fatigue Handbook - offshore steel structures*. Norges tekniske högskole published by the Tapir Publishers. Trondheim, Norway. 1985.

API-RP-2A-WSD (2000): *Recommended Practice for the Planning, Designing and Constructing - Fixed Offshore Platforms - Working Stress Design*. American Petroleum Institute, Recommended Practice, 21st Edition. Washington, USA. 12/2000.

Bignonnet, A (1987): Corrosion Fatigue of Steel in Marine Structures – A Decade of Progress. In: *Steel in Marine Structures* edited by C Nordhoek and J de Back, Elsevier, Amsterdam, The Netherlands, 1987.

Böker (2009): *Load simulation and local dynamics of support structures for offshore eind turbines*. Dissertation Thesis. Institute for Steel Construction, Leibniz Universität Hannover. Aachen: Shaker, 2009.

BSH (2007): *Standard - Design of Offshore Wind Turbines*. Federal Maritime and Hydrographic Agency (BSH). Hamburg, Germany. December 2007.

Ciamberlano F (2006): Engineering Insurance of Offshore Wind Turbines. Proceedings of the 39th IMIA Annual Conference. 12. September, Bosten, USA, 2006.

DIBt (2004): *Richtlinie für Windenergieanlagen - Einwirkungen und Standsicherheitsnachweise für Turm und Gründung*. Deutsches Institut für Bautechnik (DIBt). Berlin, Germany. March 2004 (in german)

DIN 18800-1 (2008): *Steel Structures – Part1: Design and Construction*. Normenausschuss Bauwesen (NABau) im DIN. Beuth Verlag Berlin, Germany. 11/2008

DIN 18800-2 (2008): *Steel Structures – Part2: Stability - Buckling of bars and skeletal structures*. Normenausschuss Bauwesen (NABau) im DIN. Beuth Verlag Berlin, Germany. 11/2008

DIN 18800-7 (2008): *Steel Structures- Part 7: Execution and constructor's qualification*. Normenausschuss Bauwesen (NABau) im DIN. Normenausschuss Schweißtechnik (NAS) im DIN. Beuth Verlag Berlin, Germany. 11/2008

DNV-RP-C205 (2007): *Environmental conditions and environmental loads*. Høvik, Norway. Det Norske Veritas, April 2007.

DNV-OS-J101 (2007): *Design of Offshore Wind Turbine Structures*. Høvik, Norway. Det Norske Veritas, October 2007.

EN 1993-1-1 (2005): *Eurocode 3: Design of steel structures – Part 1-1: General rules and rules for buildings*. CEN. Brussels, Belgium. 2005.

EN 1993-1-8 (2005): *Eurocode 3: Design of steel structures - Part 1-8: Design of joints*. CEN European Committee for Standardization. Brussels, Belgium. 2005.

EN 1993-1-9 (2005): *Eurocode 3: Design of steel structures - Part 1-9: Fatigue*. CEN European Committee for Standardization. Brussels, Belgium. 2005.

EN 10056-1 (1998): *Structural Steel equal or unequal leg angles - Part 1: Dimensions*. NA Eisen und Stahl (FES) im DIN, Beuth Verlag. Berlin, Germany. 10.1998.

ENV 1993-3-2 (1997): Eurocode 3: Design of steel structures - Part 3-2: Towers, masts and chimneys, CEN European Committee for Standardization. Brussels, Belgium. 1997.

Greenpeace (2000): *North Sea Offshore Wind – A powerhouse for Europe. Technical Possibilities and Ecological Considerations – Study*. Deutsches Windenergie Institut DEWI. Wilhelmshaven, Germany. 2000.

GL-Onshore Guideline (2003): *Guideline for the certification of Wind Turbines*. Germanischer Lloyd Industrial Services. Hamburg, Germany. Edition 2003 with Supplement 2004, November 2003.

GL-Offshore Guideline (2005): *Guideline for the certification of Offshore Wind Turbines*, Germanischer Lloyd Industrial Services. Hamburg, Germany. June 2005.

Hapel K-H (1990): *Festigkeitsanalyse dynamisch beanspruchter Offshore-Konstruktionen*. Vieweg-Verlag Braunschweig, Germany.

Kleineidam P (2005): *Zur Bemessung der Tragstrukturen von Offshore-Windenergieanlagen gegen Ermüdung*. Dissertation Thesis. Institute for Steel Construction, Leibniz Universität Hannover. Aachen: Shaker, 2005.

IIW (2007): *XIII-2151-07/XV-1254-07 - Recommendations for Fatigue Design of Welded Joints and Components*. International Institute of Welding IIW, edited by A. Hobbacher. Paris, France. May 2007.

IEC 61400-3 (2005) Ed. 3, Wind Turbines – Part 1: Design Requirements, International IEC 61400–1 Electrotechnical Commission (IEC). Genève, Switzerland. 2005.

IEC 61400–3 (2009) Ed. 1, Wind Turbines – Part 3: Design Requirements for Offshore Wind Turbines, International Electrotechnical Commission (IEC). Genève, Switzerland. 2009.

Mittendorf K (2006): *Hydromechanical Design Parameters and Design Loads for Offshore Wind Energy Converters*. Dissertation Thesis, Leibniz Universität Hannover. Hannover: Institute of Fluid Mechanics, 2006.

Petersen (2001): *Stahlbau - Grundlagen der Berechnung und baulichen Ausbildung von Stahlbauten*. Vieweg Verlag. Wiesbaden, Germany. 2001

Pierson W J and Moskowitz L (1964): A proposed spectral form for fully developed wind seas based on the similarity theory of S.A. Kitaigordskii, *Journal of Geophysical Research 69*: 5181 - 5190.

Radaj and Sonsino (2006): *Fatigue assessment of welded joints by local approaches. 2nd edition*. Woodhead Publishing Ltd. & Maney Publishing Ltd., Cambridge, UK. 2006.

Schaumann et al. (2007-01): *Tragstrukturen für Windenergieanlagen*. in: Stahlbaukalender 2007. Edited by U Kuhlmann. Verlag Ernst & Sohn. Berlin, Germany. 2007.

Schaumann et al. (2007-02): *Numerical Analysis of the Load Bearing Behavior of Slip Resistant Prestressed Bolt Connections with Consideration of Adhesion*. Bauingenieur. Heft 2, p. 77 - 84. Springer Verlag. Düsseldorf, Germany. 2007. (in german).

Schaumann et al. (2010): *Fatigue design for axially loaded grouted connections of offshore wind turbine support structures in deeper waters*. Proceedings of the 12th Biennial ASCE Aerospace Division International Conference. 14th -17th March, Honolulu, Hawaii. 2010.

Seidel (2001): *Zur Bemessung geschraubter Ringflanschverbindungen von Windenergieanlagen*. Dissertation Thesis. Institute for Steel Construction, Leibniz Universität Hannover. Aachen: Shaker, 2001.

Schmidt/Neuper (1997): Zum elastostatischen Tragverhalten exzentrisch gezogener L-Stöße mit vorgespannten Schrauben. Stahlbau 66, Heft 3. Ernst&Sohn. Berlin, Germany 1997.

Topics on the design of tubular steel wind turbine towers

C.C. Baniotopoulos , I. Lavassas , G. Nikolaidis , P. Zervas

Institute of Metal Structures, Aristotle University of Thessaloniki, Greece

Introduction

The wind turbine tower is mainly a simple cantiliver beam. However, its section forms a thin-walled cylindrical shell and therefore, several issues arise during the analysis such as the local buckling of the shell structure or the stress concentrations around the door opening which must be thoroughly examined. The design is governed by the extreme wind loading; earthquake loading should also be taken into account when designing the turbine tower on seismic hazardous areas.

1. Evaluating the tower shell thicknesses

1.1 General aspects

The evaluation of the shell thicknesses of the tower is performed by the plastic limit state design [LS1] and buckling limit state desingn [LS3] as they are described in EC-3-1-6.

The wind tower is a cantilever beam; in this sense, the simplest way is to perform a hand calculation, considering the tower clamped to its base. A computer analysis using a linear model can also be done. In this case the various courses of the tower are represented as linear elements. The tower can be considered either as clamped to the base, or including the foundation using a linear beam grid. By the previous type of approach the internal forces at every point of the tower are calculated. Special cases as the local buckling of the shell or the stress state at specific points, like flange positions or door opening must be examined by applying special calculation models. However, as nowadays the designer has advanced computational tools in his hands, it is now becoming easy to model the whole tower with all the details included using an integrated Finite Element model. Constructing such a model, the designer should evaluate directly the stress state at any point of the tower (including flange connections, door opening etc.)

In the paragraphs below, an analysis of the tower using the aforementioned methods is attempted. The prototype tower examined has a height of 76.15 m. The shell diameter at the base is 4.30 m and the diameter at the tower top is 3.0 m. Shell thicknesses vary from 30 mm at the bottom to 12 mm ato the top. The tower is divided into three parts connected together by bolted flanges.

1.2 Tower Loading

1.2.1 Vertical loads [G]

In a structural model, the self-weight of the shell is usually estimated directly by the FE software, as a function of the geometry and the unit mass of the steel elements. The contribution of the platforms and the ancillary equipment (ladders, cable racks etc.) to the total weight of the tower could be neglected.

The weight of the nacelle, including the blades and the rotor, is provided by the manufacturer. In the example at hand, it is assumed to be equal to: Gr=1067,00kN, having the center of gravity shifted horizontally +0,725m from the axis of the tower and vertically +0.50 to +1.00 m above the upper flange level (+76.15 m).

1.2.2 Extreme Wind loads [W]

The loads over the tower stem are calculated, for the specific dynamic characteristics and geometry of the structure, according to [EC 1-1-4]. In the example a basic wind velocity at 10m above ground of: vb=27,00 m/sec and for a terrain of category [II] is used.

The distribution of the wind forces along the height [z] of the shell is given as a function of the diameter [D], by the equations (z, D in [m], F_W in [kN/m]):

$z \leq 2{,}00\text{m} \quad : F_W = 0{,}51 \cdot D$

$z > 2{,}00\text{m} \quad : F_W = 0{,}013 \cdot \ln(20 \cdot z) \cdot [\ln(20 \cdot z) + 7] \cdot D$

where $D = -0{,}01775 \cdot z + 4{,}30266$ (variation of tower diameter along the height)

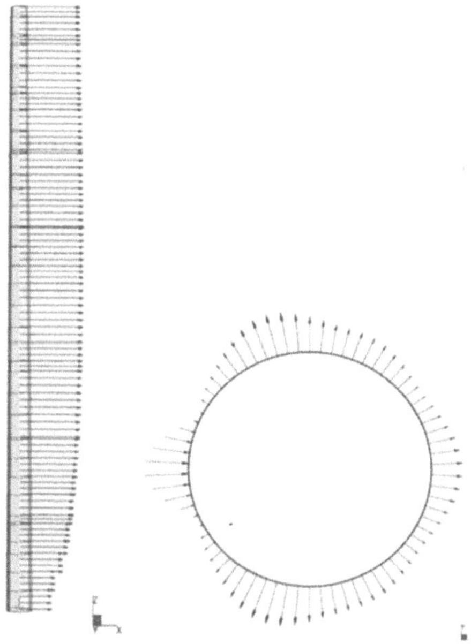

Figure 1. Wind load distribution along the tower height and around the circumference

The distribution of the wind forces along the circumference of the shell are expressed according to EC1-4:

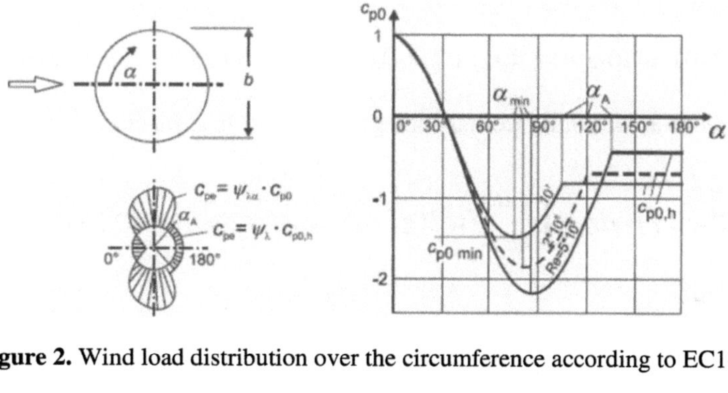

Figure 2. Wind load distribution over the circumference according to EC1-4.

1.3 Hand calculation

The tower is actually a cantilever beam (see Figure 3). Since this is a statically determinate system, the shear force and the bending moment at every point could be easily calculated by hand. Calculation of tower deformations is quite difficult to be performed by hand, due to the varying moment of inertia along the height of the tower.

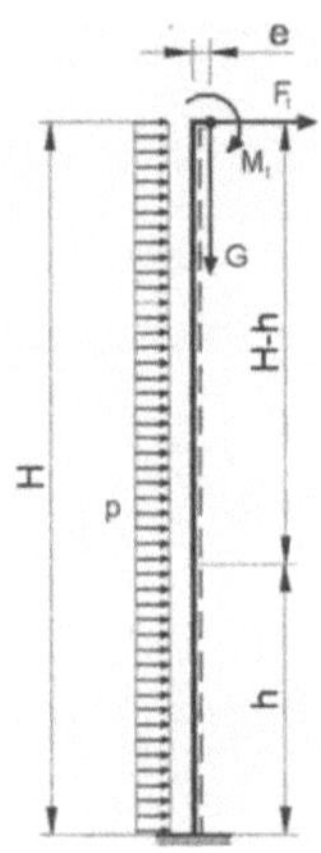

Figure 3. Simplified structural model

For the self-weight, the reactions V,M at the base are:

Load eccentricity: e_x=0,725 m, e_z=+0,50 m

V=0,0 kN

M=1.067•0,725=773,60 kNm (due to the eccentricity of the load)

The total weight of the tower (including flanges & stiffeners) is: 1.422 kN

Wind loads on the tower top are producing the following reactions:

V=598,74 kN , M=1.665,41+598,74*(76,15+e_z)=47.558,83 kNm

The corresponding wind forces over the height are (as above):

Fw1(z) = 0,51•(-0,01775•z+4,30266) , z < 2,0 m

Fw2(z) = 0,013n(20•z)•[ln(20•z) + 7]•(-0,01775•z+4,30266) , z >= 2,0 m

Base reactions are resulting from the integration of the load functions:

$$V = \int_{z=0}^{2.0} Fw1(z) \cdot dz + \int_{z=2}^{76.15} Fw2(z) \cdot dz = 4{,}37 + 298{,}19 = 302{,}56 \text{ kN}$$

$$M = \int_{z=0}^{2.0} Fw1(z) \cdot z \cdot dz + \int_{z=2}^{76.15} Fw2(z) \cdot z \cdot dz$$ =4,36+12.023=12.027,36 kNm

The total wind forces at the tower base are:

N=1422,00+1.067,00=2.489,00 kN

V=598,74+302,56=901,30 kN

M=47.558,83+12027.36=59.587 kNm

Similarly, for the load combination: [G+1.50·W] we have:

N=2489,00 kN

V=0,00+1,5• 901,3=1.351.95 kN

M=1,5•59.586,19-773,58=88.606 kNm

1.4 Developing a linear model for the tower

It is quite easy to develop such a model for the tower using linear beam elements. As the tower is divided into courses with different thickness, each course is modeled as a linear beam element having as cross-section, the cross-section of the tower at the specific height. In the case that the tower base is considered as clamped, the foundation should be designed separately using the section forces at the tower base. It is suggested to include also the foundation to the linear model as a grid of linear elements with elastic support to the ground.

1.5 Developing a complete FE model for the tower

The basic concepts on developing an analytic Finite Element model for the tower is to be described. The present structural design is based on non-linear analysis [GMNIA] for the wind loading and on spectral response analysis for the seismic loading. The engineering software used is [Strand7 / Straus7].

The overall FE tower model (see Figure 4) is composed by the shell, the intermediate flanges and the embedded to the foundation skirt. The reinforced concrete foundation is also modeled.

The structural model becomes denser in the vicinity of the flanges, the door opening and the base ring, in order to more accurately describe the local concentration of the stresses. For a more efficient use of the FE software, a cylindrical coordinate system is used.

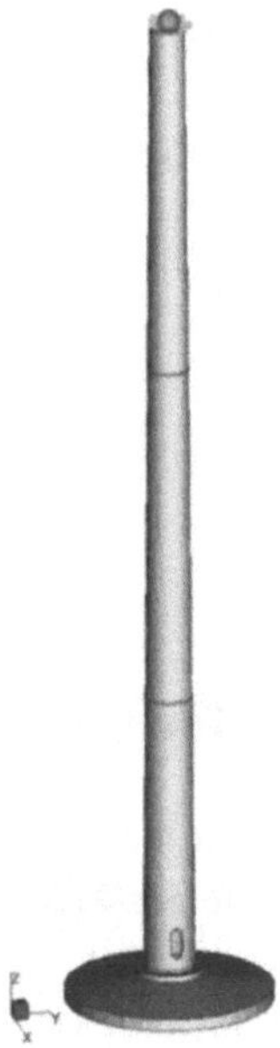

Figure 4. The FEM model

The shell is divided along the height into skirts of varying thicknesses, each of which constitutes an individual FE group. The element arrangement along the circumference is determined by the number of bolts of the connection flanges. Doing so, there is a node at each bolt position.

The intermediate flanges (see Figure 5) are modeled by the use of brick elements. The interfaces of the flanges are connected by means of frictional unilateral contact elements, active only in compression. On bolt positions, the upper node of the upper flange is attached to the lower node of the lower flange (Figure 5, points 1 & 2), via the prestressed linear elements of cable type, active only in tension. Especially for the top flange, the contribution of the nacelle to the horizontal rigidness of the section is achieved by the introduction of master – slave links, converging at the center of the circle.

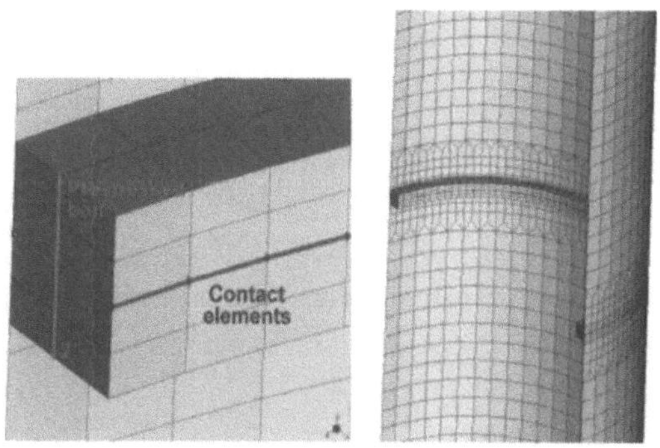

Figure 5. Detail of the flange positions

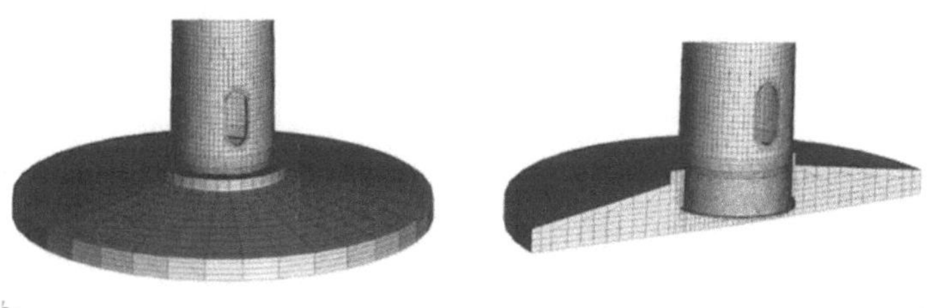

Figure 6. Details of the foundation

The foundation is modeled together with the tower body (Figure 6). The whole system is assumed to be elastically supported to the foundation base, taking account of the soil-structure interaction. The foundation has been introduced by means of brick elements, elastically supported to the ground, through unilateral contact and friction conditions.

The model will be analyzed using GMNIA. (Geometric & Material Non-linear - with global and local Imperfections - Analysis).

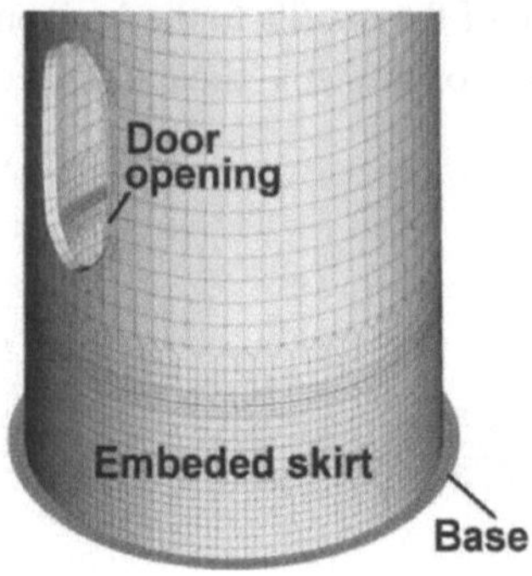

Figure 7. Detail to the door and the embeded skirt

1.6 Analysis of the tower

1.6.1 Serviceability check

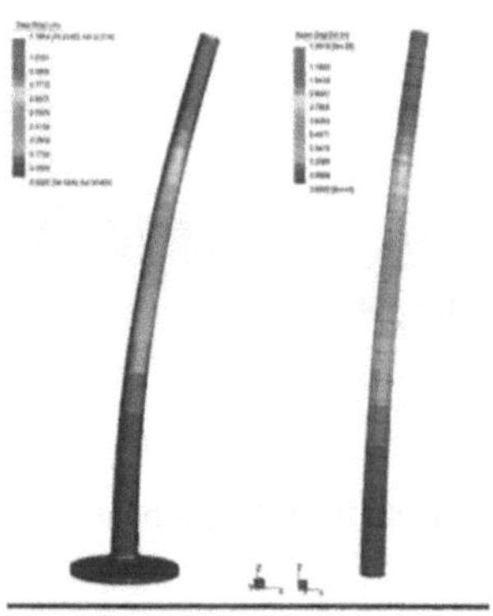

Figure 8. Tower displacements – FE and linear model

The tower is a flexible structure. The total allowed horizontal displacements on the top of the tower are usually prescribed by the manufacturer of the wind turbine in order not to damage it. In the example, for the load combination [G+W] are (see Figure 8.):

FE model: dx=1,195 m , Linear model: dx=1,391 m

On the other hand, of great importance on the SLS limit state is to check the percentage of the uplift of the foundation due to the extreme wind or seismic loading. As an example, for the above mentioned combination, an uplift of about 20% of the foundation was recorded (Figure 9), which leads to a maximum vertical displacement of 0,3mm.

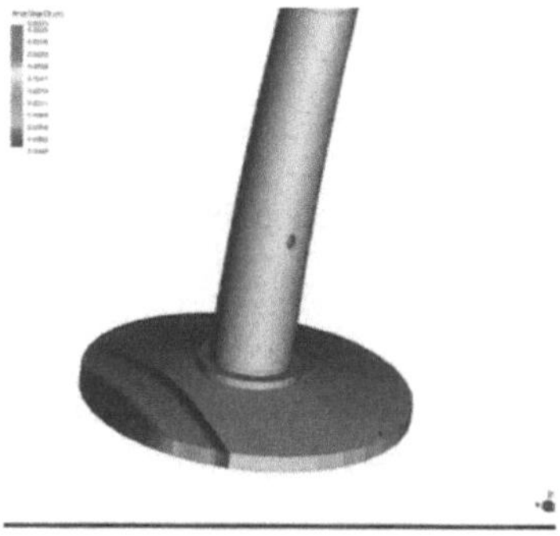

Figure 9. Uplift of the foundation

1.6.2 Section forces to the tower

The cross sectional [N], [V], [M] diagrams are presented for the Linear model (see Figure 10). At the tower base the resultant values for the load combination [G+1,50·W] are:

N=2.405,00 kN , V=1.346,74 kN, M=88.798,36 kNm

It worhts to be mentioned that the values above are a close match to the ones calculated by hand on the engineering model.

For the Finite Element model it is not feasible to obtain directly the corresponding diagrams. For the sake of comparison, the values of [N], [V], [M] have been assessed at the tower base by integration of the resulting stresses:

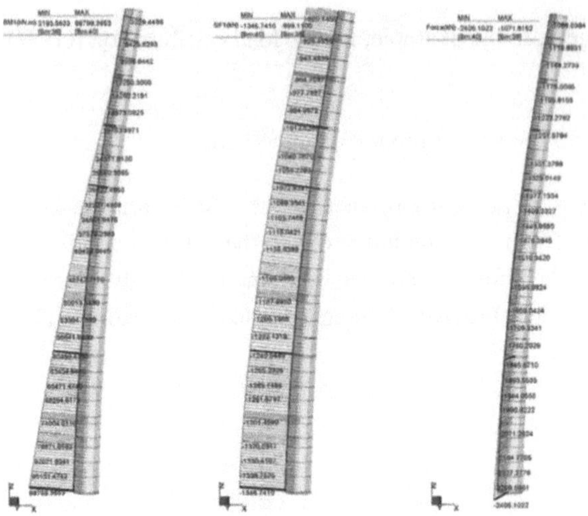

Figure 10. Section forces to the Linear model

$$N= \int_{\theta=0}^{2\bullet\pi} \sigma z \cdot t \cdot R \cdot d\theta = 2.453,40 \text{ kN}$$

$$V= \int_{\theta=0}^{2\bullet\pi} \sigma x \cdot t \cdot R \cdot d\theta = 1342.25 \text{ kN}$$

$$M= \int_{\theta=0}^{2\bullet\pi} \sigma z \cdot t \cdot x \cdot R \cdot d\theta \ = 88731.22 \text{ kNm}$$

1.6.3 Limit state design according to EC3-1-4

Having on hand the analysis results, the tower must be designed for the following limit states [LS1] (plastic) & [LS3] (buckling). For the plastic limit state, the maximum von Mises stress at any point of the tower is compared to the yield limit of the steel. Buckling limit state [LS3] will be examined in detail on Ch 3.

In our example, high strength steel is used to the shell and flanges of the tower (fy=355 MPa). According to the the Finite Element model results, the maximum von Mises stress on the tower shell has been found equal to 348 MPa, at the vicinity of the door. For the main body of the tower, the maximum von Mises stress is lower, reaching 293 MPa. It is worth mentioning that there is an almost uniform distribution along the 2/3 of the tower height.

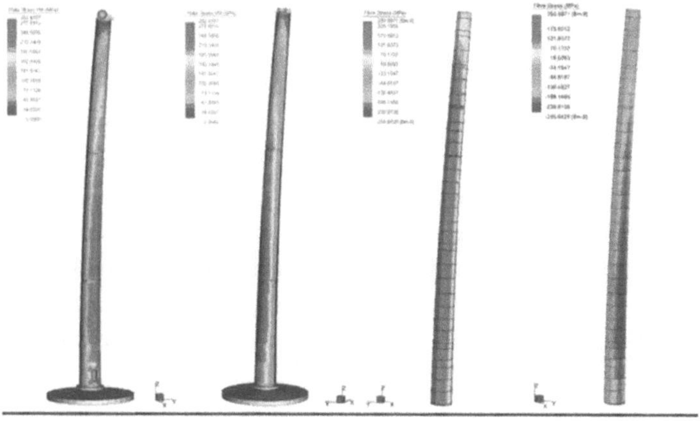

Figure 11. Vm stresses: Fe model (left), Linear model (right)

Regarding the Linear Model, the corresponding maximum stress is 251 MPa. As shown in Figure 11 there is a uniform distribution of the stresses for the two lower courses of the tower, excluding the two first at the bottom, which have been stiffened because of the presence of the door.

1.7 Conclusive remarks

The methods used for the evaluation of the shell thickness were:

- Linear static analysis (LA)
- Static GMNA & GMNIA analysis

Linear model (as well as the calculation by hand, given that the wind tower is a statically determinate structure) are sufficient for estimating the forces on the tower, but they are inadequate to acess the total stress state on the tower shell and in detail, especially because of the complicated distribution of the wind force over the tower circumference. Stress state on the tower details may be evaluated by the use of additional Finite Element models together with hand calculation or a linear model, but in this case the boundary conditions on the additional FE models need to be estimated. A complete FE model for the tower may need more computation effort, but it provides a complete stress state, when accomplished with a simplified model and hand-calculation for the check of the results.

Based on the previously presented analyses, the two lowest sections of the tower are mainly designed against the plastic limit state, having a uniform distribution of maximum von Mises stress. The lower courses (t=30 mm) seem to be ovedesigned considering the tower as a linear model, or a FE model without openings, but their dimensioning is governed by the local stress concentration on the shell around the door. The top part of the tower is mainly designed for buckling. Top course is usually stiffer in order to provide smooth transfer of the top forces from the upper flange to the tower body.

2. Seismic design

2.1 General

Wind action is usually the dominant loading for the design of a wind turbine. However, the turbine as it is a flexible structure having a big height, and the mass of the rotor system and the blades are concentrated at the top, it forms an inverse pendulum. When designing the tower for an area with seismic risc, the tower must be designed for the seismic loads as well.

2.2 Fundamental requirements

Steel towers designed to withstand the actions of the earthquake shall conform to the following European Codes:

EC 8-1: Design of structures for earthquake resistance – General rules, seismic actions and rules for buildings [10]

EC 8-6: Design of structures for earthquake resistance – Towers, masts and chimneys [11]

The relevant clauses of [GL Wind 2003 IV – Part1] [14] may also be taken into consideration.

Both requirements of [EC 8-1 §2.1], no–collapse (ultimate limit state) and damage limitation must be met. The later is considered to have been satisfied if, under a seismic action having a larger probability of occurrence than the design seismic action corresponding to the "no-collapse requirement", the displacements, as calculated according to [EC 8-1 §4.4.3], are limited, in order to prevent permanent damage of the equipment. If no specific information is available by the relevant National Annexes, the owner or the rotor supplier, a reduction factor: ν=0,5 for structures of high importance is recommended by the above mentioned Code. It must be noted though that in practice, a wind turbine tower designed according to the relevant set of Eurocodes (limitation of the 2nd order effects, competent fatigue assessment, consideration of the aeroelastic actions, adequate distance of the natural frequencies of the tower from the excitation frequencies of the rotor blades etc.), no special restrictions to the magnitude of the displacements are usually needed.

2.3 Methods of analysis

The seismic actions should be determined on the basis of the linear-elastic behavior of the structure, whereas the reference method should be the modal response spectrum analysis. As regards the alternative approaches mentioned in [EC 8-1 §4.3.3], it is remarked that:

Lateral force method of analysis in not applicable in the case of the wind towers, since the contributions from modes of vibration higher than the fundamental are always significant.

Non-linear time history (dynamic) analysis is predicated on artificial, recorded or simulated accelerograms, which are not always available, at least to the required degree of accuracy and completeness. This method though can be used supplementary as a verification of the modal response spectrum analysis and for the investigation of the response of the structural elements with unilateral behavior (preloaded base anchors, footings susceptible to local uplift etc.).

Non-linear static (pushover) analysis may be applied only for verification purposes (see [EC 8-1 §4.3.3.4.2.1].

2.4 Earthquake motion representation

The earthquake motion is represented by an elastic ground acceleration response spectrum, defined as "elastic response spectrum". As an alternative, the "design spectrum" may be used, when the capacity of the structure to dissipate energy, through mainly ductile behavior of its elements and / or other mechanisms, is taken into account. It must be noted though that the required compliance with the additional rules for dissipative behavior of [EC 8-1 §6.9] [10] concerning inverted pendulum structures is rather problematic, since the procedure indicated in [§6.5.5-1] for the estimation of the seismic action magnifying factor [Ω] makes no sense in the case of single column structures, in contradiction with the relevant provisions of the previous edition of the Code [preEN 1998-1-3:2000 §3.9-3]. At any rate, this does not seem to be a design issue, since the wind as a rule is the dominant load case [11]. This statement is demonstrated by a working example, given in the Table 1, where the relevant load combination assumptions are as follows:

Earthquake data: a=0,36, TB =0.15, T C=0,60, γI=1.40, ζ=2%

Table 1. Displacements & base forces for wind & seismic loading

No	Description	G+W	G+1.5W	G+E
1.	f,top (mm)	845		687
2.	V,Base (kN)		1,051	665
3.	M,Base (kNm)		67,652	38,080

Therefore, this fact allows the adoption of the elastic response spectrum for the seismic excitation, where no ductility requirements are involved. However, the seismic analysis should be carried out in any case, especially to areas with extreme seismic data (high risk seismic zone, weak soil, etc).

In chapters 2.5.1 & 2.5.2, the parameters involved to the above mentioned translational spectra, as given by [EC 8-1 §3.2.2], are described. Regarding the rotational component of the ground motion, [EC 8-6 §3.1] states that it should be taken into account for tall structures in regions of high seismicity, as specified by the National Annexes. The suggested by the Code field of application includes the following types of structures and site characteristics:

[§3.1]: Structures taller than 80m in regions where the peak ground acceleration times the soil factor exceeds the 25% of the gravity acceleration: $a_g \bullet S \geq 0{,}25 \bullet g$.

[§4.2.5]: Structures with height greater than five times the maximum base dimension. Obviously this criterion does not apply to pile foundations.

The method for the quantification of rotational spectra components is described in [EC 8-6 Annex A]. As a justified simplification, the vertical component [$R_z^\theta(T)$] may be disregarded, while the horizontal [$R_h^\theta(T)$] could conservatively be combined with the transitional component (SRSS procedure – see 2.6) by means of the relation:

$$\sqrt{S_e(T)^2 + \left(R_d^\theta(T) \bullet h_r\right)^2}$$ where [hr] is the hub height

With the exception of towers erected on rock or very stiff soil, the structural model of the steel shell must be built together with the one of the foundation, so that the 2nd order effects and the contribution of the soil–structure interaction are properly taken into account (see [EC 8-1 §4.3.1] & [EC 8-6 §4.2.5] [10][11]).

2.5 Response spectra

2.5.1 Elastic response spectrum

The shape of the elastic response spectrum [Se(T)] is presented in Figure 12. The horizontal translation components of the seismic action are determined by the formulas:

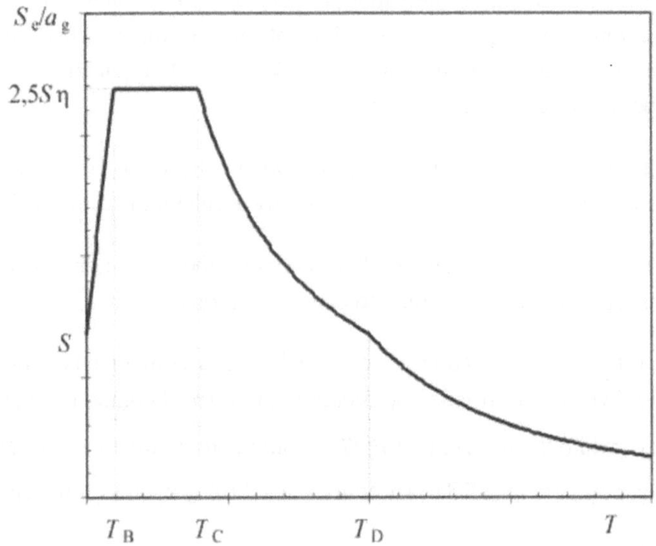

Figure 12. Elastic response spectrum

$$S_e(T) = \alpha_g \bullet S \bullet \left[1 + \frac{T}{T_B} \bullet (\eta \bullet \beta_o - 1)\right]$$ when: 0≤T<TB

$$S_e(T) = \alpha_g \bullet S \bullet \eta \bullet \beta_o$$ when: TB≤T≤TC

$$S_e(T) = \alpha_g \bullet S \bullet \eta \bullet \beta_o \bullet \left(\frac{T_C}{T}\right)$$ when: TC≤T≤TD

$$S_e(T) = \alpha_g \bullet S \bullet \eta \bullet \beta_o \bullet \left(\frac{T_C \bullet T_D}{T^2} \right) \qquad \text{when: } TD \leq T \leq 4s$$

The damping correction factor [η] and the components of the horizontal design ground acceleration [α_g] can be derived by the relations:

$$\eta = \sqrt{\frac{10}{5+\xi}} \geq 0,55 \quad \alpha_g = \gamma_I \bullet \alpha_g R$$

The values of the factors involved in the formulas, are:

α_gR: Reference peak ground acceleration on type [A] ground, specified by the National Annexes.

γ_I: Importance factor. For the electric power plants it is recommended by [EC 8-6 §4.1]: γI =1,40.

β_o: Coefficient of spectral amplification: β_0=2,50

ξ: Critical damping factor ratio. For welded steel structures the value: ζ=2% is adopted as a rule.

S: Soil factor, given in Tables [3.2] & [3.3] of [EC 8-1] for the specific ground type as classified in Table [3.1] and for the spectra type, depending on the surface–wave magnitude of the earthquakes that contribute most to the seismic hazard defined for the site.

T: Vibration period of a linear single-degree-of-freedom system.

TB, TC, TD: Characteristic values of spectrum, defined in the same way as the soil factor [S].

The vertical component of the seismic action is determined by the same expressions, with the following modifications:

S = 1,00 β_o = 3,00

TB = 0,05s TC = 0,15s TD = 1,05s

The horizontal design ground acceleration [α_g] is replaced by the vertical [α_{vg}], as given in Table [3.4] of [EC 8-1].

2.5.2 Design spectrum

The horizontal translation components of the design spectrum [Sd(T)] are defined by the following expressions:

$$S_d(T) = \alpha_g \bullet S \bullet \left[\frac{2}{3} + \frac{T}{T_B} \bullet \left(\frac{2,5}{q} - \frac{2}{3}\right)\right] \quad \text{when: } 0 \le T < TB$$

$$S_d(T) = \alpha_g \bullet S \bullet \frac{2,5}{q} \quad \text{when: } TB \le T \le TC$$

$$S_d(T) = \alpha_g \bullet S \bullet \frac{2,5}{q} \bullet \left(\frac{T_C}{T}\right) \ge \beta \bullet a_g \quad \text{when: } TC \le T \le TD$$

$$S_d(T) = \alpha_g \bullet S \bullet \frac{2,5}{q} \bullet \left(\frac{T_C \bullet T_D}{T^2}\right) \ge \beta \bullet a_g \quad \text{when: } TD \le T \le 4s$$

It is noted that:

The parameters [a_g], [S], [TC] & [TD] are as defined in §2.5.1.

The lower bound factor [β] may be taken as equal to 0,2, when not determined by the National Annex.

The behavior (shape) factor [q] can be assigned with the value: q = 1,50 (see [EC 8-6 §4.7.5, §4.10 & 6.1]), relevant to the cross–sectional Class [4] of the tower shell, according to the categorization of [EC 3-1-1 Table 5.2].

For the vertical component of the seismic action, the design spectrum is given by the same expressions, but with the following modifications:

S = 1,00

TB = 0,05s TC = 0,15s TD = 1,05s

The horizontal design ground acceleration [αg] is replaced by the vertical [α_{vg}], as given in Table [3.4] of [EC 8-1].

2.6 Combinations of seismic actions

The sum of the effective modal masses for the modes taken into account should amount to at least 90% of the total mass of the structure. The combination of the values of the seismic action effects for the governing modal responses will be assessed according to [EC 8-6 §4.3.3.3] and [EC 8-1 §4.3.3.3.2], by applying the SRSS method (square root of sum of squares), provided that the periods in any two vibration modes may be taken as independent of each other ($T_{min} \leq 0{,}9 \cdot T_{max}$). If the above term is not satisfied, more accurate procedures for the combination of the modal maxima, such as the CQC (Complete Quadratic Combination) should be adopted.

The effects of any rotational component of the ground motion may be combined, if significant, with those of the translational component via the SRSS procedure, since they are not generally in phase. Any rotational components about a horizontal direction should first be combined with those of the translational ones in the orthogonal horizontal direction.

The horizontal [EEdx], [EEdy] and the vertical [EEdz] components of the seismic excitation may be combined in respect with the stipulations of [EC 8-1 §4.3.3.5]:

$$\pm 1{,}0 \cdot EEdx \pm 0{,}3 \cdot EEdy \pm 0{,}3 \cdot EEdz$$

$$\pm 0{,}3 \cdot EEdx \pm 1{,}0 \cdot EEdy \pm 0{,}3 \cdot EEdz$$

$$\pm 0{,}3 \cdot EEdx \pm 0{,}3 \cdot EEdy \pm 1{,}0 \cdot EEdz$$

Regarding the combinations of the seismic action with other actions, it is considered that, instead of the general purpose approach of [EC 8-6 §4.5] & [EC 8-1 §3.2.4], the more specialized for wind towers provisions of [GL Wind 2003 IV – Part1 §4.4.3.3] should be applied.

2.7 Seismic design example

The tower of the previous presented example will be analyzed for seismic loading. At first, an eigenvalue analysis has to be performed. Using the results of the eigenvalue analysis, and the design spectrum, a response spectrum analysis follows for the evaluation of the stress-state of the tower for the earthquake loads.

2.7.1 Eigenvalue analysis

The governing eigenfrequencies for each model and for mass at the tower top are presented in Table 2 and Figures 13 & 14:

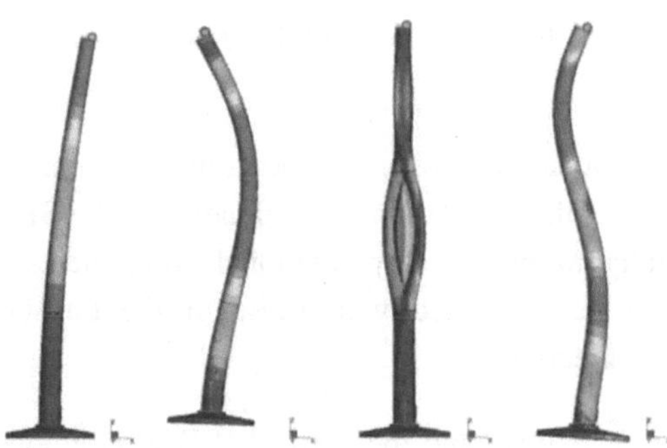

Figure 13. FE Model: Mode shapes

Figure 14. Linear Model: Mode shapes

Table 2. Eigenfrequencies (FE model & Linear model)				
Dynamic characteristics	FE Model		Linear Model	
	k=15400	k=14950	k=15400	k=14950
1st eigen frequency	0,357	0,357	0,324	0,322
3rd eigen frequency	2,820	2,822	2,626	2,610
9 th eigen frequency	7,520	7,490	7,850	7,833

2.7.2 Response Spectrum analysis

The Response Spectrum analysis for the tower example, has been performed, as presented above, for the following parameters:

a=0.24 , TB=0.10 , TC=0.40 , γI=1.40 , ζ=2%

For the assessment of the effects due to the combination of the components of the seismic action, the SRSS procedure was adopted.

The total participating mass, not considering the contribution of the foundation, is about 93% in all directions. The governing eigenvalues are as follows:

Direction X: 1st: 61,7% 3rd: 14,9% 5th: 4,9%

Direction Y: 2nd: 67,2%, 4th: 16,3%, 6th: 6,0%

The maximum seismic displacement at the top of the tower is 0,53m. The total shear [V] and overturning moment [M] at the tower base are:

V=499,33 kN , M=25.449,32 kNm

which correspond to approximately 30% of the relevant values of the extreme wind load combination.

The dominant stress components for the seismic loading are the meridional ones.

The critical loading for the tower design is the extreme wind loading. In the example above, the seismic loading leads to a stress-state of about 30% of the corresponding one extreme wind loading.

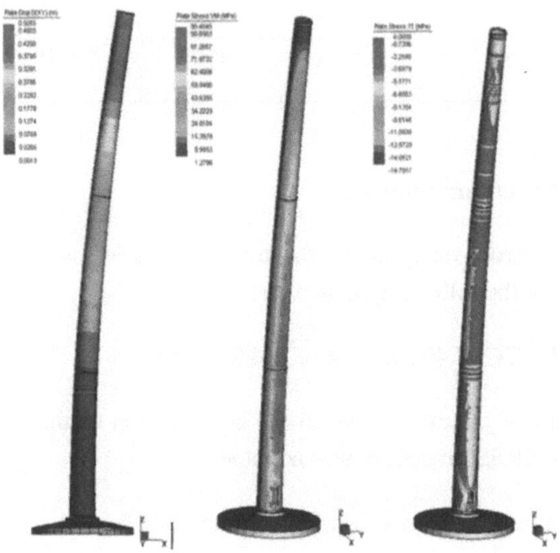

Figure 15. Displacements, von Mises stresses and compressive circumferential stresses (seismic loading – spectrum analysis)

3. Buckling capacity of the tower

3.1 Geometrical imperfections. [EC 3-1-6] specifications

The effects of geometrical imperfections on the tower are being investigated, as proposed by EC-3-1-6 by means of linear buckling analysis (LBA) and geometrical & material non-linear analysis with imperfections (GMNIA). Finally, the design of the tower by the three methods proposed by EC3-1-6 is attempted.

As proposed by EC3-1-6, there are three fabrication classes in the production:

Class A: Excellent

Class B: High

Class C: Normal

The main types of imperfections proposed by EC3-1-6 are classified as follows:

3.1.1 Out-of-roundness tolerance

The out-of-roundness tolerance is determined in terms of the parameter Ur:

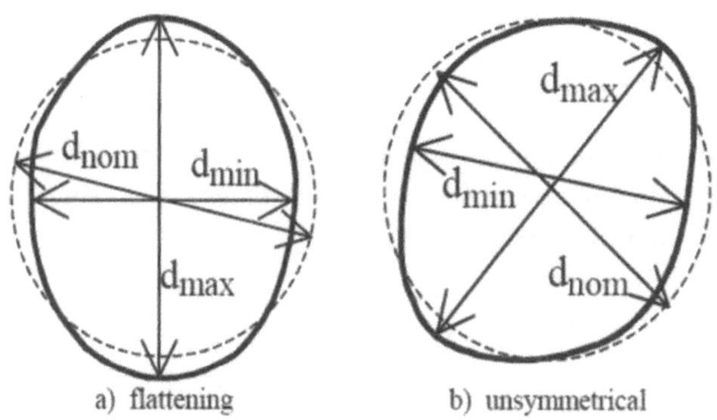

Figure 16. Out-of-roundness tolerance

$$Ur = \frac{d\max - d\min}{dnom}$$

where:

dmax and dmin are the maximum and minimum measured internal diameters dnom is the nominal internal diameter

The recommended values for the of-roundness tolerances are given in Table 3 derived by [EN 3-1-6 Table 8.1].

Table 3. Out-of-roundness tolerance parameter Ur,max

Fabrication tolerance class	Diam. range	$d \leq 0{,}50m$	$0{,}50m \leq d \leq 1{,}25m$	$1{,}25m \leq d$
	Description	Value of Ur,max		
Class A	Excellent	0,014	0,007+0,0093·1,25-d)	0,007
Class B	High	0,020	0,007+0,0133·1,25-d)	0,10
Class C	Normal	0,030	0,007+0,0300·1,25-d)	0,015

3.1.2 Dimple tolerances

A dimple measurement gauge should be used in every position for both the meridional and circumferential directions (see Figure 17.). The dimple tolerance parameter is given by the formula:

$$U_{0,\max} = \frac{\Delta_{w0}}{l_g}$$

The recommended values for the relevant fabrication classes are given in Table 4 as derived by [EN 3-1-6 Table 8.4]:

Table 4. Dimple tolerance parameter $U_{r,max}$		
Fabrication tolerance quality class	Description	Value of $U_{0,max}$
Class A	Excellent	0,006
Class B	High	0,010
Class C	Normal	0,016

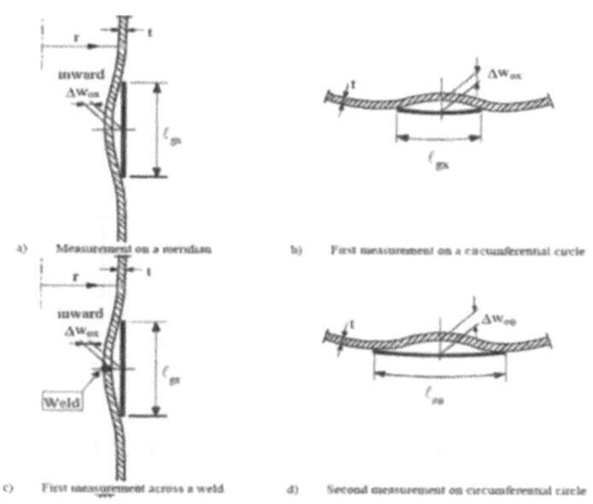

Figure 17. Dimple tolerances

3.1.3 Accidental eccentricity imperfections

Those are imperfections in which two jointed parts are placed accidentally so their theoretical mid-plane axes are not continuous as in the theoretical model, but having a specific eccentricity (see Figure 18). The values for the maximum allowable eccentricity for each fabrication class are given in Table 5, as derived by [EN 3-1-6 Table 8.2]:

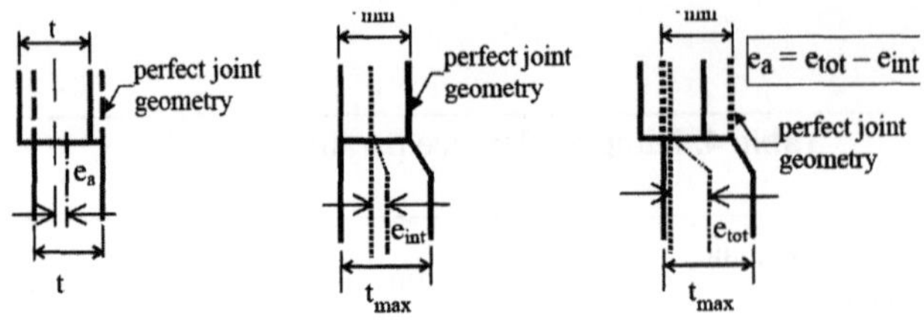

Figure 18. Accidental eccentricity and intended offset at a joint

Table 5. Accidental eccentricities $e_{a,max}$		
Fabrication tolerance quality class	Description	Maximum permitted accidental eccentricity
Class A	Excellent	$e_a \leq 2mm$
Class B	High	$e_a \leq 3mm$
Class C	Normal	$e_a \leq 4mm$

3.2 Impact of the imperfections to the resistance of the tower

The investigation of the effects of the various types of imperfections to the resistance of the tower requires the introduction of the imperfect geometry to the calculation model. To this end, the steps below have been followed :

- An analytical Finite Element model for the whole structure (perfect geometry) was built.
- Buckling analysis for the perfect geometry model was performed
- The fabrication tolerances to the Finite Element model, leading thus to an imperfect geometry model were introduced

- Buckling analysis for the imperfect geometry model was performed
- The analysis results were compared to each other

3.3 Buckling analysis for the perfect shell

A linear buckling analysis (LBA) is performed to the perfect shell for the extreme wind load combination [1,0·G+1,50·W]. The first 10 buckling eigenvalues are calculated, as presented in Figure 19. The linear buckling eigenvalues represent the factors [r_{Rcr}] determined by the expression:

$$r_{Rcr} = FRk / FEd$$

where [F_{Ed}] represents the design loads and [F_{Ed}] the characteristic buckling resistance, at the bifurcation point.

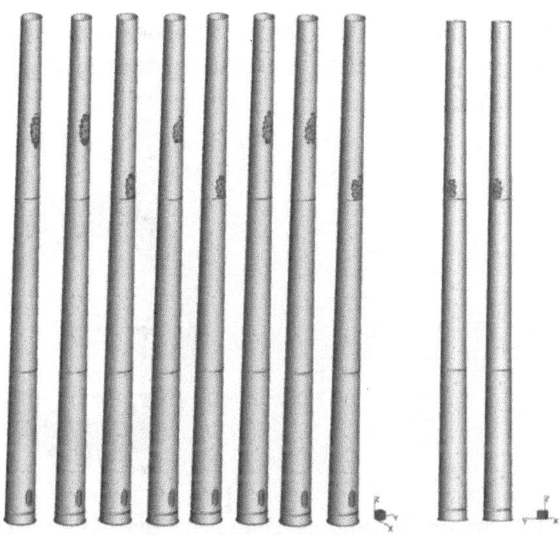

Figure 19. Eigenmodes [1]÷[5], [7], [8], [10] (positive) & [6], [9] (negative)

All buckling eigenvalues are located to the upper part of the tower. Eigenvalues [6] & [9] are negative, and therefore, they correspond to the reverse load direction.

3.4. Introduction of the imperfections to the FE model

3.4.1 Out of-roundness imperfections

The introduction of the out-of-roundness imperfections to the Finite Element model is achieved by the enforcement of the nodes of the perfect geometry structure to the desired positions, by means of appropriately defined constraints. At a next stage, the nodal displacements are added to the initial node coordinates, producing thus the imperfect shape of the shell.

For the specific case at hand, the circular cross-sections of the tower have to be ovalized, that means that they must be transformed into egg-type shapes, having the maximum diameter perpendicular to the load direction, so that the maximum radius of curvature on the expectant buckling position of the shell is attained.

The application of a sinusoidal variated constraint to all nodes of the shell has been adopted as the most suitable method to accomplish the above deformation of the cross-sections. This constraint function forces the nodes to reach the minimum axis of the section [dmin] at positions θ=0° & θ=180° and the maximum axis [dmax] at θ=90° & θ=270° (see Figure 20).

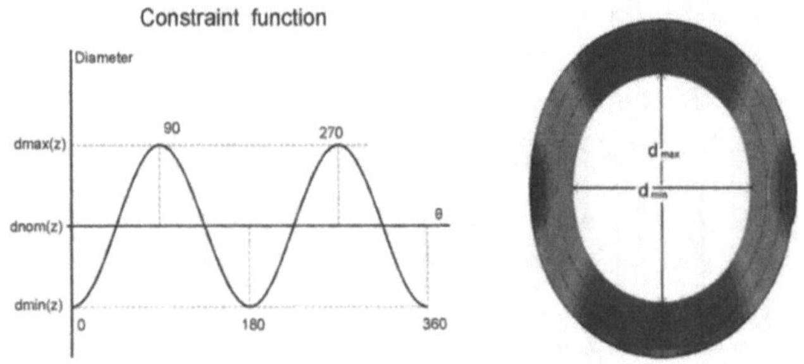

Figure 20. Constraint function and deformed shape of the tower section

Four computational models have been built for the comparison:

- Perfect shell
- Imperfect shell – Class A fabrication tolerances

- Imperfect shell – Class B fabrication tolerances
- Imperfect shell – Class C fabrication tolerances

Table 6. Buckling eigenvalues [r_{Rcr}]										
Eigenmode	[1]	[2]	[3]	[4]	[5]	[6]	[7]	[8]	[9]	[10]
Eigenvalue	3.44	3.53	3.57	3.60	3.62	-3.65	3.66	3.69	-3.72	3.73

A linear buckling analysis is next performed for the four models and the first 5 eigenvalues are compared to the one corresponding of the perfect shell. As it is demonstrated in Table 7, the buckling eigenvectors of the perfect and imperfect shell configurations are almost identical, the deviations being insignificant.

3.4.2 Eccentricity imperfections

Accidental eccentricity imperfections can be introduced to the Finite Element model by the application of a rigid offset to the elements. It is obvious that the application of this deformation to the entire tower shell is meaningless, since this will practically eliminate the imperfections. Therefore, a specific course (or a group of non-adjacent courses) must be selected for the application of the rigid offset to the elements. In the case under consideration, courses on which the first buckling eigenvalues appear for the perfect shell, are selected (see Figure 21).

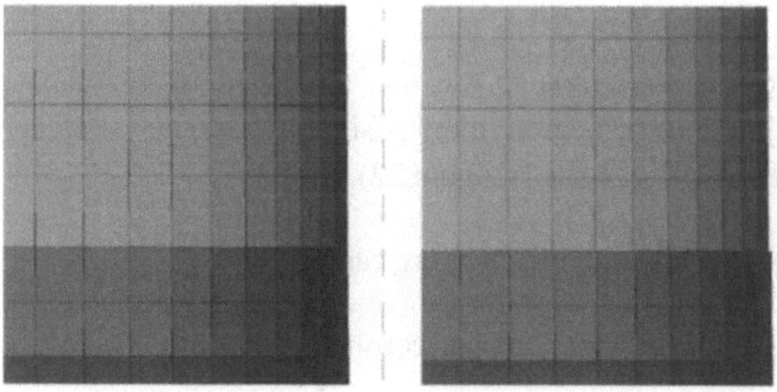

Figure 21. Perfect shell (left), Shell with accidental eccentricity (right)

As demonstrated in Table 6 there is no notable difference between the 5 first eigenvalues. By the comparison of the eigenmodes though, it is evident that, in contradiction to the restriction of the perfect model buckling within the specific course, the corresponding imperfect models present a propagation of the buckling to the neighboring courses (see Figure 23).

3.4.3 Dimple imperfections

The introduction of the dimple imperfections to the Finite Element model is a rather laborious task, due to the fact that, having by nature a random distribution, they can

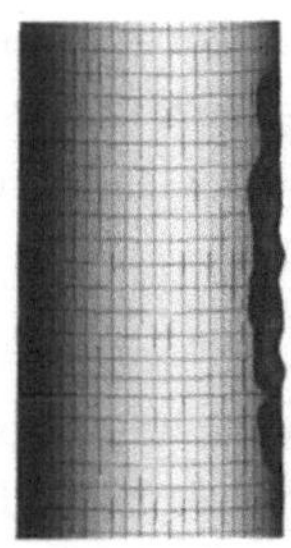

Figure 22. Dimples generated from the 1st eigenmode of the perfect shell (magnified)

appear anywhere along the tower body.

Hence, an approach analogous to the one described in §3.4.2 is performed and specifically, a dimple imperfection in the position where the 1st eigenmode occurs in the perfect shell will be generated (Figure 22).

For this scope, the following procedure are carried out:

- A buckling analysis on the perfect shell is executed
- The 1st buckling eigenmode is examined

- The length of main dimple on the 1st buckling eigenmode is measured
- The 1st buckling eigenmode is multiplied by a factor, in order to accomplish the dimple tolerance parameter suitable for each fabrication class
- Multiplied nodal displacements is added to the node coordinates
- A buckling analysis of the imperfect structure is performed

The comparison of results is given in Table 7.

Table 7. Dimple imperfections – Buckling eigenvalues [r_{Rcr}]

Eigen mode	**Perfect shell**	**Class A imperfection**	**Class B imperfection**	**Class C imperfection**
[1]	3.44	3.42	3.40	3.36
[2]	3.53	3.53	3.52	3.47
[3]	3.57	3.57	3.57	3.57
[4]	3.60	3.59	3.57	3.57
[5]	3.62	3.62	3.62	3.62

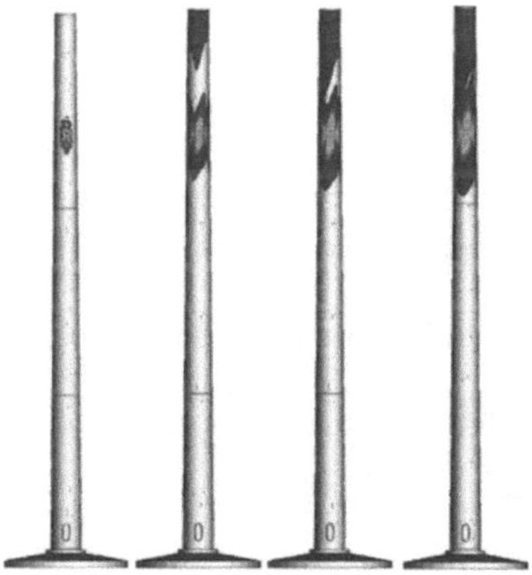

Figure 23. 1st buckling eigenmode – Perfect shell and fabrication classes A, B, C dimple imperfect shells (perfect shell & fabrication classes A,B,C)

3.5 Assessment of the limit load for the tower

The bifurcation load factor for the perfect shell is derived from the Linear Buckling Analysis (LBA) and equals to: r_{Rcr} = 3.44. Similarly, the limit load factor is determined by an iterative Geometrically and Materially Nonlinear Analysis, by means of step-by-step increments of the wind load, until the collapse of the structure. It is underlined that for the model under consideration, the unilateral contact conditions are removed and the contact of the foundation base to the ground is considered as bilateral elastic. This is an inevitable intervention, in order to prevent

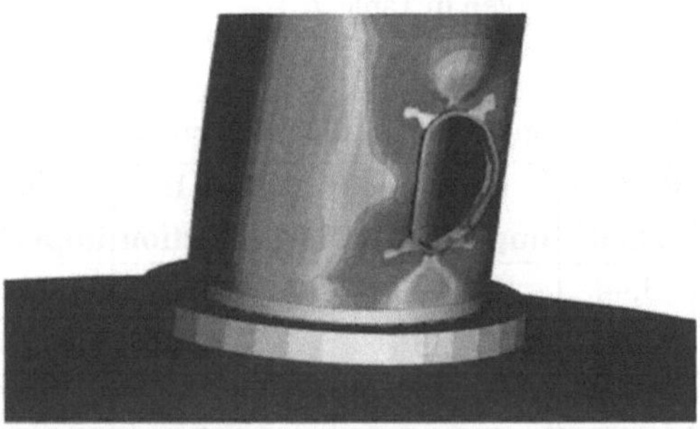

Figure 24. Shell buckling in the vicinity of the door opening (plasticized areas marked as white)

the overturning of the tower as a whole, before it reaches the buckling or plastic limit state (even for the design combination: [G+1.5·W] there is a 40% uplift of the foundation).

Having in mind that the specific approach requires a significant computational effort, only three of the various cases are investigated:

Case [1] :Shell with Class B out-of-roundness tolerances, globally to the tower

Case [2] :Shell with Class B dimple tolerances, to the location of the 1st eigenmode

Case [3] :Perfect shell

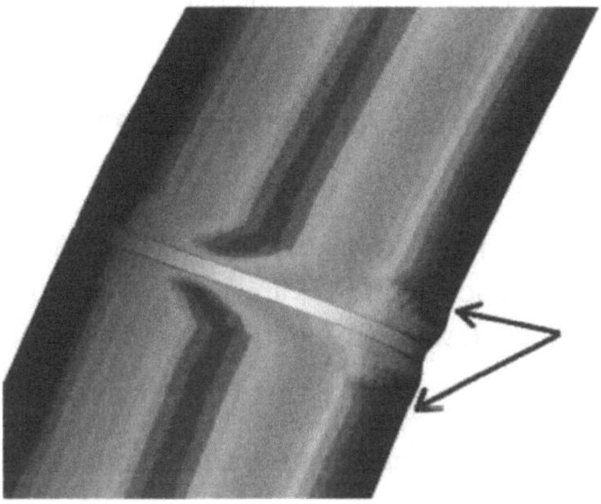

Figure 25. Shell buckling on flange position (arrows indicate the material yielding zones)

Case [1]: The procedure diverged for a characteristic resistance ratio (limit load factor): r_{Rk} = 1,95, which corresponds to the load combination: [1.0*G+2.05*W]. The failure of the structure was due to the buckling at the location of the door, induced by the excessive incremental displacements. As shown in Figures 24 , 25 the buckling of the shell takes place at the material yielding zones, when the modulus of Elasticity was significantly reduced. In other words, the shell has entered into the plastic state, before reaching the bifurcation point. In addition, shell buckling occurs at the vicinity of the flanges, due to the presence of high circumferential stresses, in conjunction with the meridional ones.

Case [2]: Despite the fact that the dimple imperfection has been introduced to the location of the 1st buckling eigenmode, the shell eventually has buckled around the door opening, for the same reason as in Case [1]. The characteristic resistance ratio was found in this case equal to: r_{Rk} = 2,05.

Case [3]: The analysis for the perfect shell results in an identical limit load factor: r_{Rk} = 2,05.

3.6 Buckling analysis according to [EC 3-1-6]

Code [EC 3-1-6] refers to three distinguished calculating procedures for the buckling assessment of the tower:

3.6.1 Design by global numerical analysis using GMNIA analysis [EC3-1-6 §8.7]

This is without doubt the most advanced design procedure suggested by the Eurocode. The imperfect elastic-plastic buckling resistance ratio [$r_{R,GMNIA}$] represents the load factor, derived as the lowest value obtained by the use of the criteria C1, C2 and C3, as follows:

Criterion C1: The maximum load factor on the load-deformation load (limit load)

Criterion C2: The bifurcation load factor, where this occurs during the loading path before reaching the limit point of the load-deformation curve

Criterion C3: The largest tolerable deformation, where this occurs during the loading path before reaching a bifurcation load or a limit load. But since no data have been supplied by the manufacturer for the investigation of [C3] and therefore this check will be omitted at this stage.

The characteristic buckling resistance ratio for the tower under consideration is found equal to: $r_{GMNIA} = 1,95$.

The calibration factor: k_{GMNIA} varies within the range: $0.8 < k_{GMNIA} < 1.2$. Since no specific data is available, the most unfavorable value is adopted: $k_{GMNIA} = 0.80$.

Substituting the values, it is verified that the design buckling resistance ratio satisfies the condition:

$$r_{Rd} = \frac{k_{GMNIA} \cdot r_{GMNIA}}{\gamma_{m1}} = \frac{0,80 \bullet 1,95}{1,1} = 1,42 > 1,00$$

3.6.2 Design by global numerical analysis using MNA and LBA analyses (as proposed by [EC3-1-6 §8.6])

The characteristic buckling resistance ratio for the perfect shell, is estimated by the [MNA] analysis: rRk = 2.05.

The overall buckling reduction factor [χov] is a function of the parameters:

$\chi_{ov} = f(\lambda_{ov}, \lambda_{ov,0}, r_{ov}, \beta_{ov}, n_{ov})$, as determined in [EC 3-1-6 Annex D §1.2.2]:

λov,0 = 0,20 βov = 0,60 nov = 1,00

$$r_{ov} = a_x = \frac{0.62}{1+(1.91/Q)^{1,44} \bullet (r/t)^{0,72}}$$

Introducing the appropriate values of the parameters, [χ] is obtained by the formulas of [EC 3-1-6 §8.5.2.2]: χov = 0,71

After substitution, it is verified that the design buckling resistance ratio satisfies the condition:

$$r_{Rd} = \frac{\chi_{ov} \bullet r_{Rk}}{\gamma_{m1}} = \frac{0,71 \bullet 2,05}{1,1} = 1,32 > 1,00$$

3.6.3 Stress design to [EC3-1-6 §8.5]

The buckling check of the shell according to the stress design method was attempted and the results of the calculations are presented in Table 8 (the combined check was not carried out).

As illustrated in this specific case, this method leads inevitably to the installation of stiffening rings. Taking for example the relevant formulas (see [EC 3-1-6 Annex D]), in the case of long cylinders:

$$\sigma_{\theta,\mathrm{Rcr}} = E \bullet \left(\frac{t}{r}\right)^2 \bullet \left[0{,}275 + 2{,}03 \bullet \left(\frac{C_\theta}{\ell}\right)^4 \bullet \frac{t^6}{r^2}\right]$$

The critical circumferential buckling stress [σ_θ,R_{cr}] is diminished for a given significant magnitude of the length between boundaries (flanges, stiffening rings or base support), resulting thus in unacceptable shell plate thicknesses.

This noteworthy deviation of the stress design method from the previously mentioned global numerical analyses should be imputed to the fact that the buckling

Table 8. Stress design

Courses	L	R_{mean}	T_{mean}	$\sigma_{\chi,Rd} / \sigma_{\chi,Ed}$	$\sigma_{\theta,Rd} / \sigma_{\theta,Ed}$	$\sigma_{\tau,Rd} / \sigma_{\tau,Ed}$
Lower	21.500	2.064	24,7	236 / 327	13,7 / 55,0	103 / 57,0
Middle	26.395	1.850	18,5	208 / 304	7,60 / 62,0	74,0 / 8,0
Upper	27.425	1.603	13,6	186 / 253	4,95 / 68,0	55,0 / 8,0

check along the boundaries is excluded only for the combined compression and not for the meridional, circumferential and shear components although, as demonstrated by the [LBA] buckling analysis, no eigenvector is reported there, even to the imperfect shell. Thus, the high intensity design stresses at the vicinity of the boundaries (see Figure 26), combined with the restricted values of the buckling resistance stresses, when no intermediate stiffeners are used, results in low strength to design load ratios of questionable reliability.

3.7 Concluding remarks

The results form the [LBA] analyses, considering elastic behavior of the structure for the perfect shell and the various types of imperfections, the influence of the imperfections, as described in [EC 3-1-6], is not significant to the buckling resistance of the tower.

On the other hand, the results from the limit load analyses indicate that the tower has been designed to reach the plastic limit load before approaching the bifurcation point. The collapse is induced by the shell buckling at the material yielding zones, where the plasticizing von Mises stresses are components of the compressive meridional and circumferential stresses.

The preferable method for the design of the tower against buckling deems to be the global numerical analysis using [LBA] and [MNA] analyses, according to [EC 3-1-6 §8.6]. In this case the imperfections are introduced indirectly, by the employment of the overall elastic imperfection factor [r_{Rov}].

The global numerical analysis using [GMNIA] analysis design according to [EC 3-1-6 §8.7] is more straightforward, but at the same time it is proved to be more tedious and requires an in-depth knowledge of the applicable imperfections and the calibration factor [k_{GMNIA}].

The stress design procedure of [EC 3-1-6 §8.5] results in rather conservative values, especially considering the circumferential stresses. The use of stiffening rings for this type of analysis is inevitable.

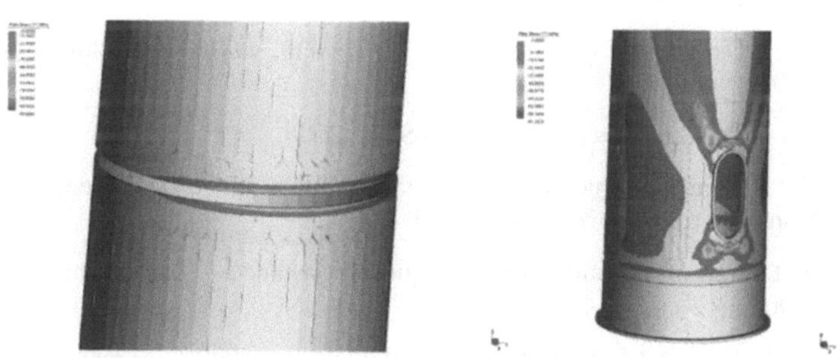

Figure 26. Compressive circumferential stresses at the vicinity of the flanges and the door opening

6. Conclusions

A wind turbine tower is designed mainly for extreme wind loading. Seismic loading must also be examined when the wind park is located in a seismic hazardous area. Although the calculation can be performed even by hand, or using simple linear computer models, an analytic FE model and advanced analysis procedures (LBA, GMNIA) give to the designer the whole picture about the stress state at every point of the structure, minimizing the assumptions that engineers need to make when transfering the actual structure into a computer model. On the other hand, the use of such advanced analyses, permits to the engineer to design the tower according to Eurocodes without the use of stiffeners, while the use of the simplified design procedure on Eurocodes leads to conservative results.

7. References

1. EN 1993-1-4: Actions on structures – General actions – Wind actions, 2005
2. EC 2-1-1: Design of concrete structures – General rules and rules for buildings, 2004
3. EN 1993-1-1: Design of steel structures – General rules and rules for buildings, 2005
4. EN 1993-1-5: Design of steel structures – Plated structural elements, 2006
5. EN 1993-1-6: Design of steel structures – General strength and stability of shell structures, 2007
6. EN 1993-1-8: Design of steel structures – Design of joints, 2005
7. EN 1993-3-1: 2006: Design of steel structures – Towers, masts and chimneys – Towers and masts, 2006
8. EN 1993-3-2: Design of steel structures – Towers, masts and chimneys – Chimneys, 2006
9. prEN 1993-4-1: Design of steel structures – Silos, 2006
10. EN 1998-1: Design of structures for earthquake resistance – General rules, seismic actions and rules for buildings, 2004
11. EN 1998-6: Design of structures for earthquake resistance – Towers, masts and chimneys, 2005
12. preEN 1998-1: Design of structures for earthquake resistance – General rules, seismic actions and rules for buildings, 2000 (withdrawn)

13. DIN 18800-4: Stuctural steelwork – Analysis of safety against buckling of shells, 1990

14. GL Wind 2003 IV – Part1: Guideline for the Certification of Wind Turbines, 2004

15. Norsok Standard N-004: Design of steel structures, 2004

16. DIBt: Guideline for Wind Energy Plants, draft version, 2004

17. ECCS-TC 13: Manual on design of Steel Structures in Seismic Zones, 1994

18. Analysis and design of a prototype of a steel 1-MW wind turbine tower - I. Lavassas , G. Nikolaidis, P.Zervas, E. Efthimiou, I.N. Doudoumis – *Engineering Structures 25 (2003)* 1097-1106

Large wind turbines in earthquake areas: structural analyses, design/construction & in-situ testing

Claudio Borri[1], Paolo Biagini[1], Enzo Marino[1]

[1] CRIACIV* c/o Department of Civil and Environmental Engineering,
University of Florence, Firenze, Italy

1 Introduction and background

The **Wind Energy Roadmap** for the European Union was published by the European Commission on Oct. 7th, 2009, in the framework of its *Communication of Financing Low Carbon Technologies*. Following its publication, the roadmap was officially presented and discussed at the Strategic Energy Technology Plan (SET-Plan) workshop, held in Stockholm on October 21st and 22nd 2009, and organised by the European Commission and the Swedish Energy Agency.

The roadmap is set to become one of the most important instruments for the development of wind power in the 2010-2020 period and will play a key-role in fighting climate change and in helping EU Member States to meet the 2020 targets identified by the new RES directive (approved in December 2008), by achieving the following goals:

- A wind energy penetration level of 20% in 2020
- Onshore wind power fully competitive in 2020
- 250.000 new skilled jobs created in the EU by the wind energy sector in the 2010 – 2020 period

The roadmap which has a total budget of € 6 bn (private and public resources) is a long term programme for increasing and coordinating the funding of wind energy R&D, so as to ensure its quick development and deployment in the EU. It focuses on the following key areas:

- New turbines and components
- Offshore technology
- Grid integration
- Resource assessment and spatial planning

* Inter-University Research Center on Building Aerodynamics and Wind Engineering.

The implementation will require a yearly investment of public and private investment of € 600 m , up from € 383 m in 2007. The roadmap is expected to be launched in 2010, after having been endorsed by the EU council. It originates from the European Wind Initiative (EWI), one of the industrial initiatives proposed by the EC in its SET-Plan, published in November 2007.

The EWI was developed by the European Wind Energy Technology Platform (TPWind) in cooperation with the European Commission and EU Member States and was therefore the result of a shared and concerted process. It was finalized in the summer of 2009 and submitted to the EC, which published as the Wind Energy Roadmap.

As requested by the SET-Plan high level Steering Group, TPWind has developed its implementation plan for the first three years (i.e. 2010 - 2012). As stated by the SET-Plan secretariat document from Dec. 14th 2009, the implementation path should cover the first 3-year period and will be revised every year, thus becoming rolling out programmes. The following elements should be included:

- Taking into account the present technology and projects financed by the EU and Member States, identifying the priority actions to move towards the objective/milestone
- Estimate the budgets, the European added-value of the actions and the risk involved by the different actions
- Identify the existing available public and private financial sources
- Identify the needed actors and potential countries to finance the actions
- Define Key Performance Indicators
- Estimate the contribution of the identified priority actions towards the 2020 objectives/milestones;
- Identify possible links with joined programmes of the European Education and Research Area (EERA).

1.1 European and world energy scenario

The worldwide energy scenario is represented in Figures 1 and 2 which show the global annual wind power installed capacity in the period 1996-2008 and the annual wind power capacity installed by region, respectively.

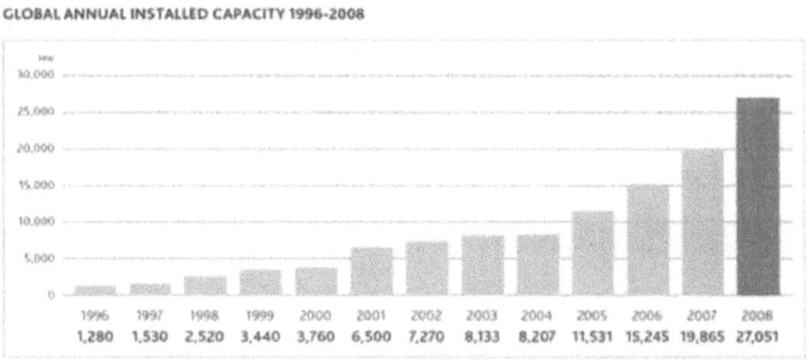

Figure 1. Global annual wind power installed capacity, 1996-2008. (Source: GWEC 2008).

Wind power is the fastest growing power generation technology in the EU with more than 35% of all new energy installations in 2008. And it is also interesting to note from Figure 2 that European wind energy installation has been leading the global installation since 2003.

Figure 3 shows that in 2008 wind energy installation was definitely dominating other energy sources. Indeed, only in this year, upon a total installed capacity of 23851 GW in Europe, approximately one third was represented by wind energy. Almost 8.9 GW of new wind turbines installed in 2008 brought European wind power generation capacity up to nearly 66 GW. Another promising sign, see Figure 4, is the diversification of the European market. 2008, in fact, saw a much more balanced expansion with not negligible contributions given by Italy, France and the UK.

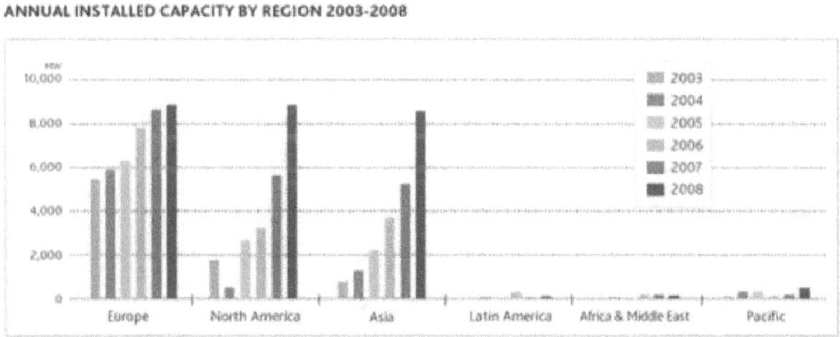

Figure 2. Annual wind power installed capacity by region, 2003-2008. (Source: GWEC 2008).

However, Germany continues to be Europe's leading market, both in terms of new and total installed capacity. To confirm this, Figure 4 shows that over 1.6 GW of new capacity was installed in 2008, brought the total German capacity up to nearly 24GW.

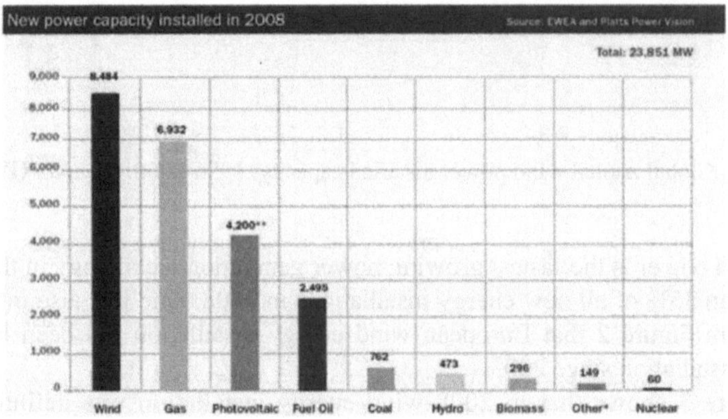

Figure 3. New power capacity installed in Europe in 2008. (Source: EWEA 2008).

As reported in the Global Wind Report (2008), it is also worth mentioning that among the growing European markets in 2008 Italy experienced a significant leap: over 1 GW of new wind turbines came on line in 2008, bringing the total installed capacity up to 3.7 GW.

A more detailed map about the cumulative installed capacity State by State is given in Figure 5

TOP 10 TOTAL INSTALLED CAPACITY 2008

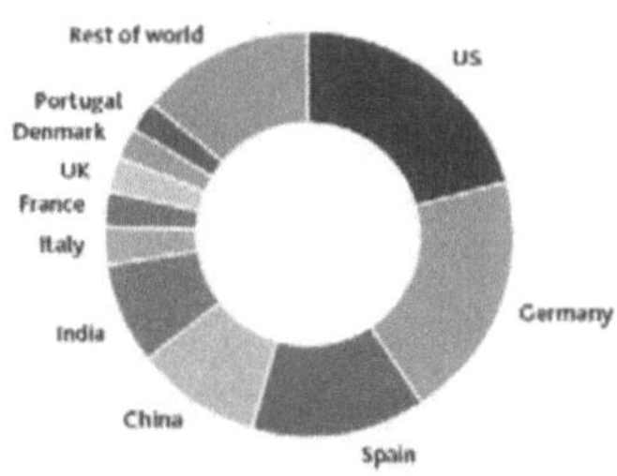

	MW	%
USA	25,170	20.8
Germany	23,903	19.8
Spain	16,754	13.9
China	12,210	10.1
India	9,645	8.0
Italy	3,736	3.1
France	3,404	2.8
UK	3,241	2.7
Denmark	3,180	2.6
Portugal	2,862	2.4
Rest of world	16,693	13.8
Total top 10	**104,104**	**86.2**
World total	**120,798**	**100.0**

TOP 10 NEW CAPACITY 2008

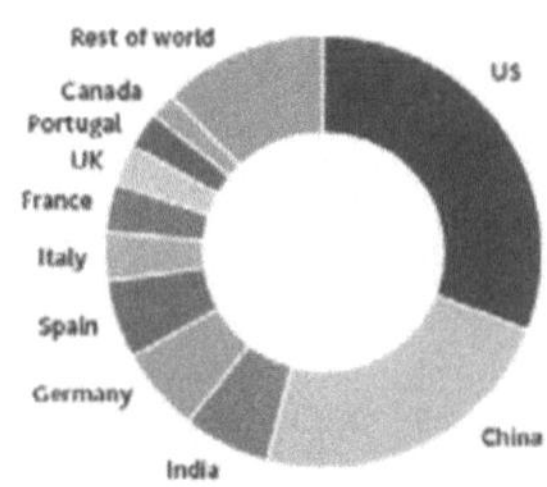

	MW	%
USA	8,358	30.9
China	6,300	23.3
India	1,800	6.7
Germany	1,665	6.2
Spain	1,609	5.9
Italy	1,010	3.7
France	950	3.5
UK	836	3.1
Portugal	712	2.6
Canada	526	1.9
Rest of world	3,285	12.2
Total top 10	**23,766**	**87.8**
World total	**27,051**	**100.0**

Figure 4. Top 10 global capacity installed, total and 2008. (Source: GWEC 2008).

Table 1. European wind power capacity.

	Installed in 2008	Cumulative, end of 2008
Total EU-27	8484MW	64 935MW
(of which Offshore)	357MW	1471MW

Figure 5. State by state cumulative installed capacity at the end of 2008. (Source: EWEA 2008).

As shown in Table 1, at the end of 2008, there were 65 GW of wind power capacity installed in the EU-27 producing 142 TWh hours of electricity which satisfies 4.2% of the whole EU electricity demand. This means that at that time offshore wind energy was able to satisfy only 0.1% of the whole EU demand. This datum makes more understandable how challenging are the targets fixed by EU shortly recalled in the next section.

In 2008 US wind industry was able to install 65 GW marking an increase in generating capacity of 50% in a single calendar year. The 2008 US growth represented about 42% of new electricity generating capacity added in the United States during the year, establishing wind as a mainstream energy source for the country, second only to natural gas in new generating capacity. US total wind generating capacity on 2008 was more than 25.17GW, producing enough electricity to power the equivalent of close to 7 million households and to meet over 1% of total US electricity demand.

1.2 Short and long term objectives

Focusing on the European situation, in the Strategic Research Agenda (SRA) - a document prepared by the Wind Technological Platform (TPWind) in 2008 - fundamental objectives in matter of wind energy development have been fixed. They are divided into:

- **Short term targets**: within 2020 reduction of greenhouse gas emission by 20% and ensure 20% of renewable energy sources in the EU
- **Long term targets**: "decarbonization", 60 - 80% reduction of the greenhouse gas emission

To meet the 2020 targets, among many other research lines, for the European Commission it is imperative to: "Double the power generation capacity of the largest wind turbines, with offshore wind as the lead application". In particular for offshore wind, the Strategic Research Agenda establishes the following special objectives to be achieved within 2030:

- More than 10% of Europe's electricity should come from offshore wind
- Make the offshore generating costs competitive with other sources of power generation
- Commercially mature technology for sites at any distance from shore with a water depth up to 50m
- Full-scale proven technology to dominate deep-water sites

Moreover, together with the above targets, five research topics have been prioritized:

- Substructures
- Assembly, installation and decommissioning
- Electrical infrastructure
- Turbines
- Operations and maintenance

With respect to on land standard design, offshore environment does introduce significant additional elements which have to be carefully considered especially in designing the support structures. Knowledge about modeling the wind and rotor aerodynamics developed for onshore sites are generally enough and do not need deep changes when moving in the offshore environment. Some adjustments are just due to the different wind characterizing offshore sites. Figure 6 gives an example of different wind shears for on- and off-shore sites, respectively.

On the contrary, for offshore plants, the concept of support structure has to be entirely rearranged. For this reason research on the substructure is always prioritized both directly, by improving the technology itself, and indirectly, that is by developing more accurate models to estimate the combined wind-waves actions. In fact, in addition to the previous research topics, the Strategic Research Agenda focuses also on the following themes:

- Development of fully integrated wind-wave-current interaction models
- Development of new substructure concepts
- Development of improved design methodologies to extend the life of structures, to reduce costs and to incorporate risk based life-cycle approaches

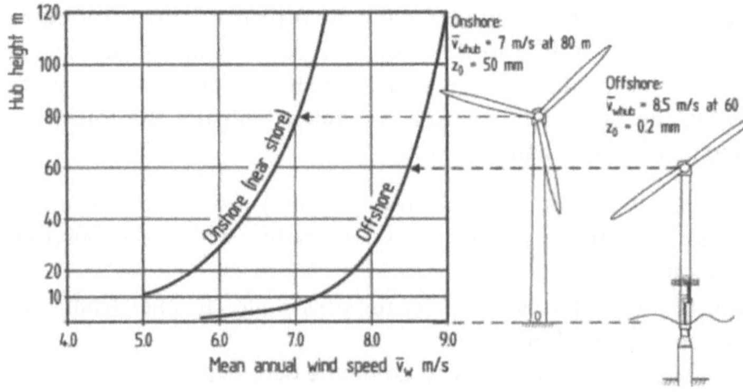

Figure 6. Logarithmic profiles of increasing wind speed with height for typical land at offshore sites. Figure from Hau (2006).

Consider that only the substructure represents approximately 25% of the whole investment and as forces (and somehow costs) used to dimension the substructure increase with the square of wind/water velocity, then it appears clear the importance of accurate models for loads prediction.

The objectives described above for wind energy development are based on the central fact that Europe has a remarkable wind potential. Figure 7 shows the map of onshore wind potential. Considering that the minimum value of the mean wind speed to make cost-effective a wind power plant is approximately 4 m/s it results that most of the European areas possess a wind energy potential.

While the offshore potential is depicted in Figure 8 which shows that in addition to the North and Baltic Seas, also some Mediterranean areas, for example between Greek and Italian coasts, the wind resource can be exploited.

1.3 General nomenclature

Before entering further the topic of this chapter, it would be useful to provide some general terminology. Figures 9 and 10 make this task much easier. Figure 9 in particular shows essential components of the upper part of the wind turbine, it depicts an onshore case with a superficial foundation. On the contrary, Figure 10 provides more details about the substructure which is defined as the structural subpart included between the sea bed and the platform.

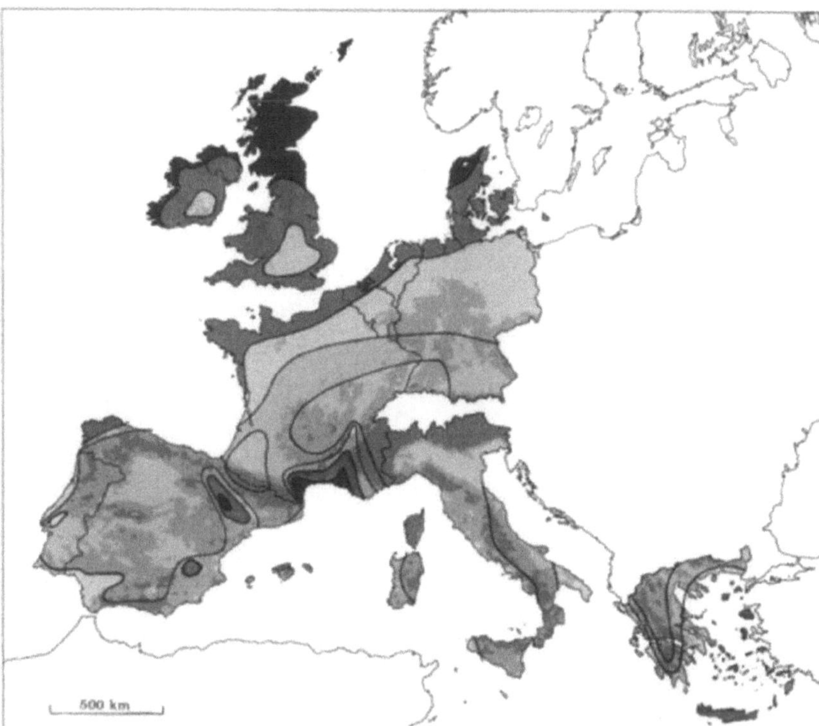

Wind resources[1] at 50 metres above ground level for five different topographic conditions										
	Sheltered terrain[2]		Open plain[3]		At a sea coast[4]		Open sea[5]		Hills and ridges[6]	
	$m s^{-1}$	Wm^{-2}	$m s^{-1}$	Wm^{-2}	$m s^{-1}$	Wm^{-2}	$m s^{-1}$	Wm^{-2}	$m s^{-1}$	Wm^{-2}
	> 6.0	> 250	> 7.5	> 500	> 8.5	> 700	> 9.0	> 800	> 11.5	> 1800
	5.0-6.0	150-250	6.5-7.5	300-500	7.0-8.5	400-700	8.0-9.0	600-800	10.0-11.5	1200-1800
	4.5-5.0	100-150	5.5-6.5	200-300	6.0-7.0	250-400	7.0-8.0	400-600	8.5-10.0	700-1200
	3.5-4.5	50-100	4.5-5.5	100-200	5.0-6.0	150-250	5.5-7.0	200-400	7.0- 8.5	400- 700
	< 3.5	< 50	< 4.5	< 100	< 5.0	< 150	< 5.5	< 200	< 7.0	< 400

Figure 7. Onshore potential. European Wind Atlas. Copyright 1989 by Risø National Laboratory, Roskilde, Denmark

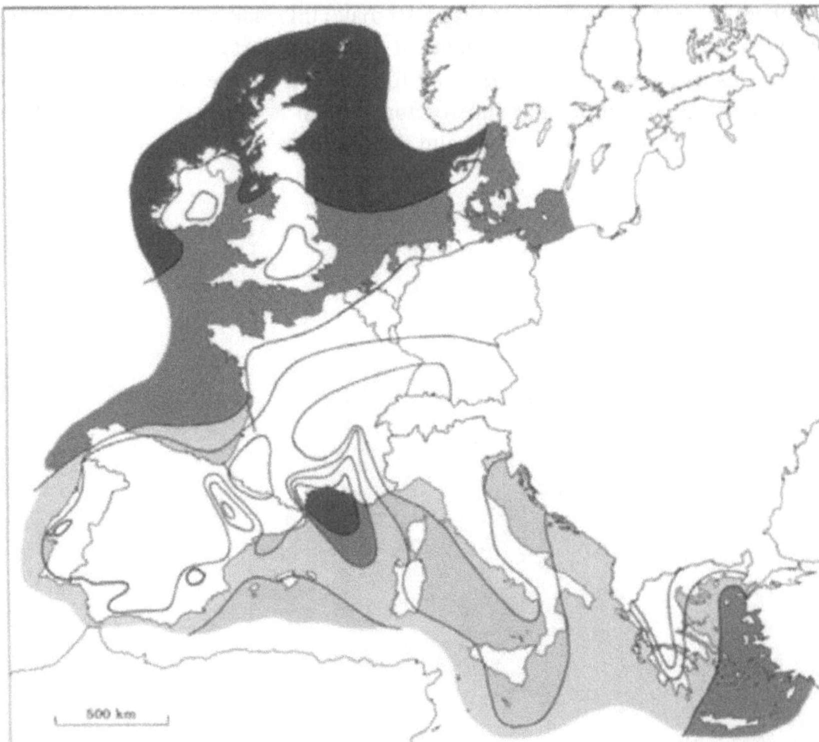

Wind resources over open sea (more than 10 km offshore) for five standard heights										
	10 m		25 m		50 m		100 m		200 m	
	$m s^{-1}$	Wm^{-2}	$m s^{-1}$	Wm^{-2}	$m s^{-1}$	Wm^{-2}	$m s^{-1}$	Wm^{-2}	$m s^{-1}$	Wm^{-2}
	> 8.0	> 600	> 8.5	> 700	> 9.0	> 800	> 10.0	> 1100	> 11.0	> 1500
	7.0-8.0	350-600	7.5-8.5	450-700	8.0-9.0	600-800	8.5-10.0	650-1100	9.5-11.0	900-1500
	6.0-7.0	250-300	6.5-7.5	300-450	7.0-8.0	400-600	7.5- 8.5	450- 650	8.0- 9.5	600- 900
	4.5-6.0	100-250	5.0-6.5	150-300	5.5-7.0	200-400	6.0- 7.5	250- 450	6.5- 8.0	300- 600
	< 4.5	< 100	< 5.0	< 150	< 5.5	< 200	< 6.0	< 250	< 6.5	< 300

Figure 8. Offshore potential. European Wind Atlas. Copyright 1989 by Risø National Laboratory, Roskilde, Denmark.

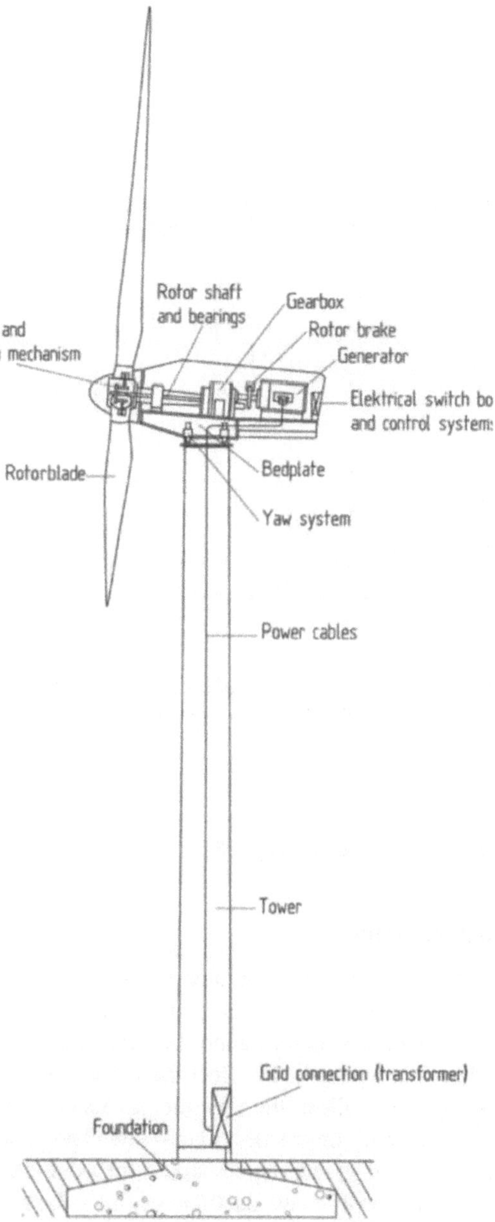

Figure 9. Main components of an horizontal axis wind turbine. Figure from Hau (2006).

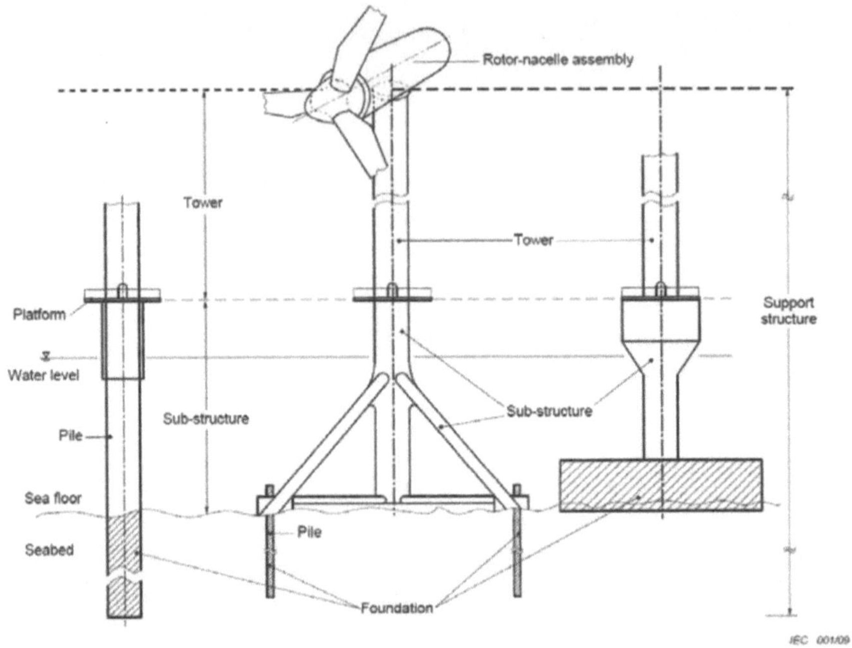

Figure 10. Main components of the support structure of an horizontal axis offshore wind turbine. Figure from IEC61400-3 [2].

2 Static and dynamic analyses

2.1 Site-Classification

Wind turbines design is mainly regulated by the international Codes such as, for example, the IEC 61400-1 or the DIBt and so on. In any case the installation of a wind turbine in a specific country cannot be performed regardless of the structural Code relative that country itself. In effect the modern structural Codes, such as for example the Eurocodes, allow the designer to take into consideration the wind actions on the structure by means of a macro-site classification. In this respect they provide a reference wind speed, depending on the position and on the orography of the site and also a wind profile law, useful to evaluate the reference wind speed at the hub height.

In general the design wind speed of the wind turbine (which has been chosen according to the IEC 61400-1 specific class) should satisfy also the wind conditions provided by the National Code for the specific site.

In applying these last instructions there is often the problem connected to the estimation of the topography coefficient, which is the responsible of the local wind speed amplification which may occur in a specific site. In this respect modern Codes permit to the designer to choose a different reference wind speed, for the specific site, in the case where a set of local measurements allow the evaluation of the actual wind conditions for the place where the WEC has to be built.

Hence, in general, the site classification is everywhere one of the most important phase in the planning of a wind park, because, based on a short or long term wind speed measurement campaign, it allows the following steps:

- Assessment of the specific wind conditions and determination of the energy yield for a planned wind farm
- Choice of the most suitable wind turbine for the planned wind farm
- Optimization of the WECs position inside the wind farm area
- Evaluation of the extreme wind condition in the WECs positions
- Check of the design parameters relative to the chosen WEC with respect to the measured site conditions and to the National Code instructions

The wind resource measurement can be performed by means of the installation, on site, of one or more masts on which a couple of anemometer are mounted together with a data recorder. The main purpose is the correlation of the data set provided by the anemometer on site with the long term measurement provided by some meteorological station present in the vicinity of the wind park area.

In order to achieve this task it is necessary that the measurement period on site would be at least 6 months, even if the more data is possible to obtain from the site measurements the more confidence there will be in the correlation.

Data collected on site should be correlated, by considering 12 30° degree sectors, with the measurement obtained, for the same period, at the meteorological station. In this way it is possible to evaluate a correction factor to be applied to the long term measurement in order to estimate the wind resource properties with a certain return period. Moreover, more sophisticated techniques are able to improve this process by taking into consideration of the site orography and of the terrain roughness (e.g., see Figure 11).

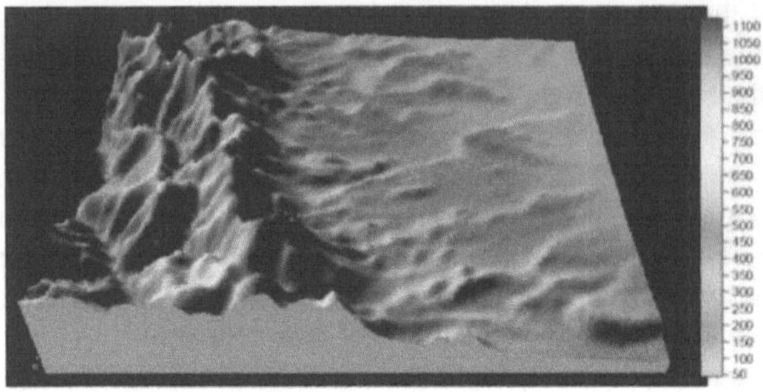

Figure 11. Digital model of the terrain.

These methods, such as for example WASP or CFD methods, allow the estimation of the wind resource at the actual position of the wind turbine. The on site measurements will allow also the estimation of the reference wind speed at the hub height. In any case, especially when the on site measurements cover a short time period, it is necessary to make use of some well known long term distribution and applying them a few appropriate correction factors in order to consider the uncertainties in the measurements. This last analysis should lead to the determination of a table, for the whole wind park, useful to the characterization every WEC position from a point of view of the structural design (cfr Table II.1 below).

Table 2. Typical classification of the WECs positions.

WEC ID	V_{ref} [m/s]	V_{ave} [m/s]	I [%]	σI	I_{15} [%]	Class [IEC]
W01	38.5	6.2	10.3	3.2	15.1	II_A

The table above should permit to initially estimate the WEC type suitable for the specific site and the V_{ref} parameter can be used to compare the classification provided both from the IEC 61400-1 and from the National Code instructions. In general, provided that the reference wind speed obtained from measurements on site should be the most reliable datum, then every design process and consideration should be based on it.

2.2 Modeling for static and dynamic analysis

Together with the general motivations regarding the global need in boosting wind energy production, nowadays research on (offshore) wind turbines is also desirable because it is crucial to get more and more reliable simulations.

Wind turbines are very sophisticated system and can be only analyzed by adopting multi-physic models. There are four coupled disciplines involved: aerodynamics, structural dynamics, hydrodynamics, soil dynamics. Each of them relates to the rotor, the tower, the substructure and the foundation, respectively. Figure 12 tries to sketch this concept, indeed, moving from right to left, the four isolated subjects are applied to the four main parts of an offshore wind turbine and thus coupled into a unique system which should render the reality as much as possible. The more general case of offshore wind energy converters is here discussed as the same concepts apply also to onshore installations with the obvious simplification of dropping the hydrodynamic model.

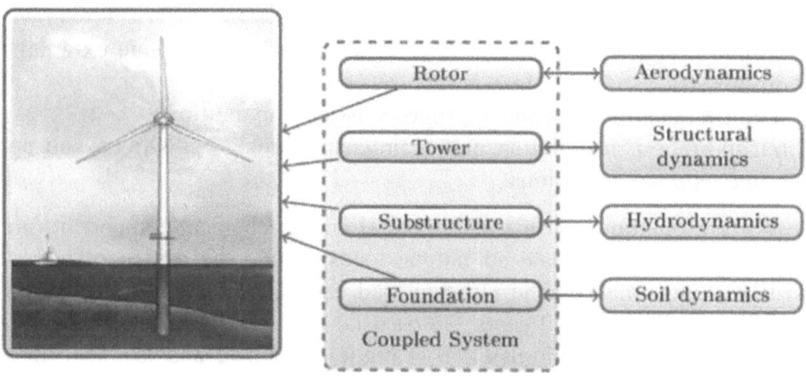

Figure 12. Multi-physic coupled system.

The current standard technique to analyze and design offshore wind turbines starts by collecting the system and environmental variables so that it is possible to provide input data to set up the numerical model. Environmental analysis has the primary goal of assessing design loads.

2.3 Design loads

Loads acting on wind turbines mostly stem from:

- rotor aerodynamics;
- gravitational and inertial forces;
- control system;

Modern turbine rotor blades are getting larger and larger and this gives rise to an increase of the dimensions of all others structural components (e.g. the tower height). Larger dimensions cause, in their turn, an augmentation of dead weight (gravitational loading), inertial forces and last but not least the effects of unsteady turbulent wind becomes more and more evident. Today some standard load assumptions are provided by IEC61400-1 and IEC61400-3, where wind turbines are divided in four classes with respect to the design wind conditions. Parameters indentifying the classes are: the reference wind speed and the expected value of the turbulence intensity. Much attention must be paid to the reference wind speed, in fact if a turbine belongs to a specific wind turbine class with a reference wind speed V_{ref}, it is designed to withstand climates for which the extreme 10 min average wind speed with a recurrence period of 50 years at the hub height is lower than or equal to V_{ref}. Design Load Case (DLC) are subsequently defined by combining:

- normal design situations and appropriate normal or extreme external conditions
- fault design situations and appropriate external conditions
- transportation, installation and maintenance design situations and appropriate external conditions

It is worth pointing out that all structural and mechanical components are required to resist both ultimate ad fatigue loads. While the design of tower and foundation is governed by the ultimate load cases, the rotor and the blades are designed by the fatigue load cases. To each DLC it is assigned a specific type of analysis denoted by U (Ultimate), F (Fatigue). Ultimate analysis are moreover distinguished in Normal (N) or Abnormal (A) and consequently partial safety factors (psf) are assigned as 1.35 for N, 1.1 for A situations. All fatigue design situations assume 1.0 as psf. The latter is rather a sensitive issue in the current design practice because IEC psf could be different compared with those imposed by national standards.

Once the DLCs are known the next step deals with the computation of internal forces. Depending upon the analysis to be carried out different load inputs are needed (wind speed time history or turbulence spectrum), then, after aerodynamic simulations coupled with structural solvers the system response is found and all the structural verifications can be performed.

As already mentioned wind turbines design is mainly regulated by the international Codes such as, for example, the IEC 61400-1 or the DIBt and so on. In any case the installation of a wind turbine in a specific country cannot be performed regardless of the structural Code relative that country itself. In effect modern structural Codes, such as for example the Eurocodes, allow the designer to take into consideration of the wind actions on the structure by means of a macro-site classification. In this respect they provide a reference wind speed, depending on the position and on the orography of the site and also a wind profile law, useful to evaluate the reference wind speed at the hub height

In addition to loads above discussed for onshore wind turbines, the offshore environment provides forces stemming from:

- Waves
- Currents
- Tides

An idea about the ultimate (U) and fatigue (F) load conditions that an offshore wind turbine is required to withstand is given in Figure 13.

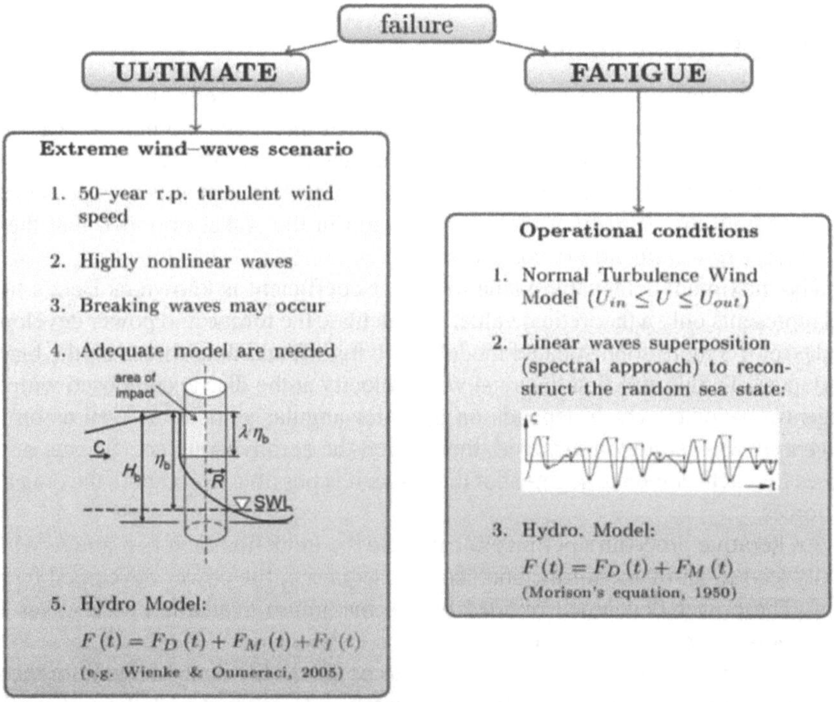

Figure 13. The two main failure conditions that offshore wind turbines may experience.

Fatigue limit state conditions are usually reached when the turbine is in operation: a turbulent wind model is reproduced with mean value falling between the cut-in and cut-out wind speed limits. It is reasonable to assume that such wind conditions generate irregular sea state so that waves can be assumed linear and their superposition is allowed by using standard spectra. In this case, in fact, hydrodynamic forces are computed by means of Morison's equation.

On the contrary, for ultimate failure conditions more representative model are required in order to capture the nonlinear contribution of extreme waves. Phenomenal sea state generated by 50-year return period wind speed may contain some very steep waves which can break against wind turbine substructures. In these extreme circumstances adequate models capable of capturing the impulsive contribution of such loads have to be employed.

According to IEC61400-1 and IEC61400-3, other loads such as wake loads, impact loads, ice loads, etc., may occur and shall be included. However in this context the focus is on wind and wave loads.

2.4 Aerodynamic loads

The primary goal of wind turbines is to subtract kinetic energy from the wind to transform it first into mechanical energy and then into electrical energy. The conversion of the wind kinetic energy into mechanical energy takes place when the air flows through the rotor disc. Given the upstream air flow velocity it is possible to calculate both the velocity at the rotor disc and in the wake, provided that the so called *axial flow induction factor* a is known.

The maximum achievable value of power coefficient is known as Betz's limit and represents only a theoretical value. To calculate the torque and power developed by the rotor a more sophisticated model involving lift and drag forces on the blades is adopted. To this aim first the axial wind velocity at the disc is composed with the tangential velocity which depends on the rotor angular velocity as well as on the *tangential flow induction factor* a', then, given the aerodynamic coefficients of the blades for each elemental segment of the blade, it is possible to compute the drag and lift forces.

An iterative procedure permits to calculate the induction factors a and a' which finally lead to know the torque and, as a consequence, the power developed by the rotor. The power developed divided by the maximum available power gives the expression of the power coefficient C_p.

The method above outlined is referred to as Blade Element Momentum theory (BEM) and it requires two corrections which take into account both the real number of blades and the case when the momentum theory is no longer allowed.

Aerodynamic solvers (as the case of AeroDyn, developed at the National Renewable Energy Laboratory (NREL), Colorado) implement also the so called Generalized Dynamic Wake (GDW) model to compute the aerodynamic forces. This theory, also known as the Method of Acceleration Potential, is based on a solution of Laplace's equation for potential problems and exhibits some advantages with respect to the BEM theory as a more general distribution of pressure across the rotor plane is obtained. Details about the GDW theory are available in Burton et al. (2001) and Moriarty and Hansen (2005).

In this context, however, only some key details about the BEM theory are provided. Thus, starting with the stream tube concept, see Burton et al. (2001), the total force acting at the rotor disc (the thrust) is given as follows

$$F_d = 2\rho A_d U_\infty^2 a\,(1-a) \tag{1}$$

where ρ is the air density, A_d is the disc area, U_∞ denotes the far upstream wind velocity and a is the *axial induction factor.*

Thus the power developed is $P_{yield} = F_d U_d = 2\rho A_d U_\infty^3 a\,(1-a)^2$ where the velocity at the rotor disc is given as $U_d = U_\infty\,(1-a)$. It is straightforward now to derive the power coefficient defined as the ratio between the extracted power and the total available power as follows

$$C_p = \frac{P_{yield}}{\frac{1}{2}\rho U_\infty^3 A_d} = 4a\,(1-a)^2 \tag{2}$$

The above coefficient is maximum when $a = 1/3$, therefore the maximum power coefficient is $C_{p\max} = 0.593$. The latter is known as Betz's limit and represents an ideal value which proves that the maximum exploitable energy theoretically equals 60 % of the available power.

Information about the extractible power as well as about the total axial force exerted on the rotor disc when a far field wind velocity U_∞ is given, although useful, are not enough to compute the aerodynamic forces acting on the blades. As already said, to this aim the BEM theory is applied as follows.

Given an elemental blade element, whose cross-section is sketched in Figure 14, an air particle past the rotor disc has a tangential velocity $U_t = 2a'\Omega R$ where Ω is the angular velocity of the rotor, while a' is the *tangential flow induction factor.* Throughout the rotor disc thickness the tangential velocity of an air particle varies from zero (upstream) to $2a'\Omega r$ (downstream), so that in the middle the tangential velocity is $a'\Omega r$. The latter, together with the tangential velocity of the blades, results in a net tangential flow experienced by the blade element equals to $\Omega r + a'\Omega r = \Omega r\,(1+a')$, therefore the total relative velocity W experienced by the blade is the following

$$W = \sqrt{U_\infty^2\,(1-a)^2 + \Omega^2 r^2\,(1+a')^2} \tag{3}$$

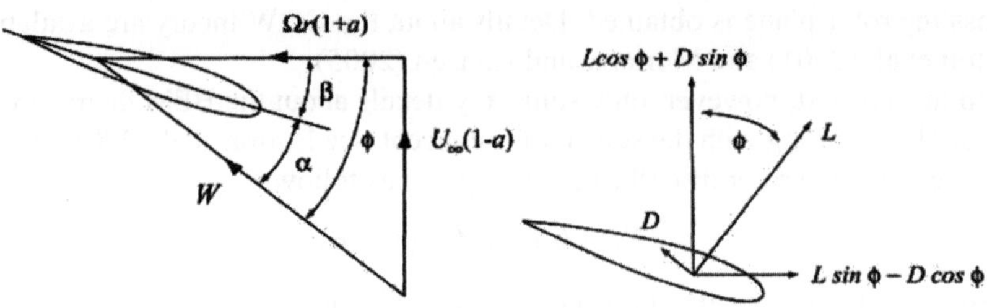

Figure 14. Velocity and forces on a blade element. Wind Energy Handbook, Burton et al. (2001).

Drag and Lift forces

Once the total velocity W of air particles past the single blade element is known, the forces induced on the blade element can be computed by using the aerodynamic properties of the blades. Namely, by means of the drag and lift coefficients, C_D and C_L, respectively, the aerodynamic forces acting on a blade element are

$$\delta D = \frac{1}{2}\rho W^2 c C_D \delta r \tag{4}$$

$$\delta L = \frac{1}{2}\rho W^2 c C_L \delta r \tag{5}$$

where δr denotes an elemental ring belonging to the rotor plane and c is the blade chord. Drag and lift forces are then projected on the normal and tangential plane with respect to the rotor disc so that the thrust and the torque induced by the blade element can be derived. For further details see Burton et al. (2001), Hansen (2008).

2.5 Wave loads

Sea waves are traditionally described by both a deterministic approach and a probabilistic model with respective advantages and disadvantages. For instance, the spectral approach permits to describe a random sea but has the drawback that only linear wave can be represented and, as a consequence, only forces stemming from linear wave theories can be derived. On the contrary, some nonlinearity magnitude can be taken into account when deterministic monochromatic wave are used. In other words, the two commonly used approaches implement respectively either

- Regular nonlinear waves, or
- Irregular linear waves;

Fortunately, in most cases the nature of oceans can be very well described by the superposition of linear regular waves, and this makes the spectral approach a tool extraordinarily effective. In fact, especially for fatigue loading condition, this approach fits fairly well the nature of the actions.

Contrarily, for ultimate failure conditions more representative models are required in order to capture the fully nonlinear contribution due to extreme (possibly breaking) waves. Figure 15 shows the two main approaches which are linked each other due to the possibility of superposing single linear waves according to the spectral formulation.

Nowadays, capabilities of modern computers permit to simulate fully nonlinear waves without penalizing the total simulation time, thus, whenever the fully nonlinear behavior of waves plays a dominant role in designing offshore structures, it seems to be opportune to adopt a direct numerical solution of the water waves governing equation without introducing any a priori hypotheses, where "a priori hypotheses" refers to the magnitude of nonlinearity, rather than to the assumptions on the fluid and the flow type.

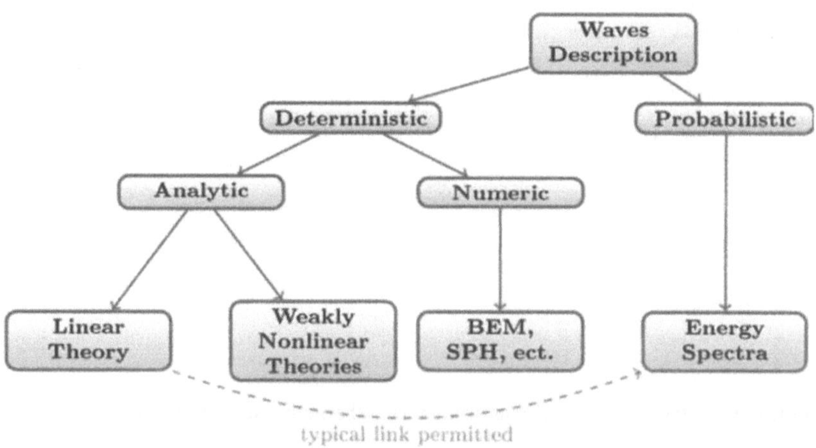

Figure 15. Traditional scheme adopted for describing ocean waves.

Analytical theories, such as Stokes-2nd and 5th order, fall in the weakly nonlinear group. In fact what distinguishes weak nonlinearity to the high (or strong) nonlinearity is essentially the asymmetry with respect to the vertical axis of the free surface. In the weakly nonlinear case only asymmetries with respect to the horizontal axis are reproducible: crests become narrower and the trough get wider.

Figure 16 presents the conceptual scheme followed when simulating an offshore wind turbine. The environmental analysis provides input data both for the hydro-aero-elastic solver. Then, during the simulation, the solver calls (at each time step) both the aerodynamic and hydrodynamic solvers to get all the forces acting on the system. Such an approach covers most of the Design Load Conditions recommended by Standards (e.g. IEC61400-3), but cannot take into account extreme sea states which can give rise to fully nonlinear waves breaking against the substructures.

According to section 6.4.1.8 of IEC61400-3, impact loads associated with plunging breakers have to be taken into account because, as demonstrated in Marino (2010), they are crucial when the goal is evaluating the structural safety during extreme environmental conditions.

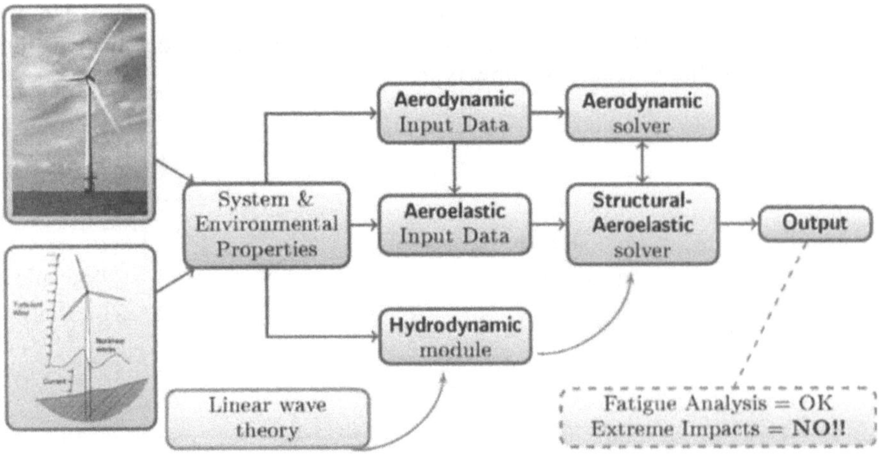

Figure 16. Offshore wind turbines simulation scheme implementing linear waves only.

The integration of a fully nonlinear numerical solution of gravity waves into the more general multi-physics framework characterizing the design of offshore wind turbines seems to be not yet a common practice. Figure 17 improves the scheme in Figure 16 by adding a fully nonlinear waves simulator.

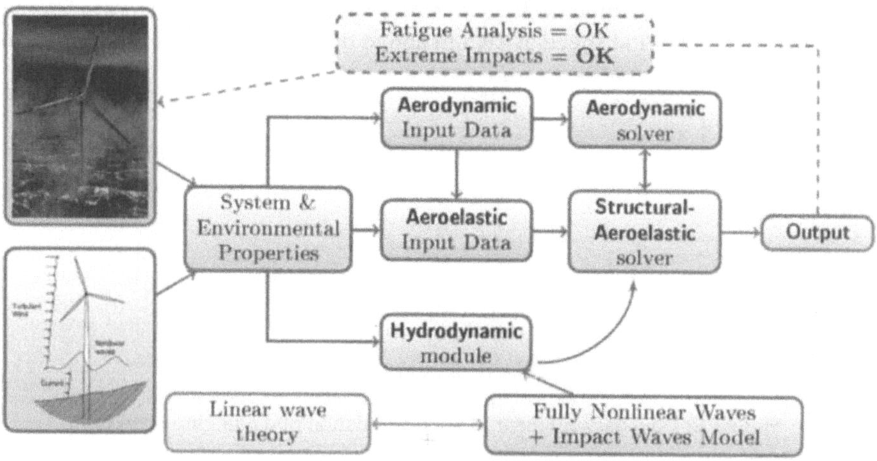

Figure 17. Advanced simulation scheme capable of capturing both fatigue state of failure and ultimate limit states stemming from extreme wind-waves actions.

The fully nonlinear wave model

The hypotheses of an incompressible and inviscid fluid and irrotational flow allow the description of fully nonlinear water waves by means of a potential model. That is, at a fixed time t, the velocity field for each point belonging to the domain $\Omega(p,t)$ can be expressed through a velocity potential $\phi(p,t)$ as follows

$$\bar{v}(p,t) = \nabla\phi(p,t) \quad \forall p,t \in \Omega(p,t) \tag{6}$$

One of the first contribution addressing the numerical solution of the above time depending Laplace's equation was due to Longuet-Higgins and Cokelet (1976) who introduced for the first time the Mixed Eulerian-Lagrangian (MEL) approach to describe such a free surface problem. Subsequently Dold (1992) proposed a new and time-effective procedure to integrate in time the dynamic and kinematic boundary conditions on the free surface. Excellent results are obtained dicretizing in space Eq. (6) by implementing the higher-order Boundary Element Method (BEM), see, Grilli et al. (1989), Grilli and Svedsen (1990), Nakayama (1983, 1990), Marino et al. (2010).

As an example, the capability of boundary element method-based fully nonlinear water waves simulator is shown in Figure 18 where a violent plunging breaker occurring during a storm is reproduced. Such an extreme storm is characterized by $U_{19.5} = 31.69$ m/s; $H_s = 11.5$ m; , $T_p = 10.60$ s, where $U_{19.5}$ is 50-year return period mean wind speed at 19.5 m above the sea level, H_s is the significant wave height and T_p the peak spectral period.

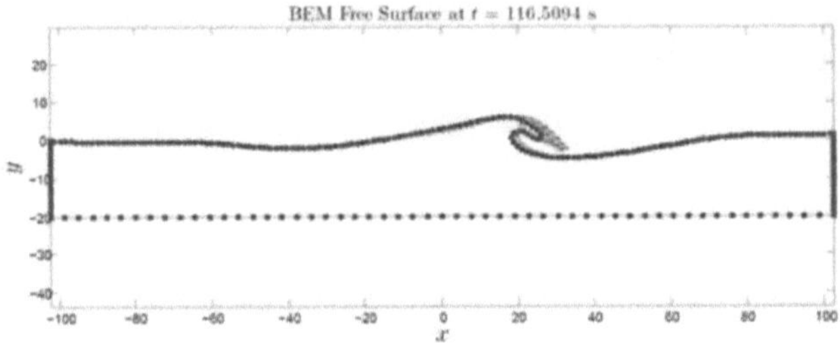

Figure 18.Boundary Element Method simulation of a plunging breaker.

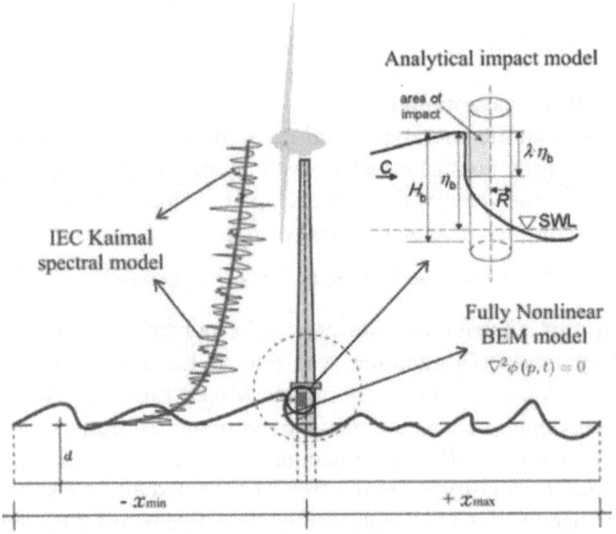

Figure 19. The coupling of the three main models involved in the simulation scheme proposed in Figure 17. Wind: IEC Kaimal turbulence model; Waves: fully nonlinear Boundary Element Method coupled with an analytical impact model.

A fully coupled model is idealized in Figure 19 where an extreme turbulent wind induces highly nonlinear sea waves (described by means of Boundary Element

Method model) which can break against the structure generating extremely dangerous impact loads computed by implementing the analytical model of Wienke and Oumeraci (2005). For further details on such a coupled model, refer to Marino et al. (2010).

3 Design and Verification

Only very recently (in 2007) R.C. towered structures for large wind turbines (>1.5 MW) have been widely employed for the realization of high-production on-shore wind farms (up to 100 MW). The R.C. towers are usually made out by concrete elements (steel, high-resistance R.C. conically shaped segments) which are prefabricated and then erected and pre-stressed on-site. This particular technology, never used in Southern Europe before, has demanded an adjustment of the original design to the Italian codes. A deep analyses of the aeroelastic loads and wind loads are needed, while considering the coexistence of the seismic action (in earthquake prone areas). The study was carried out for different heights of the towers as well as for different kinds of the foundation blocks, which can be direct or pile sustained. The steel-concrete mixed system revealed consistent economic advantages with respect to the traditional steel-only system, even if particular care has to be taken into account during the realization phase.

3.1 The ENERCON E-82

The turbine, each of the maximum power of 2MW, consisting of a block of rotor type (diameter of 82 meters) installed on top of wind tower. The hub height is 78-138 m. In Italy such machines are supported by mixed steel pre-stressed R.C. towers (hub heights of 84 m and 98 m). The lower part of the tower (+0.000 until +57.390/+68.838 m) is made of pre-cast concrete segments, while the upper part (+57.390/+68.838 until +83.300/+97.100 m) is made of steel. The outer base diameter is 6.369 m (for the tower 84 m WTG) and 7.500 m (for the tower 98 m WTG) (see Figure 20).

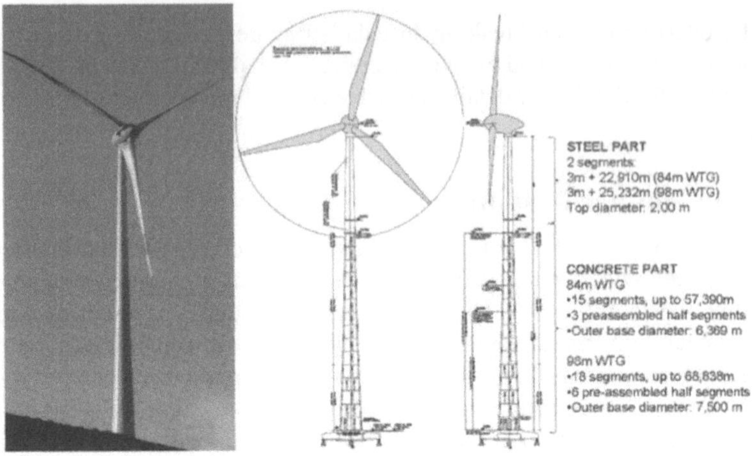

Figure 20. A Typical 2MW wind turbine tower.

Concrete tower is pre-stressed with 32 or 36 strands, depending on the performance requirements, anchored to the foundation crown at the base and to the first steel segment on the top. Segments are built with high strength concrete, from C45/55 up to C70/85. The material qualification follows the common rules as for the usual building construction, even if the production is prefabricated (and in quality control).

The foundations are made by means of a monolithic concrete block. The circular foundation can be built in 3 different configurations: Direct foundations without buoyancy (diameter = 14.2 m); Direct foundations with buoyancy (diameter = 16.8 m); Piled foundations (12 to 16 piles ø =1.0 ÷ 1.2 m, foundations diameter = 15.6 to 15.8 m). The foundations crown hosts also the anchorage devices for the pre-stressing system (see Figure 21).

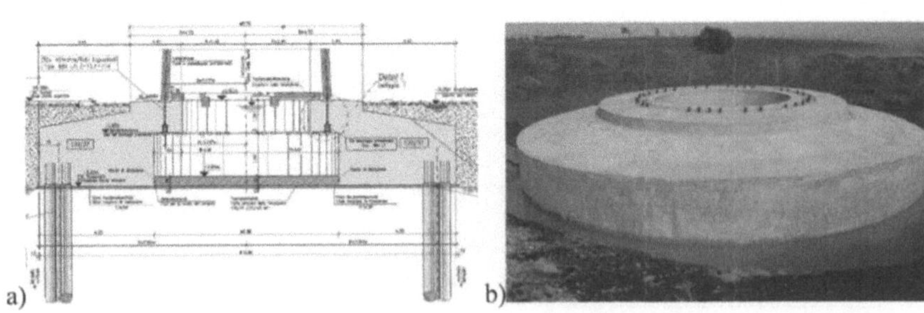

Figure 21. Foundations: a) Scheme of piled foundations; b) View of foundations.

3.2 FE Model

Both supporting structure and foundation structure have been analysed by means of a FEM model (see Figure 22).

The tower has been modeled by using beam elements with 6 degrees of freedom at each node. The FE model of the structure is implemented to check the natural frequencies of the tower and to evaluate the effects due to the seismic action. The eigenmodes are the same as in a cantilever beam. In general two FE models have been analyzed: the first one on elastic foundation while the second one on rigid restraint. In every case, the number of the eigenmodes which are considered in the analysis has to guarantee that the percentage of the excited mass is at least the 85% in every direction. With regards to the torsional effects due to the eccentricity of the rotor mass, the participating of mass is always high. As expected the differences between the model on rigid foundation and on elastic foundation are not so evident with respect to the internal forces. Anyway, as regards the evaluation of the seismic effects on the structure, the foundation is considered rigid, being this one on the safe side. Even if it is not required by the IEC 61400-1, the ratio between the frequency of the rotor and the frequency of the structure has been checked, in fact the following limetations are required by the German code DIBt:

$$3\frac{f_{R2}}{f_{T2}} \leq 0.95 \qquad 3\frac{f_{R2}}{f_{T1}} \geq 1.05 \qquad \frac{f_{R2}}{f_{T1}} \leq 0.95 \tag{7}$$

Where f_{R2} is the second frequency of the rotor, f_{T1} and f_{T2} are the first and second frequency of the tower.

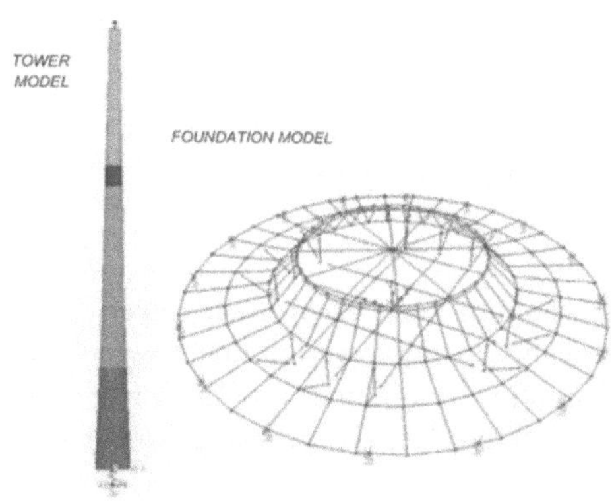

Figure 22. Models FEM.

The foundations has been modeled by using beam elements in a three-dimensional configuration: 4 rings elements of different diameter and three radial beams. The basic constraints are different depending on the type of foundation: Winkler bed springs simulating the elasticity of the soil (for direct foundations); 12-16 joint springs simulating the vertical stiffness of the piled foundation.

3.3 Actions on the structure

The design and verification of the E-82 wind turbines structures (Towered support and Foundations) have been made according to the new Italian Code (D.M. 14.01.2008) and the Euro Codes (where the Italian Code lacks of information). Structural analyses in Earthquake prone areas are needed to check the full compliance of dimensioning criteria and design loads with the National Codes. National specifications (standards) are needed for large wind turbines and their carrying structures. As an example, for the towers design in Italy, the recourse to the following regulations has been needed:

- IEC 61400-1 3rd Ed. :2005, Wind turbines – Part 1: Design requirements;
- DIBt, "Richtlinie für Windenergieanlagen", 2004, Berlin;
- D.M. 14.01.2008, Norme tecniche per le costruzioni;
- Circ. 02.02.2009 n.617, Istruzioni per l'applicazione delle nuove norme tecniche per le costruzioni;
- Euro Codes (EC).

The following elementary load conditions have been taken into account: self weight and permanent loads, imperfections due to the non-verticality of the tower, wind and seismic actions.

Self weight and permanent loads. The self weight and the permanent loads due to equipment inside the tower and the rotor system were calculated through the numerical model.

Imperfections due to the non-verticality of the tower. The analysis includes the effect due to non-verticality of the tower structure, which has been assumed as a deviation equal to 5 mm/m. In addition, one has to consider a deviation of the foundation structures equal to 40 mm in the base of foundations diameter.

Wind action. The wind action (load of the rotor and distributed load along the tower) is modeled referring to a specific code (IEC 61400-1:2005), being the Italian code D.M. 14.01.2008 not sufficiently detailed for such a structure. A comparison between the two codes is then needed, in order to identify the Italian sites where the wind profile - calculated according to the IEC code - is not lower than the one calculated according to the Italian code. In any case the recourse to a local site condition assessment is always strictly necessary. For the 82m rotor blade tower design in Italy is strongly recommended a special consideration about the reference wind speed at hub height according to the IEC 61400, Italian code (D.M. 14.01.2008) and to the site wind report. According to these Codes, the wind velocity profile follows a log-law depending on the height on the terrain. The mean wind velocity at a height z depends on the terrain roughness and orography and on the basic wind velocity (see Figure 23). The peak wind velocity includes gust effects through the exposure factor.

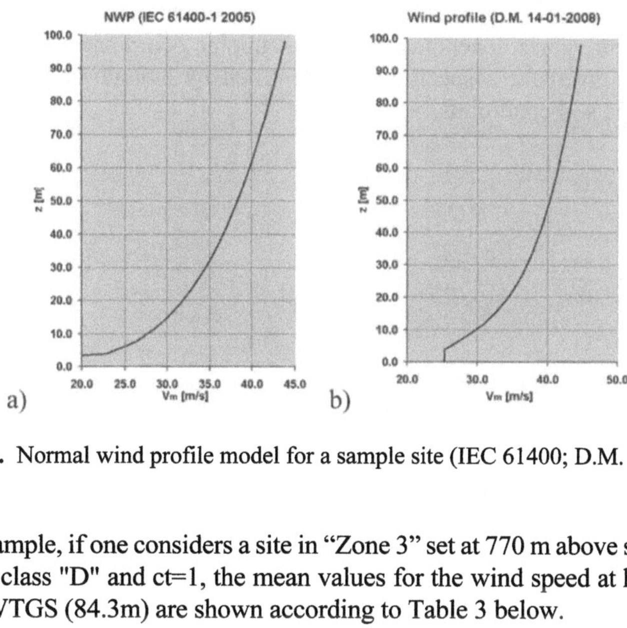

Figure 23. Normal wind profile model for a sample site (IEC 61400; D.M. 14.01.2008).

For example, if one considers a site in "Zone 3" set at 770 m above sea level with roughness class "D" and ct=1, the mean values for the wind speed at hub height of the E-82 WTGS (84.3m) are shown according to Table 3 below.

Table 3. Average wind speed at hub height.

	$V_{m(84.3)}$ [m/s]
D.M. 14.01.2008	43.68
UNI EN 1991-1-4	43.68
IEC 61400 (class IIA)	42.50

Following the values from Table III.1, the design wind reference speed Vd must be greater than 43.68 m/s. In any case it is also necessary to refer to the wind report results, that are based on measurements on site through a reconstruction of the wind velocity field "VWR" in a given site. The evaluation of the topographic coefficient according to the various Codes is often not adequate to represent the win field in a specific site. For this reason it is strongly recommended that the wind turbine design wind speed will verify also that Vd > VWR.

Seismic action. The seismic action is represented through the elastic response spectra and the design response spectra at the ultimate limit state. The seismic action on the tower has been described by the ground acceleration response spectra according to national Code (D.M. 14.01.2008). The national territory is classified with a micro zonation (see Figure 24) in which the spectral parameters (a_g, F_0, Tc*) assume different values depending on the geographic coordinates of the reference site. They are given for a specific site (identified by latitude and longitude) for horizontal, rigid soil conditions.

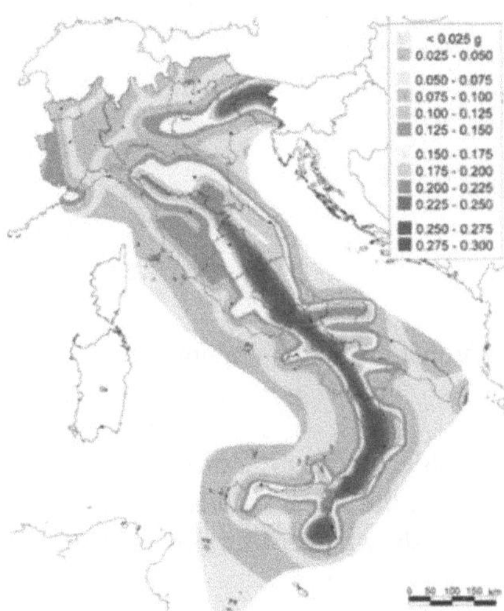

Figure 24. Micro Zonation of the national site according to D.M. 14.01.2008.

For the calculation of the seismic action, the first step consists in the identification of the three parameters (a_g, F_0, Tc*) for the examined site. They are given for the 10.751 nodes of the referring reticulum and for the islands. For different sites interpolation is allowed. The second step aims the definition of the referring period of the structure, in order to compute – in the third step of the design procedure – the response spectrum of the seismic action. The latest one depends on the amplifications due to the stratigraphic and topographic conditions of the site. The behavior factor for the wind turbines has been assumed to be q = 1.5 (inverted pendulum system).

Assuming a specific site in high seismicity area, as in the south Italy, in the following Figure 25 one can see the corresponding design spectrum for the horizontal component according to the D.M. 14.01.2008.

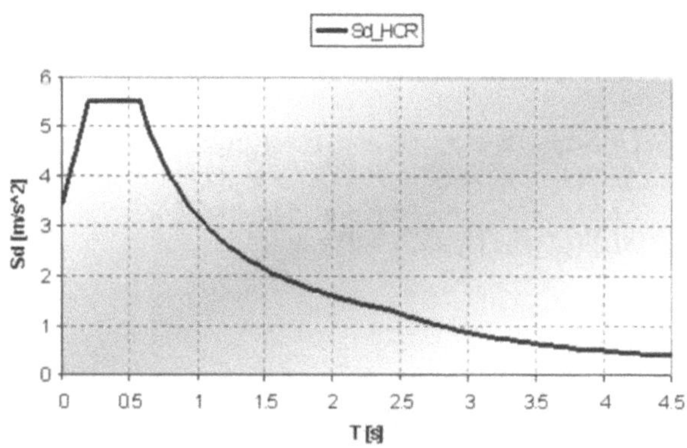

Figure 25. Horizontal design response spectrum for a specific site in Italy.

In conclusion, there are a certain number of wind condition, provided by the manufacturer and that one can substantially group as follows:

- Wind IEC (with safety factor) – ULS conditions;
- Wind IEC (without safety factor) – SLS characteristic conditions;
- Wind IEC (without safety factor) – SLS frequent conditions;
- Wind IEC (without safety factor) – SLS quasi-permanent conditions;
- Wind Seismic = Wind load to be superposed to seismic action (for the ULS);

while other 24 elementary load conditions are seismic conditions which are divided in three groups as follow:

- Seismic_X = ± Ex ± 0.3*Ey ± 0.3*Ez;
- Seismic_Y = ± Ey ± 0.3*Ex ± 0.3*Ez;
- Seismic_Z = ± Ez ± 0.3*Ex ± 0.3*Ey;

Where the Ex and Ey are the horizontal component of the design seismic spectrum, and Ez is the vertical component.

At last, the remaining tree conditions are:

- PP = self weight of the tower and rotor system;
- PERM = permanent loads;
- ΔM = imperfections due to the non-verticality of the tower;

Moreover the *thermal effect* has to be carefully considered for the design of transversal reinforcement in the concrete part of the tower.

The elementary load conditions are combined by partial safety factors, according to the D.M. 14.01.2008 and the Euro Codes (UNI EN 1990 – 2006) (see Figure 26). The combinations of actions to be taken into account in the relevant design situations should be appropriate for the serviceability requirements and performance criteria being verified.

$$\begin{aligned}&ULS: \gamma_{G1}G_1+\gamma_{G2}G_2+\gamma_P P+\gamma_{Q1}Q_{k1}+\gamma_{Qi}\psi_{0i}Q_{ki}\\&SLS(R.): G_1+G_2+P+Q_{k1}+\psi_{0i}Q_{ki}\\&SLS(F.): G_1+G_2+P+\psi_{11}Q_{k1}+\psi_{2i}Q_{ki}\\&SLS(Q.p.): G_1+G_2+P+\psi_{2i}Q_{ki}\end{aligned} \quad (8)$$

- G_i = permanent actions;
- Q_{ki} = variable actions;
- γ_{Gi} / γ_{Qi} = partial factor for actions
- ψ_{0i} / ψ_{1i} / ψ_{2i} = factor for combination value of a variable action;

		SLS			ULS	
		Characteristic	Frequent	Quasi - Permanent	Wind	Wind + Earthquake
γ_G	Fav.	1,00	1,00	1,00	0,90	1,00
	Unfav.	1,00	1,00	1,00	1,35	1,00
γ_P	Fav.	1,00	1,00	1,00	1,00	1,00
	Unfav.	1,00	1,00	1,00	1,00	1,00
γ_q Wind		1,00	1,00	1,00	1,10 1,35	Constructor data
Earthquake		-	-	-	-	1,00

Figure 26. Partial factors for actions of the tower.

According to the D.M. 14.01.2008 and the Euro Codes, the earthquake and wind load are considered independently of each other, in fact, both standards has proved (for seismic combination):

$$ULS(E): G_1 + G_2 + P + \psi_{2i} Q_{ki} \tag{9}$$

- E = seismic load;
- ψ_{2i} = factor for combination value of a variable action (= 0 for wind load);

Moreover, since these structures are designed to be continuously subjected to the wind action, it is unsustainable to not include a combination of wind and earthquake loads. The IEC 61400-1 requires the overlap of the seismic action with a wind load obtained by assuming a *loads during normal power production by averaging over the lifetime* and considering even the eventuality of the system shutdown.

According to the new Italian Code (D.M. 14.01.2008) and the Euro Codes, for the structures design refers to the method of limit states.

3.4 Towered support Design – Concrete part

The towered support design needs verifications of concrete part: ULS for normal+bending actions and torsion+shear strength; SLS for stresses limitation (in concrete, pre-stressing cables and ordinary reinforcements); SLS for limit state of cracking and shrinkage and finally FLS (ultimate limit state of fatigue) in concrete and pre-stressing cables. In the following the verification on the main components of the concrete part of the towered support will be shown.

Post-tensioned cables. There are immediate (cfr Eq. 10) and time-dependent losses (cfr Eq. 11). Immediate losses are due to friction and also caused by the devices displacements. Time-dependent losses are caused by concrete deformation (due to fluage and shrinkage) and steel relaxation.

$$\Delta P_{attr} = P_{\max}(1 - e^{-\mu(\theta + Kx)}) \tag{10}$$

$$\Delta P_{c+s+r} = A_p \Delta\sigma_{p,c+s+r} = A_p \frac{\varepsilon_{cs} E_p + 0.8\Delta\sigma_{pr} + \frac{E_p}{E_{cm}} \varphi(t,t_0) \cdot \sigma_{c,Qp}}{1 + \frac{E_p}{E_{cm}} \frac{A_p}{A_c} \left(1 + \frac{A_c}{I_c} z_{cp}^2\right) [1 + 0.8\varphi(t,t_0)]} \tag{11}$$

Concrete at ULS (N+M). Results of the design for bending and axial force are reported in the interaction diagrams N-M. The resistance domain of each of the sections examined is traced with reference to actual coordinate couples (N,M) obtained from the results (see Figure 27). They take into account differences in the cross-sections due to the openings in the first segment and in the middle segment.

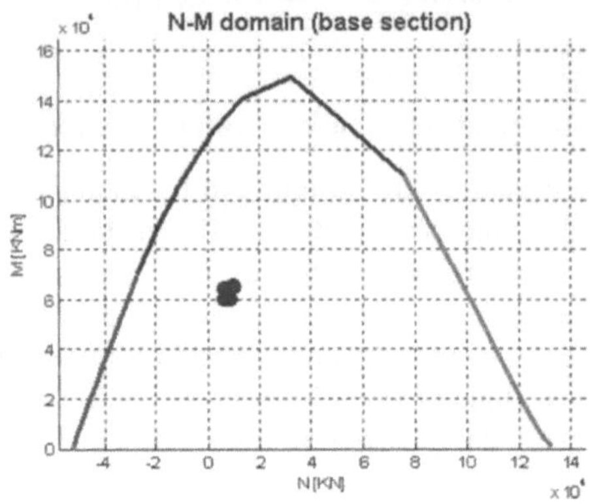

Figure 27. Example for the resistance domain N-M (base section).

Concrete at ULS (V+T). For shear and torsion verification, two different methods are followed: "Stress design" and "Resistant design". In the first case the check of the principal stresses in the concrete has to be performed and compared to the mean bending concrete tensile strength f_{ctm}. The calculation is computed in four points (see Figure 28). A minimum amount of reinforcement for shear and torsion is also calculated.

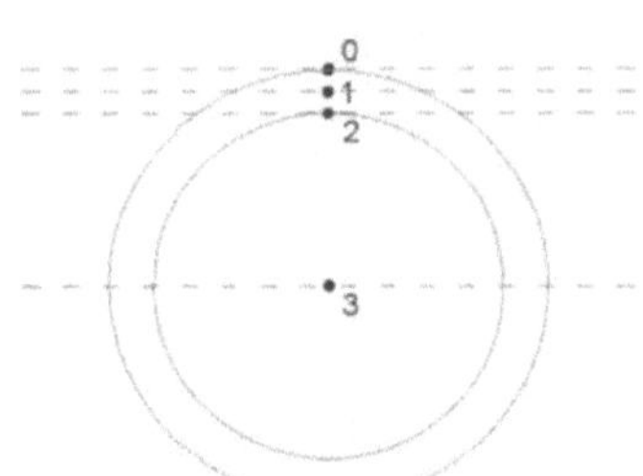

Figure 28. Four points of section verification (shear and torsion design).

In the second case, the D.M. 14.01.2008 requires the control of the ratio between the shear resistance and the design shear force resulting from the combinations of actions. For the shear resistance it is more appropriate to consider a small portion of the circular hollow section as defined in the Figure 29.

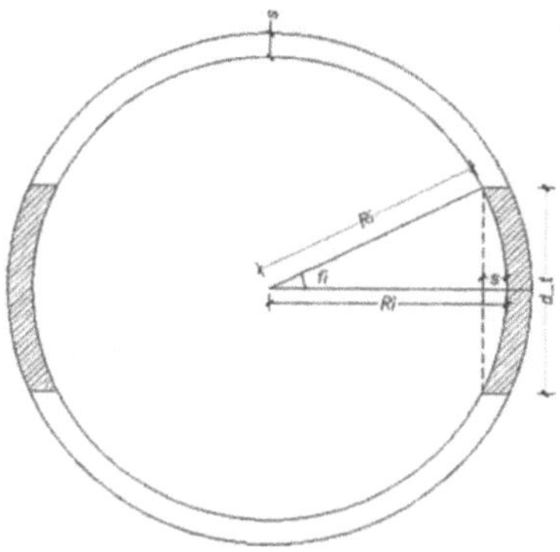

Figure 29. Shear resistant section.

where the effective section depth *d_t* is calculated depending on the concrete thickness *s* of the section, as:

$$d\_t = 2 \cdot R_i \cdot sen(f_i)\,, \quad f_i = a\cos\left(\frac{R_i - s}{R_i}\right) \tag{12}$$

Moreover a combination of the torsional and shear effects has ro be considered according to the following equation:

$$\frac{T_{Ed}}{T_{Rd}} + \frac{V_{Ed}}{V_{Rd}} \leq 1 \tag{13}$$

Where T_{ed} and V_{ed} are the torsional and shear force,while T_{Rcd} and V_{Rcd} are the torsional and shear resistance.

Concrete at SLS (Stresses limitation). The concrete design in the serviceability limit states includes the characteristic, the frequent and the quasi-permanent combinations of actions. Concrete normal stresses are then calculated and compared to the limitations required by the codes:

$$\begin{aligned} &R.: \sigma_{c,c} \leq 0.6 \cdot f_{ck}, \sigma_{c,t} \leq f_{ctm}/1.2 \\ &F.: \sigma_c \leq 0 \\ &Q.p.: \sigma_{c,c} \leq 0.45 \cdot f_{ck} \end{aligned} \tag{14}$$

Under the rare combination of actions there is also a tensile stresses limitation in the pre-stressing reinforcement:

$$R.: \sigma_p \leq \min\left(0.75 \cdot f_{ptk}; 0.85 \cdot f_{p0,1k}\right) \tag{15}$$

Concrete at SLS (Limit state of cracking). The limit state of cracking requires the calculation of minimum reinforcement areas (cfr 7.3 EC2) and the calculation of cracks width (cfr 7.3 EC2 and cfr C4.1.2.2.4.6 Circ. 2009). The first calculation, which aims the evaluation of the amount of transversal reinforcement due to cracks, takes into account of two phases: in the first one (after 4 days) cracks are induced by shrinkage effects, while in the second one (> 28 days) cracks are induced by serviceability loads. The calculation of the cracks width is mainly due to the temperature gradient. For the control of crack width, the limitation of the D.M. 14.01.2008 are more strict than EC2 one (see Figure 30).

UNI EN 1992

ENVIRONMENTAL CONDITIONS	REINFORCEMENT ORDINARY and PRESTRESSING (Post-tensioned)
XC2, XC3, XC4	Quasi-permanent
XD1, XD2, XS1, XS2, XS3	0,3

W_{lim} DM < EC

D.M. 14.01.2008

ENVIRONMENTAL CONDITIONS		COMBINATIONS OF ACTIONS	REINFORCEMENT					
			PRESTRESSING			ORDINARY		
			Limit State	w_d [mm]		Limit State	w_d [mm]	
b	Aggressive	Frequent	Crack width	≤	0.2	Crack width	≤	0.3
		Quasi-permanent	Decompression	-		Crack width	≤	0.2

Figure 30. Confrontation of limit crack width (EC2-D.M. 2008).

Concrete at FLS (Limit state of fatigue). The design according to the ultimate limit states of fatigue in the concrete relies on the given values of bending moments under the assigned number of loads cycles. The total bending moment in the structures is then given by the superposition of bending moments due to loads cycles and bending moments due to other acting loads, which do not cause fatigue in the structure. The fatigue design according to "CEB - FIP Model Code 1990: Design code" consists in the evaluation of the fatigue damage according to the Palmgren-Miner summation:

$$\textit{(CEB-FIP Model Code cfr 2.1.7, 6.7.4)} \qquad D_{Ed} = \sum_j \frac{n(\Delta\sigma_i)}{N(\Delta\sigma_i)} < 1 \qquad (16)$$

Where $n(\Delta\sigma_i)$ is the number of acting stress cycles at a given stress level and stress range; $N(\Delta\sigma_i)$ is the number of resisting stress cycle.

There is also another control, which is required by UNI EN 1992-1-1: 2006, based on the calculation of normal stresses in the concrete:

$$\textit{(EC2 cfr 6.8.7)} \qquad E_{cd,\max} + 0.43\sqrt{1 - R_{equ}} \le 1 \qquad (17)$$

The design of the pre-stressed reinforcement under fatigue loads is based on the evaluation of the fatigue damage according to the Palmgren-Miner summation with S-N curves for reinforcing steel (cfr 6.7.4 CEB - FIP Model Code 1990, cfr 6.8.4 EC2). The control is needed at knots where the anchorage of the strands is located.

Base concrete segment. The base segment needs a detailed analysis since there are openings for inspection and ventilation. It is modeled through two-dimensional shell finite elements in order to achieve a more detailed representation of the stresses distribution around the openings in the concrete surface (see Figure 31).

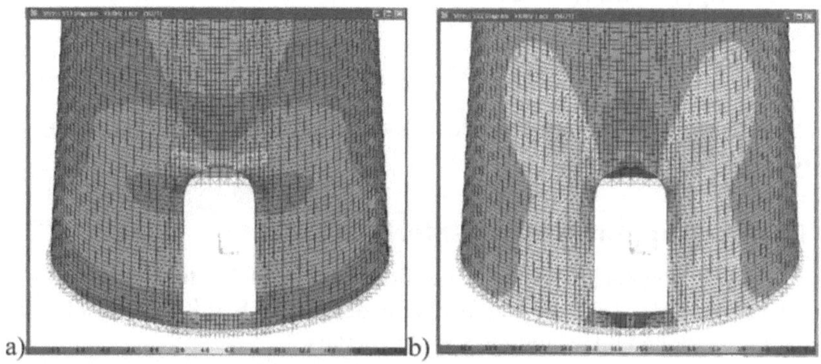

Figure 31. Stresses distribution around the openings in the concrete shell: a) Membrane tensions (S11) b) Circumferential tensions (S22).

This detailed analysis allowed the design of the minimum steel reinforcement to be put in the door lintel (see Figure 32).

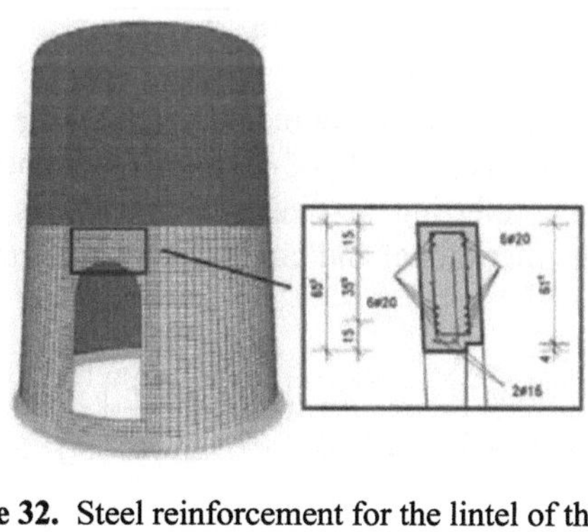

Figure 32. Steel reinforcement for the lintel of the door.

Transversal reinforcement in the concrete part. The amount of reinforcement (A_{s_Env}) is the maximum value between: minimum area of reinforcement at the ULS for shear + torsional effect (wind and/or earthquake load) and for bending effect (temperature); minimum area of reinforcement for Code requirements; minimum area of reinforcement for shrinkage and cracking. The effects of wind, earthquake and temperature are combined by partial safety factors according to the D.M. 14.01.2008 (A_{s_ULS}) (cfr Eq. 18), and then enveloped with the results of the Italian Code requirements (A_{s_code}) and of the shrinkage and cracking effects (A_{s_FLS}) (cfr Eq. 19):

$$A_{s\_ULS} = \max\left(1.5\cdot\Delta T + 0.6\cdot Wind; 1.5\cdot 0.6\cdot\Delta T + 1\cdot Wind\right) \tag{18}$$

$$A_{s\_Env} = envelope\left(A_{s\_ULS}; A_{s\_code}; A_{s\_FLS}\right) \tag{19}$$

With these results we calculate the amount of transversal reinforcement in the concrete part.

3.5 Towered support Design – Steel part

The towered support design needs the verification even of the steel part: ULS for the internal actions (N+M+V+T) and buckling design; FLS (ultimate limit state of fatigue) for the bolted knots and ULS for the details.

Below are the main verification for the steel part of the towered support:

Steel at ULS (strength and buckling). The steel design strength at the ultimate limit states refers to D.M. 14.01.2008 and UNI EN 1993-1-1. The stresses calculation in the shell is computed according to the theory of the global beam as in the lower and upper parts of the shell stiffening rings have been introduced. In the strength design the maximum stress is calculated according to Von Mises criteria:

$$\sigma = \sqrt{\sigma^2_{(N,M)} + 3(\tau_V^2 + \tau_T^2)} \tag{20}$$

There are also horizontal bending moments given by the wind action, stressing the shell wall according to UNI EN 1993-3-2 (for the Chimneys).

The buckling design, referring to steel shell structures (UNI EN 1993-1-6), takes into account of the meridional and circumferential normal stresses and of the shear ones as well:

$$\left[\left(\frac{\sigma_{xEd}}{\sigma_{xRd}}\right)^{k_x} + \left(\frac{\sigma_{\theta Ed}}{\sigma_{\theta Rd}}\right)^{k_\theta} + \left(\frac{\tau_{x\theta Ed}}{\tau_{x\theta Rd}}\right)^{k_\tau}\right] \le 1 \tag{21}$$

Steel at FLS (Limit state of fatigue). The steel fatigue failure is checked according to the Italian Code Circolare 02.02.2009 and to UNI EN 1993-1-9:2005. The fatigue design consists in the evaluation of the fatigue damage according to the Palmgren-Miner summation (cfr. Eq. 16).

The fatigue strength for nominal stress ranges is represented by S-N curves which correspond to typical detail categories. Each detail category is designated by a number which represents the reference value for the fatigue strength at 2 million cycles (see Figure 33).

The analyzed knots for the steel part of the tower are classified according to the Tables of the detail category in the Euro and Italian Codes.

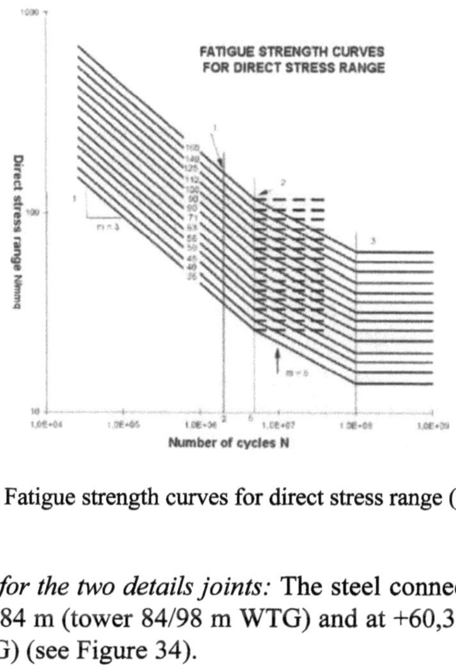

Figure 33. Fatigue strength curves for direct stress range (S-N curves).

- *Steel at ULS for the two details joints:* The steel connections are located at +57.390/+68.84 m (tower 84/98 m WTG) and at +60,390/+71.84 m (tower 84/98 m WTG) (see Figure 34).

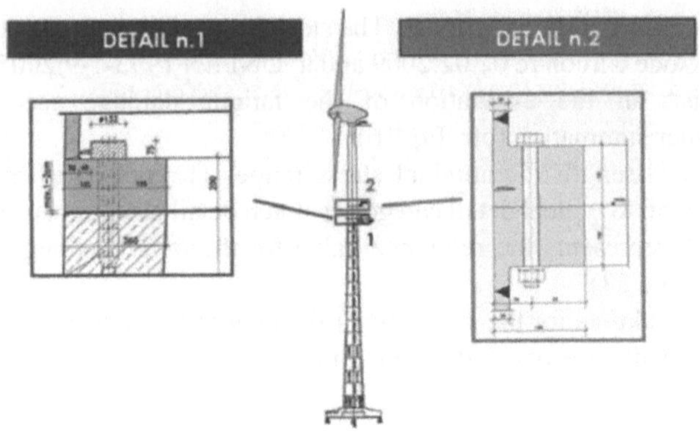

Figure 34. View of the two details joints of the tower witch are calculated.

The first one connects the concrete and the steel parts of the tower and the needed controls are: the verification of the contact pressure on concrete and the steel flange design for bending and torsional stresses.
The second connection requires the verification of bolts for shear, traction, creep and fatigue.

3.6 Foundations Design.

According to D.M. 14.01.2008, for the foundations design the seismic actions to be taken into account should be considered as the design resistance of the overhanging elements but they need not exceed the action effects corresponding to the response of the structure under the seismic design situation inherent to the assumption of an elastic behaviour (q = 1). The design values on the foundations need also not exceed the action effects from the analysis of the design seismic action amplified with a overstrength factor γ_{Rd}, taken as being equal to 1.1 for the ductility class (CD) "B" and 1.3 for CD"A":

$$E_{Ed} = R_{Ed} \qquad E_{Ed} = E_E(q=1) \qquad E_{Ed} \le \gamma_{Rd} E_E \tag{22}$$

In the above equation E_{Ed} is the design value of the seismic action effects on the foundations, R_{Ed} is the design resistance of the overhanging elements, $E_E(q=1)$ is the seismic action effect corresponding to the response spectrum with the elastic behaviour, γ_{Rd} is the overstrength factor, E_E is the action effect from the analysis relative to the design seismic action.

- The verifications of foundation elements (ring and radial) are:
- ULS for normal + bending strength and shear strength;
- SLS for stresses limitation (in concrete and ordinary reinforcements);

- SLS for limit state of cracking;
- FLS (ultimate limit state of fatigue) in ordinary reinforcements;
- Strut & Tie model for the verification of the console zone;
- Load concentration for the piled foundations;
- Concrete slab verification for the hydrostatic thrust in the direct foundation with buoyancy.

Below are the verification for the main elements of the foundations.

Ring n.2. The ring n. 2 is the more stressed element for ULS, SLS and FLS design of the foundation (see Figure 35).

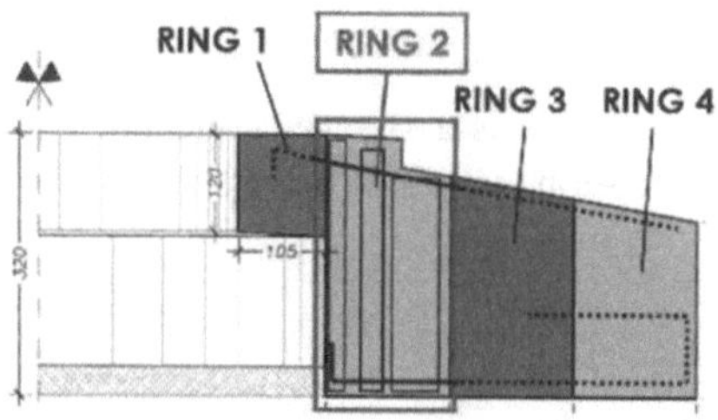

Figure 35. View of the section of foundation – Ring n. 2.

At ULS, results of the design for bending and axial force are reported in the interaction diagrams N-M. With respect to the shear strength, both Eurocode and the D.M. 14.01.2008 require a control based on the comparison between the shear force, resulting from the combinations of actions, and the design resistance shear witch is calculated in presence of transversal reinforcement.

At SLS, the concrete design for the stresses limitation includes the characteristic, the frequent and the quasi-permanent combinations of actions. Concrete normal stresses are then calculated and compared to the limitations required by the Euro and Italian Codes (cfr Eq. 14). Under the rare combination of actions there is also a tensile stresses limitation in the reinforcement:

$$R.: \sigma_s \leq 0.8 \cdot f_{yk} \tag{23}$$

The limit state of cracking requires the calculation of cracks width (cfr 7.3 EC2 and cfr C4.1.2.2.4.6 Circ. 2009). For the control of crack width, the limitation of the D.M. 14.01.2008 are more strict than EC2 (see Figure 30). Under the frequent and the quasi-permanent combinations of actions, the limit of crack width (D.M. 2008) for the aggressive condition and ordinary reinforcement are:

$$\begin{aligned} F.: w_d \leq w_2 = 0.3mm \\ Q.p.: w_d \leq w_1 = 0.2mm \end{aligned} \tag{24}$$

The design according to the ultimate limit states of fatigue (FLS) is carried out in the section in which the phenomenon of fatigue produces the greatest effects, that is at the lower reinforcement of the ring 2. The design is performed on the given values of the bending moments depending on the number of loads cycles and on normal stresses due to other acting loads, which do not imply fatigue in the structure. The fatigue design according to UNI EN 1992-1-1: 2006 and to CEB - FIP Model Code 1990, consists in the evaluation of the fatigue damage following to the Palmgren-Miner summation with S-N curves for ordinary reinforcement steel (cfr Eq. 16).

Ring n.1. In the ring n. 1 there is a stresses concentration in the contact area between tower and the foundation (see Figure 36).

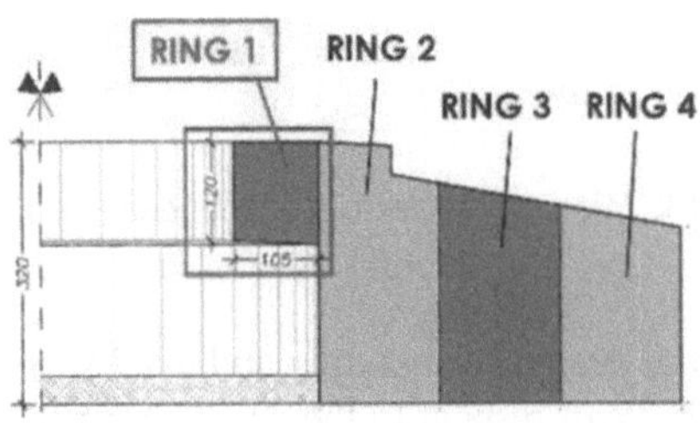

Figure 36. View of the section of foundation – Ring n. 1.

As regards to the design of the ring n. 1 it is possible to consider the Strut & Tie model for the verification of the strength of the lower and upper reinforcement (see Figure 37). These verification is made according to Euro Codes (EC 2) and to Italian Code (Circolare 02.02.2009).

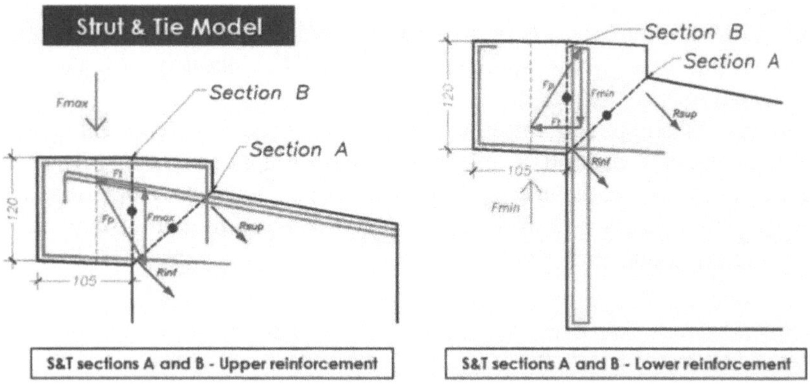

Figure 37. Strut & Tie model for the lower and upper reinforcement of the ring n. 1.

Even for the ring n. 1 one has to check the fatigue limit state design (FLS) following the UNI EN 1992-1-1: 2006 and CEB-FIP Model Code 1990 requirements for the ordinary reinforcement (cfr Eq. 16).

Ring n.4. For the piled foundations there is a stresses concentration in the contact area between the foundation and the pile in the ring n. 4. (see Figure 38).

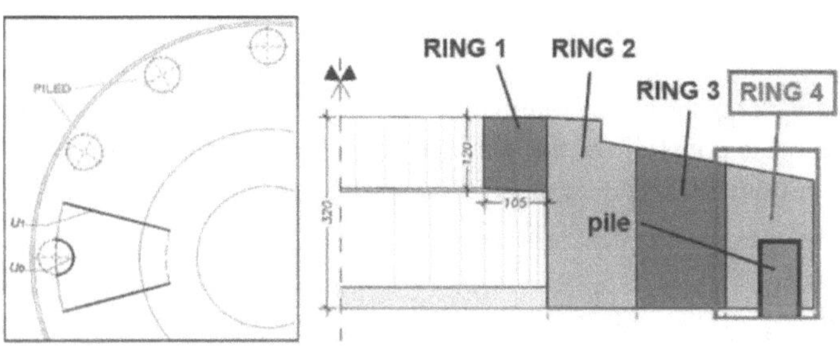

Figure 38. View of the section of foundation – Ring n. 4.

According to UNI EN 1992-1-1: 2006 the stress concentration at the interface between pile and foundation has to be checked.

3.7 Soil – structure Interaction

The FE tower model has been analyzed in two configurations: the first one with a perfect restraint at the base (joint fully restraint), while the second one with an elastic spring at the base (joint springs), simulating the elasticity of the system soil-foundation (see Figure 39).

The shallow and piled foundations are modeled by springs representing the soil behavior in the first case and the pile group behavior in the second one. The springs are located in a node at the interface level between the foundation and the soil. This node is rigidly connected to the basis of the tower. It means that the circular foundation ring is considered as rigid.

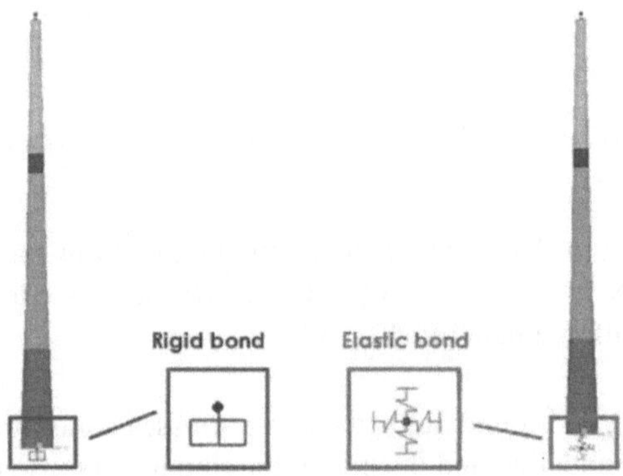

Figure 39. Basic constraints in the FE model of the tower.

The analysis of the direct foundation has been made by a finite elements model by using, as basic constraints, the winkler bed springs representing the behaviour of the soil (see Figure 40). The values for the springs of the direct foundation is variable and it depends on the characteristic of the soil:

$$K_{vert} = 3000 \div 21000 kN / m^3 \quad (25)$$

The FE model of the piled foundation has as many joint springs as the number of piles supporting the structure (see Figure 40). The values for the springs of the paled foundation is also variable depending on the characteristics of the soil:

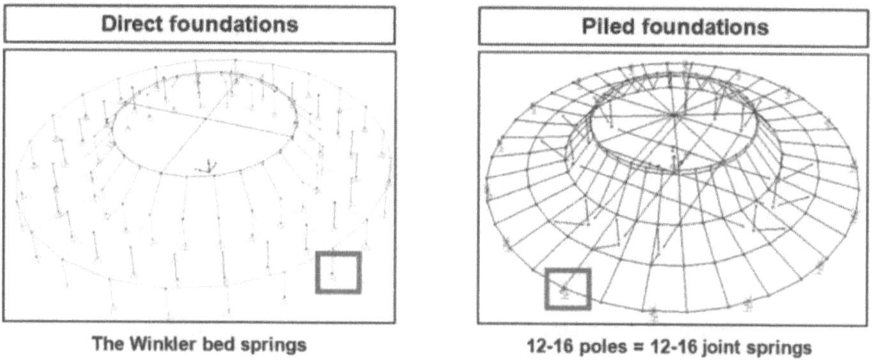

Figure 40. Basic constraints for the FE model of direct and piled foundations.

It is important to note that elastic springs have to be considered as not resistant to traction forces in the case of shallow foundation. This will allow to correctly represent the loss of contact between the foundation and the soil during extreme actions.

4 Technology & Construction

In the present chapter the main phases of the mixed steel and prestressed concrete towers construction will be shown and discussed. The installation of such a type of structure can be divided in four subsequent phases:

- realization of the foundation system;
- pre-assembling of the base segments at the ground level;
- assembling of the concrete part of the tower (including the first 3 m steel segment and prestressing phase);
- erection of the 25 meters steel section and of the rotor.

4.1 Foundations

The concrete foundation system can be installed on a group of piles or directly on a poor concrete layer casted on a regularized soil surface In the first case the number of the piles can vary from 12 to 16 (see Figure 41).

a)  b) 

Figure 41. Foundation piles a) and basement reinforcement b).

After the realization of the group of piles the lean concrete layer is casted and then is possible to start with the assembling of the basement reinforcement as shown in Figure IV.1b. The concrete casting for the basement has to be done in one time in order to prevent the cracking induced by a differential shrinkage. The final result for the basement is shown in Figure 42 before to start the lateral backfilling.

Figure 42. Foundation block before lateral backfilling.

4.2 Tower construction

Once completed the foundation it is possible to start the tower erection. This process is divided in three phases: the first one is related to the base segments assembling (see Figure 43a), the second one consist in the erection and positioning of the concrete segments (see Figure 43b and Figure 44a) until the first steel one and in the prestressing procedure, while in the last one the final steel segment and the hub with the rotor will be installed.

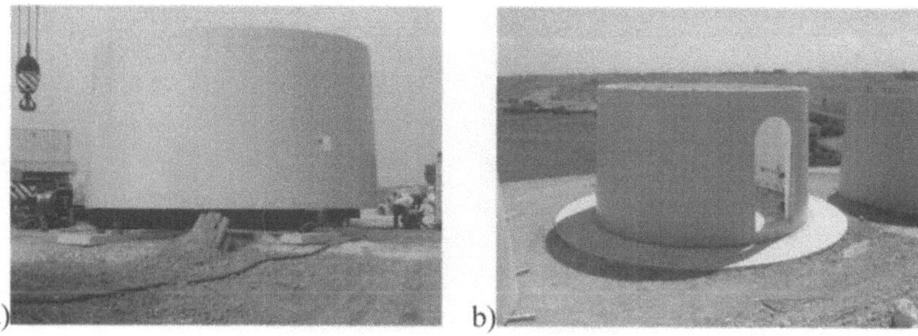

a) b)

Figure 43. Pre-assembling of the base segments.

The concrete segments can be fixed together by means of a special mortar with high tensile strength in order to guarantee the stability of the tower until the prestressing procedure has completed. Prestressing cables are introduced in the concrete thickness from the top of the tower (see Figure 44b) where the dead anchorage has to be placed. The active anchorage and hence the prestressing procedure has to be performed inside the foundation block.

a) b)

Figure 44. Installation of a concrete segment a) and of the 3m steel segment b).

Once the prestressing procedure has been completed, the cables injection has to be immediately done and then we can proceed to the third phase which can be completed in three times. In the first one the 25 m steel segment is bolted to the lower part of the tower (see Figure 45).

Figure 45. Installation of the 25m steel segment.

In the second one the whole nacelle, made by the machine house and the power generator, has to be installed and bolted to the top flange of the tower (see Figure 46a). Finally the rotor has to be assembled at ground level and then uplifted in front of the power generator (see Figure 46b).

a) b)

Figure 46. Installation of the nacelle a) and of the rotor b).

4.3 Logistics

The installation of a mixed steel and prestressed concrete tower put in evidence some difficulties that have to faced and well planned before the starting of the building site.

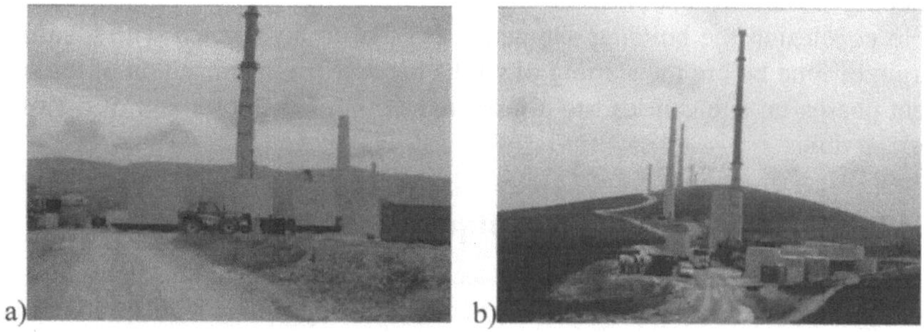

Figure 47. Installation of the nacelle a) and of the rotor b).

In effect a lot of spaces have to be found on site in order to stock the huge amount of material mainly represented by the concrete segment, which have to be positioned, if possible, around the tower to be built (see Figure 47).

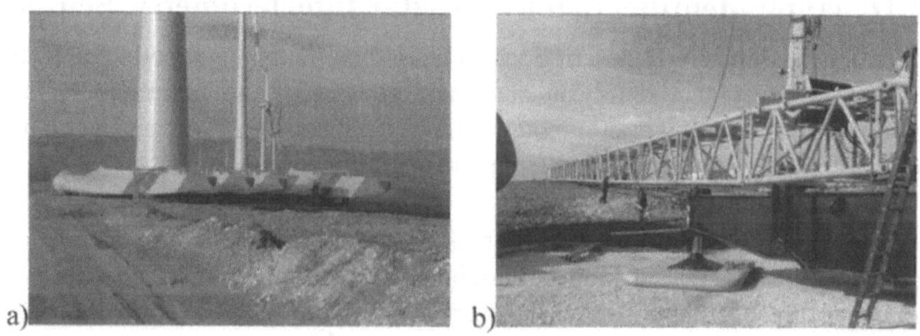

Figure 48. Installation of the nacelle a) and of the rotor b).

Moreover it is important to make the building site roads easily accessible by the trucks bringing the blades, the steel segments and any other material exceeding the standard dimensions. In effect a lot of this material is coming by ship and once in the harbor area it has to arrive on site as soonest as possible. This means to arrange a lot of free areas, of thousands of square meters, where to leave the material.

Another important aspect to be taken into account is the in site movement of the cranes like the one shown in Figure 48b, which are made of several framed segments and having a total length of more than hundred meters. In effect as many towers can be erected at the same time, it is necessary to find some places where to place the lattice segments constituting the cranes that have to erect the upper part of the tower, including the nacelle and the rotor.

In conclusion the building site activities have to be planned with a sufficient margin of time before the starting of works because the optimization of the subsequent phases on site can lead to a great reduction of the costs and of the overall working time.

5 Monitoring & Testing techniques

Non-intrusive and remote monitoring techniques and sensing are nowadays well developed and increasingly adopted in the sector of dynamic identification of large structures, industrial plants, monumental buildings as well as in cultural heritage masterpieces in order to preserve the integrity of the analysed system as well as to save its safety and resistance. In addition to these advantages, the time and costs of the measurement campaigns are definitely competitive and therefore widely accepted and employed.

5.1 Dynamic Identification by a Radar Interferometric Sensor

In this first section, a brief description of the used instrument is reported. It is able to perform a non-contact detection and measurement of the deformations of the structure under test. It works remotely offering a great improvement over conventional techniques regarding installation and operation simplicity.

Measurement of Dynamic Characteristics of Structures. The estimation of dynamic characteristics of industrial/architectonic structures is based on the measurement of the oscillations induced by artificial (hydraulic jacks, percussions, etc.) or environmentally induced (wind, vehicular traffic, etc.) causes on the structure itself.

Currently, the mostly employed technique is based on the use of a set of accelerometers installed on the structure. They are only able to detect the acceleration at installation points in one or more directions.

However, in a lot of situations, this method can result in a complex implementation as it is necessary to install the sensors directly on the structure, often in unreachable places, and then to collect the information using a transmission cabled network.

All these operations can become very cumbersome especially in the case of structures with limited accessibility (for example towers or chimneys). Furthermore in some cases, the structures need to be shut down (out of service).

The radar interferometric sensor (interferometer). The used prototype of interferometric sensor (see Figure 49) is a portable device (weight 12 Kg) which is temporarily installed, over a tripod, near the structure under test. The compact size and the decomposability of the main components are important features that enables the instrument to be a true portable equipment. Its physical aspect is very similar to a camcorder but it is able to image the scenario and to measure the radial displacements of all the object located in its view cone.

Figure 49. Prototype of the interferometric sensor.

The instrument is power supplied by a battery pack which enables the user to operate continuously for a maximum of 5 – 6 hours. The system is controlled by software installed in a regular notebook whose hard disk is used to store all the acquired data. The instrument is connected to the notebook through a USB 2.0 link.

System performances depend on the selected configuration but on the operating conditions too. However, generally speaking, the instrument can be used to measure displacements of objects within a maximum distance of 2km, with an high accuracy (better than 0.1mm) and a selectable sampling frequency between 10Hz and 100

Instrumental set-up. In order to detect the dynamic behavior of different sections of a whole structure, the instrument is temporarily installed near it in such a way that all the interesting zones of the structure itself are visible and different parts appear from the interferometer at different distances. An installation scheme, used in case of vertical structures, is reported in Figure 50.

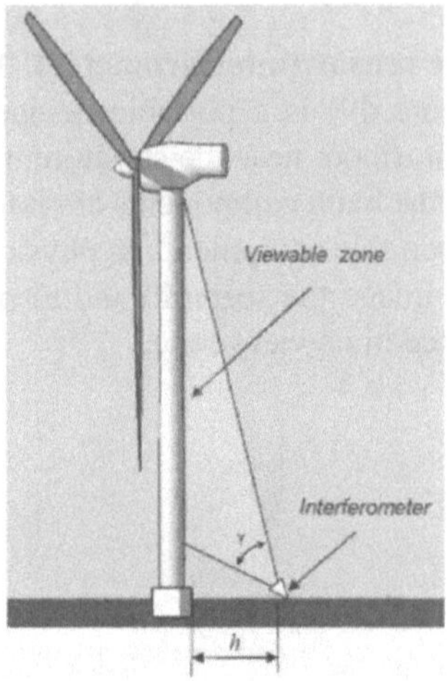

Figure 50. Positioning of the interferometric sensor.

The correct instrument positioning lies on the determining its optimal distance *h* from the structure under test. It depends on the structure effective dimension and the angular aperture γ of the view cone (13° in this case).

If general information on oscillation frequencies of a structure are only needed, the distance parameter is only important in order to obtain a good S/N. The structure can be measured from great distance but, in this case, no correct displacement measurement is achievable. This is what one refers to as "Landscape measurement" mode.

Performing measurements. The measurement is usually performed installing the instrument, as described before. After its configuration and set-up, the measurement is started recording the raw data in the hard disk of notebook.

During the measurement time, which technically has no temporal limit, the instrument acquires continuously the structural deformations with the configured sampling frequency. Being a continuous acquisition, all the acquired data will contain the deformations information of all the measurement cycle.

During processing time, the displacements of all horizontal sections of the structure are calculated. A section is distinguished from the others in function of its distance from the instrument with the system resolution (about 40cm). Depending on the structure dimension it is so possible to obtain a large number of measurement points. These points can be considered as virtual sensors installed on the structure itself from which it is possible to obtain the desired displacements information.

Therefore, it is possible to perform subsequent processing able to reconstruct the temporal evolution of the structural deformation map or calculate the oscillation spectrum of all the sections. In case of landscape measurements, the structure is visible as a single point from which it is possible to obtain the oscillation spectrum only.

5.2 A typical measurement campaign

The measurement results reported in this document are related to the wind generators of a wind farm located in East Calabria, Southern Italy (a set of 8 towers out of 48; height at hub: 98m; 2 MW and 82m rotor turbines). A picture of a typical wind turbine is shown in Figure 51.

Figure 51. A 2 MW wind turbine (98m height, 82m).

Description and Purpose of the Measurements. The measurements on the wind generators of the above set were mainly aimed to identify the natural oscillation frequency of the towers and their modal shapes.

Each measurement was performed using an interferometric sensor and lasted for about 20 ÷ 30 minutes. The sensor has recorded all the displacement map of the towers, in digital form, on a notebook hard disk. The towers displacements were excited by the natural wind and/or by using a “startup/emergency stop” sequence of the turbine of the towers (see Table 4). The emergency stop command was normally given once the rotor has reached a speed of 8 ÷ 18 rpm.

Table 4. Radar tests load conditions.

Tower Code	Measurement Number	Excitation
CR01	2	Wind + Startup/Emergency Stop
CR02	1	Wind Only
CR03	1	Wind + Startup/Emergency Stop
CR04	1	Wind Only
CR05	1	Wind + Startup/Emergency Stop
CR06	1	Wind Only
N06	1	Wind Only
N07	2	Wind + Startup/Emergency Stop

All towers of the above set of the site have been measured; half of the towers were measured by using both the startup/emergency stop sequence and the wind excitation, while the second half were tested by the natural wind excitation only. A landscape mode was not employed here.

Processing of data. All data recorded on the notebook hard disk have been processed in order to extract the oscillation spectrum of the towers and theirs modal shapes. All identification data are documented (tower numbering, measurement date, wind conditions, turbine rotor conditions, sampling frequency and a picture of the monitored tower) as well as all measurements results (accelerations spectrum of the oscillations and a list of the main oscillation frequencies detected).

The spectrum of oscillations of each tower has been calculated using a time span, of the total measurement time, selected in an interference free / high oscillation conditions (the movements of the turbines, which were rotating freely, are an example of measurements interference). The oscillations were caused by a sudden braking during the exercise of the turbine. For any tower, a given and suitable time period (window) for the spectral analysis has been selected, which was not the same for all the towers.

However, this was always greater than 100 sec, in order to obtain a frequency resolution higher than 0.01Hz and a very low noise level. It has been considered that performing a spectral analysis over a greater time period was not necessary, as the obtained frequency resolution can considered suitable for the scope of this report. The main frequencies, listed by analyzing the spectral function diagram, have been obtained by looking at all the maxima and discarding all peaks lower than 5% of the greatest value.

The main oscillation modes of the towers are obtained through a FDD method (Frequency Domain Decomposition). To apply this method, a selected number of target points, spread over a vertical line (generatrix) of tower, are selected. For each one of the targets, the FFD (Fast Fourier Transform) of the recorded displacements (over the same time span) is computed. Then, for any frequency, the cross-correlation matrix is computed and decomposed by using the SVD (Singular Value Decomposition) process. This decomposition gives single values as well as their vectors: the first ones represent the spectral response of the various oscillation modes, while the second ones deliver the associated modal shape.

References

Brebbia, C.A., Dominguez, J. (1998). Boundary Elements, An Introductory Course. WIT Press, Boston, Southampton.

Burton, T., Sharpe, D., Jenkins, N., and Bossanyi, E. (2001). Wind Energy Handbook, John Wiley & Sons.

Cointe, R. (1990). Numerical simulation of a wave channel. Engineering Analysis with Boundary Elements 7(4), 167-177.

Dold, J.W. (1992). An efficient surface-integral algorithm applied to unsteady gravity waves. Journal of Computational Physics 103, 90-115.

Grilli, S.T., Svendsen, I.A. (1990). Corner problems and global accuracy in the boundary element solution of nonlinear wave flows. Engineering Analysis with Boundary Elements 7(4), 178-195.

Grilli, S.T., Skourup, J., Svedsen I.A. (1989). An efficient boundary element method for nonlinear water waves. Engineering Analysis with Boundary Elements 6(2), 97-107.

Hansen, M. O. L. (2008). Aerodynamics of Wind Turbines. Earthscan.

Hau, E. (2006). Wind Turbines: Fundamentals, Technologies, Application, Economics. Springer.

Jonkman, J. M., 2005. Buhl Jr., M.L. 2005. FAST User's Guide, NREL/EL-500-29798. Golden, Colorado: National Renewable Energy Laboratory.

Jonkman, J. Butterfield, S. Musial, W. Scott, G. 2009. Definition of a 5-MW reference wind turbine for offshore system development. Technical Report, NREL.

Longuet-Higgins, M.S., Cokelet, E.D. 1976. The deformation of steep surface waves on water, I. A numerical method of computation. Proceedings of the Royal Society of London. Series A, Mathematical and Physical Sciences, 350(1660):1-26.

Machane, R., Canot, E. 1997. High-order schemes in boundary element methods for transient non-linear free surface problems. International Journal for Numerical Methods in Fluids 24(10), 1049-1072.

Marino, E. (2010). An Integrated Nonlinear Wind-Waves Model for Offshore Wind Turbines. PhD Thesis.

Marino, E., Borri, C., Peil, U. (2010). Offshore wind turbines: a wind-fully nonlinear waves integrated model. Submitted to The Fifth International Symposium on Computational Wind Engineering (CWE2010).

Moriarty, P. J., Hansen, A. C. (2005). Aerodyn theory manual. Technical report, NREL/EL-500-36881.

Morison, J. R., O'Brien, M.P., Johnson, J.W., Schaaf, S.A. 1950. The force exerted by surface wave on piles. Petroleum Transactions (American Institute of Mining Engineers) 154, 189-149.

Nakayama, T. 1983. Boundary element analysis of nonlinear water wave problems. International Journal for Numerical Methods in Engineering 19(7), 953-970.

Nakayama, T. 1990. A computational method for simulating transient motions of an incompressible inviscid fluid with a free surface. International Journal for Numerical Methods in Fluids 10(6), 683-695.

Peil, U., Corte, C. 2005. Numerical simulation of breaking wave load on offshore wind turbines. J. Napstrek & C. Fisher (eds); ITAM AS CR, Prague.

Peregrine, D.H., Dold, J.W. 1986. An efficient boundary integral method for steep unsteady water waves. Numerical Methods for Fluid Dynamics II, 671-679.

Tsai, W., Yue, D.K.P. 1996. Computation of nonlinear free-surface flows. Annual Review of Fluid Mechanics 28,249-278.

Wang, P. Yitao, Y. Tulin, M.P. 1995. An efficient numerical tank for nonlinear water waves, based on the multi-subdomain approach with BEM. International Journal for Numerical Methods in Fluids 20, 1315-1336.

Wienke, J., Oumeraci, H. (2005). Breaking wave impact force on a vertical and inclined slender pile-theoretical and large-scale model investigations. Coastal Engineering 52, 435-462.

Wind energy statistics. (2008). European Wind Energy Association (EWEA).

Global Wind Report (2008). Technical report, Global Wind Energy Council (GWEC).

IEC 61400-1, 3rd ed. 2005. International Standards. Wind turbines Part 1:Design requirements.

Norme Tecniche per le Costruzioni (2008). Ministero delle Infrastrutture, Italy, 14.01.2008.

UNI EN 1990:2006 (2006). Eurocode 1: Basis of structural design.

UNI EN 1992-1-1:2005 (2005). Eurocode 2: Design of concrete structures - Part 1-1: General rules and rules for buildings.

DIBt (2004). Richtlinie für Windenergieanlagen, Deutsches Institut für Bautechnik.

ERRATA CORRIGE

CONTENTS

MIX
Papier aus verantwortungsvollen Quellen
Paper from responsible sources
FSC® C105338

If you have any concerns about our products,
you can contact us on
ProductSafety@springernature.com

In case Publisher is established outside the EU,
the EU authorized representative is:
Springer Nature Customer Service Center GmbH
Europaplatz 3, 69115 Heidelberg, Germany

Printed by Libri Plureos GmbH
in Hamburg, Germany